ACCESS TO MEDICINE IN THE GLOBAL ECONOMY

Access to Medicine in the Global Economy:

INTERNATIONAL AGREEMENTS ON PATENTS AND RELATED RIGHTS

Cynthia M. Ho

OXFORD
UNIVERSITY PRESS

OXFORD
UNIVERSITY PRESS

Oxford University Press, Inc., publishes works that further Oxford University's objective of excellence in research, scholarship, and education.

Oxford New York
Auckland Cape Town Dar es Salaam Hong Kong Karachi Kuala Lumpur Madrid Melbourne Mexico City Nairobi New Delhi Shanghai Taipei Toronto

With offices in
Argentina Austria Brazil Chile Czech Republic France Greece Guatemala Hungary Italy Japan Poland Portugal Singapore South Korea Switzerland Thailand Turkey Ukraine Vietnam

Published by Oxford University Press, Inc.
198 Madison Avenue, New York, New York 10016

Library of Congress Cataloging-in-Publication Data

Ho, Cynthia M.
Access to medicine in the global economy : international agreements on patents and related rights / Cynthia M. Ho.
p. cm.
Includes bibliographical references and index.
ISBN 978-0-19-539012-4 ((hardback) : alk. paper)
1. Drugs—Patents. I. Title.
K1519.D78H6 2011
346.04'86—dc22 2010048464

---

2 3 4 5 6 7 8 9

Printed in the United States of America on acid-free paper

Note to Readers

This publication is designed to provide accurate and authoritative information in regard to the subject matter covered. It is based upon sources believed to be accurate and reliable and is intended to be current as of the time it was written. It is sold with the understanding that the publisher is not engaged in rendering legal, accounting, or other professional services. If legal advice or other expert assistance is required, the services of a competent professional person should be sought. Also, to confirm that the information has not been affected or changed by recent developments, traditional legal research techniques should be used, including checking primary sources where appropriate.

(Based on the Declaration of Principles jointly adopted by a Committee of the American Bar Association and a Committee of Publishers and Associations.)

Table of Contents

List of Figures and Tables vii
Preface ix
Acknowledgments xi
Introduction xiii

PART ONE | BACKGROUND

1. *From Conception to Commercial Success* 3

2. *A Limitation on the Patent Right to Exclude: An Introduction to International Exhaustion* 35

PART TWO | THE CURRENT FRAMEWORK

3. *An Introduction to TRIPS* 55

4. *Freedom under TRIPS: India as an Example* 89

5. *Compulsory Licensing under TRIPS: An Introduction* 125

6. *Compulsory License Case Study: An Introduction to Competing Patent Perspectives* 157

7. *Complicated Compulsory Licenses: The Waiver/Article 31bis "Solution"* 195

PART THREE | THE EVOLVING FRAMEWORK

8. *An Overview of "TRIPS-Plus" Standards* 223

9. *Beyond Patents: Protecting Drugs through Regulatory Laws* 253

10. Suspensions of In-Transit "Generic Drugs": A Case Study of Competing Perspectives 285

11. A History of Access to Medicine through the Lens of Patent Perspectives 325

12. Looking Toward the Future 355

APPENDICES

Appendix A: Selected TRIPS Provisions 375

Appendix B: Declaration on the TRIPS Agreement and Public Health 382

Appendix C: Draft Ministerial Declaration (Developed Countries) 384

Appendix D: Draft Ministerial Declaration (Developing Countries) 386

Appendix E: Glossary of Terms 390

INDEX 399

List of Figures and Tables

Figures

1. Drug Development Process 8
2. Patentability Analysis 18
3. Patent versus Marketing Timelines 24
4. Domestic Exhaustion 39
5. International Exhaustion—Look to Importing Country (B's) 41
6. "Global" Example 42
7. TRIPS Patent Term—Scenario 1 69
8. TRIPS Patent Term—Scenario 2 70
9. Indian Opposition Periods 100

Tables

1. Patent vs. Marketing Approval Requirements 23
2. International Exhaustion—Definitions of Related Terms 44
3. Are Imported Goods Permissible?—Summary of Related Terms 44
4. International Exhaustion—Sample Countries 49
5. "Unfair Commercial Use"—Possible Implications 81
6. TRIPS Compliance Dates 84
7. Indian Patent Rights Mailbox vs. Non-Mailbox Patent Remedies 114

Preface

WHEN I FIRST began practicing law, I never imagined that I would write this book. My first job was as a patent attorney; in that role, I often represented the interests of large companies, including multinational drug corporations. At the time, I did not seriously question the assumption held by such entities that patents provide a critical incentive to development of new medical treatments. However, since then I have developed a more nuanced view; although I still believe that these companies perform an important role in developing new drugs, patents do not always promote development of the most socially beneficial ones. Moreover, even drugs that are socially beneficial may be of little use to those who cannot afford to purchase them.

The issue of how patents impact medicine has increased in significance within the last decade. Just as I was transitioning from private practice to academia, a landmark international agreement (TRIPS) was concluded that requires most countries to provide patents. Significantly, TRIPS required a fundamental shift in the laws of many countries that previously excluded drugs from patentability. The requirement to patent drugs obviously has implications on the cost of drugs and whether citizens in poor countries have access to them.

The conclusion of TRIPS has increased attention on how patents impact access to medicine, but this new focus has not always led to productive results. Discussions concerning the impact on access to medicine often degenerate into finger-pointing. Patent-owning companies are often vilified as greedy corporations that place profits above people while those who advocate greater access to drugs are accused of stealing private property. These accusations seem to be based on deeply held views about the role of patents (referred to in the book as "competing patent perspectives"). On one hand, patents are seen as a tool to promote innovation, and as such, they can (and should) be modified. On the other hand, patents are viewed as an important property right that should seldom be subject to exceptions, especially considering its limited term. As explained in this book, these competing views have led to confusion and obfuscation of the law. I was inspired to write this book in part to clarify widely prevalent misconceptions, as reflected in both reports from the popular press and in some academic publications.

This book has two principal goals. The first is to provide a clear explanation of the current international infrastructure that requires most nations to provide patent and related rights regarding drugs. A second and complementary goal is to explain how competing patent perspectives play a thus far unacknowledged role in promoting distortion and confusion.

I believe this book can make a unique contribution, even in an area where there is a wealth of existing resources. The primary audience I want to reach includes those who are interested (but not yet experts) on the current issues, especially those who are not familiar with either intellectual property or patent law. In addition, I believe my book may be a resource for those who have some prior knowledge of the issues discussed. I realize it is a challenge for a book to adequately speak to both audiences; however, the book in fact does so. It is designed to explain things using only a minimum of jargon (although necessary terms are defined in the Glossary) visual diagrams and "frequently asked questions" also help to reinforce what might otherwise be complicated material. At the same time, the book provides new legal analysis as well as a new theory of patent perspectives that can provide an important step toward possible solutions through better understanding.

Although this book makes no attempt to provide a grand solution to the long-standing question of how to balance patents and access to medicine, it should promote an eventual solution by clearing away some confusion and facilitating a better understanding of competing views of patents. I strongly believe the best results will emerge once there is both a broad-based understanding of the legal issues and a greater understanding and acceptance of different views. The views of some may be resistant to change, but that does not mean it is not worth trying to do so. It is in that spirit that this book was written.

Cynthia M. Ho
November 12, 2010

Acknowledgments

THIS BOOK WOULD not exist without the assistance and support of a number of individuals and institutions. Ironically, I am grateful to a major publisher of legal books for rejecting an earlier proposal of a textbook on this issue for lack of an "adequate" market. That led me to reconceptualize who might benefit from a book that aimed to explain how international laws impact patents. I then realized that a monograph rather than a textbook could be used in a number of different academic environments, as well as a resource to researchers and policy makers. In addition, reaching a broader audience would be very desirable for this important topic of how laws impact global access to medicine. Needless to say, I am grateful to Oxford (and the anonymous reviewers of my initial proposal) for believing that such a book would be useful. I am similarly grateful to Loyola University of Chicago for providing institutional support to enable me to write it.

In terms of the substance of the book, I have many individuals to thank for providing comments on one or more chapters. I am particularly grateful that some well-known scholars—including some that I have yet to have the pleasure of meeting in person—were so generous with their time. At the risk of omitting someone, I thank the following individuals for their helpful feedback and support: Shamnad Basheer, Vince Chiappetta, Colleen Chien, Carlos Correa, Peter Drahos, Erika George, Daniel Gervais, Joe Grant, Leah Chan Grinvald, Joan Krause, Chunlin Leonhard, Valbona Muzaka, Tu Nguyen, Kevin Outterson, Jerome Reichman, Henning Grosse Ruse-Khan, and Nadia Sawicki. In addition, I owe thanks to a number of people who helped answer specific questions about chapters, including Margo Bagley, Sam Brunson, Thomas Jaeger, Molly Beutz Land, and Ed Lee. I am also thankful to some of these indivduals (and others) for their scholarship, which helped inform my own understanding. Finally, although there are too many students to name in person, I am nonetheless thankful for all the students who read and commented on early draft chapters that I used in class.

I am also grateful for feedback on earlier articles or presentations that evolved into the current chapters. I am especially thankful to Graeme Dinwoodie for his consistent support and encouragement of my work in the area of access to medicine; his support was especially important in the publication of "A New World Order for Addressing Patent

Rights and Public Health," 82 Chicago Kent Law Review 1469 (2007), which formed an important foundation for several subsequent chapters of this book. In particular, I thank Greg Vetter and his colleagues at the University of Houston for inviting me to present an article at its symposium on international aspects of intellectual property this opportunity provided me the opportunity to publish "Unveiling Competing Patent Perspectives," 46 Houston Law Review 1047 (2009), from which I adapted Chapter 6. Similarly, I am grateful to Deakin University and the University of Hyderabad for inviting me to a conference where I was able to present the draft chapter on in-transit patent infringement. I particularly appreciated the supportive comments of Dean Coldicott, Owain Williams, and Vivienne Eggers. In addition, others who provided helpful comments and support on earlier articles that evolved into chapters include Nancy Kim, Susan Kuo, Sean Pager, and Angie Upchurch.

A few individuals deserve special recognition. To begin with, I am very grateful to L. Song Richardson for her helpful feedback and support during the early stages of my writing process. I am also very grateful to Jacqui Lipton, not only for all her substantive feedback, but for bearing with me and responding to the many questions I had about writing my first-ever monograph. I also appreciate the insight and encouragement of veteran authors Cynthia Lee and Alex Tsesis. I am also very thankful for Maggie Chon, not only for her seemingly unfailing support of this project, but for encouraging me to provide draft chapters of my book to her class that resulted in some very helpful feedback. Moreover, I am enormously impressed by and grateful to Peter Yu for reading and providing comments on the entire manuscript. That was not a task I had the audacity to ask of him; however, Peter kindly offered to do so, and I could not turn down his generous offer to essentially do a single-handed peer review.

There is also a small army of individuals who assisted me with the individual elements of the book. I am thankful to Chris Nemes for helping to create the many figures. In addition, I very much thank Patricia Scott, librarian extraordinaire, who always cheerfully responded to my requests to find sometimes obscure references—and at times seemed to work magic to make them appear. I'm also grateful to her for enlisting additional support, especially with foreign language materials from Julliene Grant. I owe a huge debt to a number of student assistants who over a number of years helped provide important research support for this book from its initial conception to current form. At an early stage, I was fortunate to have excellent help from Anna Barriero Megan Simpson and LeighAnne Thompson. More recently, I am grateful to a number of students for not only researching specific issues but also checking the accuracy of the cited sources. I especially thank Kathleen Sullivan Klein, Kate McNamara, Melissa Miltonberger, and Brad Snyder. Brad and Melissa deserve special recognition for helpful comments on the text of the entire book.

Last, but, certainly not least, I am very grateful to my family and friends for putting up with me while I was working on this book. I could not have done it without the consistent and constant support of those who often seemed to have a stronger belief in the book than I did myself.

Introduction

PATENTS ARE FREQUENTLY praised or reviled for their impact on public health. Companies that develop and market patented drugs hail patents as essential to developing drugs that benefit all of society. Conversely, consumer advocates emphasize that patents limit competition and result in inflated prices that frequently impair access to essential and life-saving drugs.

Although some continue to debate whether patents adequately promote innovation in light of short-term access costs, some of the debate is moot because of international laws that *mandate* patents. In particular, all countries that are members of the World Trade Organization (WTO) are *required* to grant patents pursuant to the WTO side agreement known as the Agreement on Trade Related Aspects of Intellectual Property, more commonly known as TRIPS. Admittedly, countries can elect not to participate in the WTO. However, in a global economy, that would be a poor choice for most nations. Indeed, the fact that well over 150 countries across a wide economic spectrum are WTO members suggests that most countries consider participation important.

There is no serious question that TRIPS—and patents—are here to stay.[1] Nonetheless, TRIPS presents a dramatic change to the international landscape. Historically, each nation decided how to best balance promoting long-term innovation versus short-term access. Over the centuries, nations have come to differing conclusions based on differing priorities. A decision to provide patents is not simply an issue of whether a country is rich or poor. Some industrialized countries declined to provide patents on pharmaceutical products until the early 1990s.[2]

[1] TRIPS has been strongly criticized by some. However, because TRIPS is linked to the WTO, nations are unlikely to abandon the WTO forum even if they dislike related agreements, such as TRIPS. Similarly, although some critics of patents would prefer they not exist, that is not realistic in a world where patents are not only a business reality, but mandatory pursuant to TRIPS.

[2] Spain and Portugal declined to grant patents to pharmaceutical products until 1992.

Today, most developing countries do not have the luxury to decide whether to provide patents on drugs. TRIPS mandates patent protection—including on drugs—for WTO member countries.[3] Such mandatory patent protection is likely to significantly impact access to medicine in countries where resources are already limited. As patent protection is no longer optional, the key question is how much discretion remains for nations to promote maximum access to drugs. A nation's ability to promote access to drugs is especially of importance to low-income countries in which the cost of drugs may be a significant barrier to obtaining access to medicine; drugs constitute a much larger percentage of budgets for individuals in poor countries than in wealthy ones.[4] Ironically, individuals from wealthier countries often pay lower prices for their drugs because their governments both impose price controls on drugs, and often have insurance that further subsidizes their out of pocket expenses. Neither of these things are typically true for individuals in poor countries. Accordingly, their out of pocket expenses for medications often far exceed those in wealthier countries.[5]

This book focuses on international agreements regarding patents and related rights because although many issues impact access to medicine, these are generally poorly understood and deserving of separate treatment.[6] Thus far, those with a vested interest–such as patent-owning drug companies–have largely controlled the development of international patent law, including TRIPS.[7] Moreover, such companies have continued to play an essential role in developing subsequent international agreements that provide more protection to drugs through stronger patent laws as well as regulatory laws that help protect exclusivity of drugs beyond patent rights.

Access to medicine impacts all citizens of the world. Knowledge about how drugs are protected from competition should not be limited to only a few, let alone those with a singular and slanted perspective. Although some people may not have problems in

---

[3] Although least-developed countries need not provide patents on drugs yet, all developing countries that are members of TRIPS must currently provide such protection.

[4] Developing countries can spend up to 60% of health care budgets on medicines whereas OECD countries spend about 18%. WHO, *World Medicines Situation* (2004); OECD, *Drug Spending in OECD countries up by nearly a third since 1998*, (Aug. 6, 1998), http://www.oecd.org/document/25/0,2340,en_2649_201185_34967193_1_1_1_1,00.html.

[5] In developing countries, up to 90% of the population pays for drugs out of pocket, making medicines the largest expenditure after food. A. Cameron et al, *Medicine prices, Availability and affordability in 36 Developing and Middle-income Countries: A Secondary Analysis*, 353 LANCET 240 (Jan. 17, 2009), citing WHO, Equitable Access to Essential Medicines: A Framework for Collective Action (2004), 8 WHO POLICY PERSPECTIVES ON MEDICINE 1 (2004).

[6] This book focuses on how patent and related rights increase the cost of drugs and thus limit access to medicine. Other factors that can play a role in the price of drugs (such as domestic tariffs and taxes) will not be addressed here. Similarly, this book does not attempt to address whether improved local medical infrastructure or the use of competition laws can improve access to drugs.

[7] *E.g.*, SUSAN K. SELL, PRIVATE POWER, PUBLIC LAW (2003); PETER DRAHOS WITH JOHN BRAITHWAITE, INFORMATION FEUDALISM: WHO OWNS THE KNOWLEDGE ECONOMY (2002).

accessing necessary drugs, global needs impact all. For example, given that many wealthy countries are involved in buying drugs for poorer countries, the cost of drugs also impacts those countries.

This book focuses on the impact of key international agreements on access to patented drugs in the existing commercial reality. It does not focus on either alternative methods beyond the patent system to promote innovation or whether the dominant corporate model for developing new drugs needs to be overhauled. These issues are noted in the book, but cannot be addressed in depth while simultaneously explaining the complex international landscape. In addition, this book assumes that despite proposals for reform, the present reality will most likely continue in the near future such that a better understanding of the laws that impact this reality is useful. In particular, this book assumes pharmaceutical companies will continue to not only make drugs, but seek to maintain and further expand their ability to provide stronger protection of such drugs in the global arena. Accordingly, the book begins with an introduction to how drugs are currently developed and marketed to set the stage for why companies are invested in the existing laws that promote this infrastructure.

In addition to explaining the relevant international laws, this book provides insight on why debates on access to medicine are not only heated, but can result in the promulgation of incorrect information. As explained in more detail in Chapter 6, this book posits that the foundation for this problem may be a spectrum of views on patents benchmarked by two distinct and seemingly irreconcilable perspectives. At one extreme, patents are seen as a mere privilege granted by a nation so that they are inherently subject to limitations to accommodate other societal goals, such as access to medicine. At the other extreme, patents are viewed as an exceptionally strong property right that should seldom (if ever) be subject to exceptions because these rights are presumed critical to promoting innovation. Adherents of each view may believe the other is necessarily incorrect. However, in reality, their strong views may make them incapable of seeing—let alone tolerating—an alternative vision. This phenomenon, as well as the existence of competing patent perspectives, is consistent with social science literature concerning how people process information. In essence, research shows that people will maintain preexisting beliefs despite evidence to the contrary.

Several chapters of the book use these patent perspectives to provide a more nuanced explanation of controversies concerning patents and access to medicine. Although the perspectives are not intended to completely explain controversies, they can provide an enriched understanding. As will be discussed, these perspectives can help explain why not only patent owners, but even those with an interest in promoting access to medicine may be equally susceptible to distorting TRIPS requirements. The influence of perspectives may make those on both sides of the debate vulnerable to overstatements, which undermines credibility, degrades the possibility of productive dialogue, and may engender confusion amongst the broader public. Accordingly, recognition of the existence and impact of these views is just as important as understanding the actual laws because these views have a strong impact on how international laws are developed and interpreted.

Although the issues involving access to medicine are many and ever-evolving, this book provides a useful introductory resource. In particular, it should be accessible to a broad range of readers with an interest in how laws can impact the availability of lower-cost generic drugs. For example, policy makers, academics, and students from a range of disciplines could benefit from a better understanding of the way international laws impact access to medicine. Although discussion of these laws tends to be limited to those with a background in either patents or international trade, there are clearly implications for public health as well as public policy. Hopefully, this book will help promote not only a better understanding of existing laws, but more balanced future discussions.

The book is divided into three parts. The first provides background information concerning the scientific, legal, and commercial implications of drug development that are relevant to any discussion of patents. The second provides an overview of the current international legal framework, beginning with an introduction to TRIPS. Finally, the third section focuses on the evolving legal framework, including agreements and events subsequent to TRIPS that impact access to medicine. Although each section builds upon other sections, the chapters can be read out of order. For example, the first section provides background information that may not be necessary to those already familiar with how drugs are developed, regulated, or legally protected from competition. To maximize the utility of this book as a resource, individual chapters refer to related chapters. In addition, the Appendix provides a glossary with explanations of terms and common abbreviations. The Appendix also reprints selected provisions of TRIPS that are discussed in the text, as well as a few other important WTO texts.

Part I

Chapter 1 provides an overview of how drugs are brought to market. Although there are many nuances to this process, the goal of this chapter is to focus on themes important to understanding subsequent chapters. This chapter explains how patented versus generic drugs are made, approved for sale, and marketed. It also explains how various laws may protect a drug in the marketplace; the discussion includes, but is not limited to patents. This chapter also focuses on commonalities among the laws of different countries so as to provide a basic understanding of the commercial and legal realities involved with making drugs.

Chapter 2 builds upon the basic patent concepts explained in Chapter 1 to explain an important yet confusing doctrine referred to as "international exhaustion," recognized by some countries. Patent-owning pharmaceutical companies are strongly opposed to this doctrine because it challenges their desired business model of segregating markets to maximize the profit in each. This doctrine impacts whether drugs first sold by the patent owner in one country may be imported by another into a second country without violating patent laws.[8] A country that recognizes international exhaustion may be able to import cheaper versions of drugs. However, as will be explained in subsequent chapters, countries may be pressured to reject this principle.

Part II

The chapters in Part II explore what is required under TRIPS. In particular, Chapter 3 introduces the overall TRIPS framework that is the foundation of all international laws regarding patents. Subsequent chapters assume knowledge of the basic TRIPS framework and provide more detail on key issues involving controversies about the extent to which TRIPS impacts access to medicine. For example, Chapter 4 explains how domestic patent laws can crafted to promote access to medicine. The remaining chapters focus on different issues involving compulsory licensing.

Chapter 3 provides the foundation of the current international framework: an overview of TRIPS. Although subsequent chapters discus specific issues, this one offers an overview of how TRIPS was concluded and the key provisions relevant to access to drugs. This chapter discusses what must be patented and the extent of patent enforcement rights. Additionally, it addresses the controversial issue of what protection must be provided for clinical data submitted in connection with obtaining approval to market a new drug.

Chapter 4 explains how nations may promote access to medicine consistent with TRIPS, using India's patent laws as an example. As a WTO member, India was required to provide more extensive patent protection under TRIPS than it had in the recent past. However, India's laws were adopted with the goal of continuing to promote access to medicine, as well as to sustain its strong generic drug industry. Accordingly, an examination of India's laws helps to illustrate how nations may attempt to craft patent laws consistent with TRIPS that still promote access to medicine.

Chapter 5 explains the basic requirements for a compulsory license—an exception from patent rights that is permissible under TRIPS, yet highly contested by patent-owning drug companies.[9] This chapter is important not only to lay the foundation for understanding subsequent chapters, but because it dispels common misconceptions frequently propagated by interested parties and the popular press.

---

[8] International exhaustion undermines pharmaceutical industry profits because drugs sold at lower prices in developing markets can be imported into wealthier markets where the imports undercut the higher-priced drugs.

[9] Where applicable, a government can impose a compulsory license on a patent, which enables the government (or an entity approved by the government) to make the patented drug, subject only to payment of a government-determined royalty. The patent is not invalidated, but the compulsory license prevents the patent owner from reaping its usual patent premiums for drugs sold under the license.

Chapter 6 builds upon Chapter 5 and also provides a complementary examination of compulsory licenses. Chapter 6 is the first chapter to introduce competing patent perspectives; it provides a case study of Thailand's recent compulsory licenses to explain some of the confusion and controversy concerning them. In particular, the lens of competing perspectives is used to provide an enriched understanding of why there has been so much confusion concerning compulsory licensing. In addition, this introduction to competing patent perspectives constructs a framework for subsequent chapters that revisit how such perspectives have played a role in issues involving access to medicine.

Chapter 7 also builds upon Chapter 5 to explain a complicated and narrow exception to compulsory licenses that is intended to help provide low-cost drugs to developing countries. This chapter also provides an overview of how the exception works and evaluates its utility with a discussion of the single instance where it has been used.

Part III

The last part discusses the evolving global landscape. Chapter 8 begins with an overview of international laws subsequent to TRIPS that further restrict national discretion to tailor patent laws to promote access to medicine. In general, these laws require more protection than does TRIPS, such that they are called "TRIPS-Plus" laws. Subsequent chapters provide more detail on TRIPS-Plus laws. For example, Chapter 9 explains regulatory laws that provide complementary protection to patent laws by limiting competition from generic manufacturers. In particular, Chapter 9 focuses on explaining "data exclusivity" and "patent linkage"—protection provided in the regulatory laws of some countries that may create additional hurdles to the entry of generic drugs. Chapter 10 explains a TRIPS-Plus law in the EU that limits the ability of countries to internationally trade generic drugs. Chapter 10 also provides a case study of the impact of patent perspectives in shaping interpretations of TRIPS; it complements the earlier case study of compulsory licenses in Chapter 6.

The last two chapters revisit the patent perspectives as a lens to better understand how the international framework has evolved since the conclusion of TRIPS. Chapter 11 uses these perspectives to gain insight on international developments since the conclusion of TRIPS. Although some developments are discussed in other chapters, this one uses the perspectives to provide a cohesive method of understanding the overall development of international law and policy. Chapter 12 outlines anticipated issues that will likely increase the existing tensions between providing patent (and related) protection versus promoting current access to medicine. The chapter primarily focuses on proposed solutions. The competing perspectives are also used to explain the genesis for these same proposals as well as the likelihood they will be well received. Although the ultimate outcome of the conflict over access to medicine remains unknown, this chapter attempts to frame current proposals in the context of possible future developments.

I Background

1 From Conception to Commercial Success

This chapter provides a snapshot of drug discovery as well as an overview of the laws applying to the development of drugs. It is not intended to provide a comprehensive discussion of regulatory approval in all nations; rather, it focuses on major issues relevant to patent and patent-like exclusivity that are discussed in the remainder of the book. This chapter answers some key questions, including:

- How are drugs developed?
- What must a company establish to legally sell a drug?
- What laws help companies maintain market exclusivity and profitability?

I. Overview 4
A. Drug Types 5
B. Drug Manufacturers 5
II. From Conception to Market Approval 6
A. Drug Development 6
1. New Drug Development Trajectories 6
2. Generic Drug Development 12
B. Marketing Approval 13
1. New Drugs: Regulatory Approval 14
2. Generic Drugs: Regulatory Approval 15
III. Achieving and Maintaining Commercial Success 16
A. Legal Toolbox of Protection 16
1. Patents 16
2. Trademarks 20
3. Regulatory Protection 21
B. The Intersection of Patent and Regulatory Approval 21
1. New Drugs: Timing and Effective Term of Protection 21
2. Legal Barriers to Generic Drugs 24
C. Market (and Sales) Strategies 26
1. New Drug Manufacturers Business Model 26
2. Generic Drug Manufacturers Business Model 32
IV. Conclusion 34

I. Overview

Before discussing how drugs are developed, it is important to clarify which drugs this chapter addresses. What consumers consider to be drugs may include both conventional drugs (often sold in pill form) and biologics (often injected). The conventional drugs— referred to as "small molecule compounds" by scientists—currently comprise the majority of the marketplace and are the focus of this chapter. Although biologics are a growing market and can treat the same conditions as conventional drugs, they have a very different scientific and regulatory path to the marketplace; thus, for the sake of simplicity they are excluded from this overview, although they are discussed further in Chapter 12.

A. DRUG TYPES

Even within the class of conventional drugs, there is an important further differentiation of drugs. In particular, conventional drugs may be either *new drugs*[1] (typically patented) or *generic drugs* (unpatented, or drugs whose patent term has expired). New drugs often have a very long development process; the total time from conception to commercialization can easily take a decade or more.[2] In addition, the process is uncertain; companies report only one in ten thousand initially tested compounds eventually reach the marketplace. Generic drugs, by contrast, are low-cost versions of drugs previously developed by another company. There is little risk in developing drugs that have already been shown to be safe and effective. Thus, a relatively minimal investment of time and money is needed to manufacture and commercialize generic drugs.[3]

What constitutes a generic drug merits some brief clarification as there is often great confusion concerning what this term means. Although the word *generic* is often used to refer to an unpatented drug, a generic drug is also one that is "bioequivalent" to a previously approved drug; this means the generic is considered not only safe and effective, but a therapeutic substitute for the original. Accordingly, generic drugs are not fake; they are legitimate drugs that provide a safe and often lower-cost alternative to a previously patented drug. Substandard drugs are a problem in the global marketplace, but approved generic drugs are not part of it.

B. DRUG MANUFACTURERS

Large multinationals with substantial resources typically bring to market new drugs as they are able to engage in the extensive—and expensive—clinical testing necessary to secure regulatory approval to sell new drugs. Some examples include Pfizer, Glaxo Smith Kline, and Roche. Although these are the primary actors that do clinical testing critical to ultimate commercialization, the path to commercialization may involve others—especially at earlier stages of development. For example, a discovery by a university or government

---

[1] The more technical name for new drugs is either *new molecular entities* or *new chemical entities*.

[2] *E.g.*, GOV'T ACCOUNTABILITY OFFICE, GAO-07-49, NEW DRUG DEVELOPMENT: SCIENCE, BUSINESS, REGULATORY, AND INTELLECTUAL PROPERTY ISSUES CITED AS HAMPERING DRUG DEVELOPMENT EFFORTS 6, 8 (2006); *Research & Development*, INTERNATIONAL FEDERATION OF PHARMACEUTICAL MANUFACTURERS AND ASSOCIATIONS, http://ifpma.org/issues/index.php?id=421; THE ASSOCIATION OF THE BRITISH PHARMACEUTICAL INDUSTRY, THE DEVELOPMENT OF MEDICINES 1 (2005), http://www.abpi.org.uk/%2Fpublications%2Fbriefings%2FDev_Medicines.pdf.

[3] However, the right to produce a generic drug sometimes generates substantial litigation expense—especially in the United States where the first company to successfully challenge a patent as invalid is awarded a six-month period as the exclusive generic. *E.g.* Maurice Ross, *Leveling the Playing Field—The Role of Venture Capital in Hatch-Waxman Patent Litigation*, 79 PAT. TRADEMARK AND COPYRIGHT J. (BNA) 730 (April 9, 2010); *see also* Federal Food, Drug, and Cosmetic Act, 21 U.S.C. § 355(j)(2)(B) (2006) (providing generic exclusivity).

agency may be further refined and tested by such a company. Alternatively, a smaller company may initially discover a drug later marketed by a large pharmaceutical firm; in fact, this is a common path to development of new drugs today.[4]

Although new drugs are brought to market by large companies with the financial resources for necessary clinical testing, a much smaller investment of capital is needed to bring a generic drug to market as a generic drug manufacturer need only do limited testing to show its proposed compound is bioequivalent to a previously approved drug. Given that generic drug development is less capital-intensive, small companies as well as companies in developing countries can succeed in the generic drug business. Some of the large companies do make generic versions of their own drugs. However, for simplicity, this chapter will use the term *generic drug manufacturer* to refer to companies that primarily make generic versions of drugs developed by others so as to highlight distinctions in manufacturing and marketing models.

II. From Conception to Market Approval

This section explains how new and generic drugs are scientifically developed, as well as what steps must take place before they may be sold to consumers. For both types of drugs, the process begins with scientific drug development; if successful, this culminates in approval by a national agency to market (i.e., sell) the drug.[5] The approval is referred to as either marketing or regulatory approval as regulatory laws often govern whether a drug may be marketed. Although this section discusses the development process for both new and generic drugs, more attention is given to new drugs because they have a longer path to commercialization.

A. DRUG DEVELOPMENT

1. New Drug Development Trajectories

Before explaining the development process, it is important to clarify that there are different types of drugs with different development trajectories that may all appear "new" to a

[4] Some have suggested new drugs may also be developed "virtually" by scientists and others who want to collaborate and use "open source" methods instead of patents—especially for treatments that are unlikely to be highly profitable, yet nonetheless socially desirable. *E.g.*, Bernard Munos, *Can Open Source R&D Reinvigorate Drug Discovery?*, NATURE REV. DRUG DISCOVERY (2006)), *available at* http://www.nature.com/nrd/journal/v5/n9/pdf/nrd2131.pdf; *see also* Amy Kapczynski et al., *Addressing Global Health Inequities: An Open Licensing Approach for University Innovations*, 20 BERKELEY TECH. L.J. 1031 (2005). However, large pharmaceutical companies (either alone or in conjunction with other companies) remain the dominant method of producing most new drugs today.

[5] The national agency may be referred to as a "National Drug Regulatory Agency" (NDRA).

consumer. The longest development path with the most rigorous and lengthy regulatory approval is for a *new molecular entity* (NME) *or new chemical entity* (NCE). Both of these terms refer to a new chemical compound rather than a combination of previously approved compounds or a new use of a known compound. This section first outlines the development of an NME, but later discusses how other variations are developed.

It is critical to understand that different types of drugs are referred to as "new" because the popular press (as well as the pharmaceutical industry) often broadly refer to all newly marketed drugs in the same way—even though they have very different development paths.[6] When drug companies note that drug discovery is long, risky, and expensive, they are actually only referring to NMEs. For NMEs, only a fraction of initially tested compounds make it through all of the necessary stages to final commercialization. The drug industry suggests that only one in eight or ten thousand lab-tested compounds eventually reach the marketplace.[7] For ultimately successful drugs, it may take another decade of testing to develop the necessary data required to seek marketing approval.[8] This additional decade is after several years of laboratory research.[9] A graphical illustration of this process is shown in Figure 1.

Development of an NME is very expensive, with estimates ranging from $200 million to over a billion dollars.[10] Although there are substantial disputes concerning the exact cost due to estimates being based on proprietary information that cannot be independently

---

[6] *E.g.*, MARCIA ANGELL, THE TRUTH ABOUT THE DRUG COMPANIES: HOW THEY DECEIVE US AND WHAT TO DO ABOUT IT 41-42 (2004) (noting that the media seem to accept reports from the drug industry on the cost of drug development, even though the studies focus only on NME).

[7] *See, e.g.*, *Innovation*, PHARMACEUTICAL RESEARCH AND MANUFACTURERS OF AMERICA ["PhRMA"], http://www.phrma.org/innovation/ (one in ten thousand); *Research & Development*, INTERNATIONAL FEDERATION OF PHARMACEUTICAL MANUFACTURERS AND ASSOCIATIONS, http://ifpma.org/issues/index.php?id=421 (one in eight thousand).

[8] *E.g.*, Salomeh Keyhani et al., *Are Development Times for Pharmaceuticals Increasing or Decreasing?*, 25 HEALTH AFFAIRS 461, 463 (2006) (reporting an average of 5 years of clinical testing, with a range of 1.4 to 14.6 years); GOV'T ACCOUNTABILITY OFFICE, *supra* note 2, at 6-7, (suggesting total time for discovery and development is 10–15 years, with an average of 6.5 years for basic discovery and preclinical testing in animals, followed by 7 years of clinical tests in humans).

[9] The pharmaceutical industry suggests that of the initial 10,000 compounds screened, only 250 enter preclinical testing, five enter human clinical trials, and only one is approved for sale. *E.g.*, PHARMACEUTICAL RESEARCH AND MANUFACTURERS OF AMERICA, SPECIAL 301 SUBMISSION 2010, at 16, *available at* http://keionline.org/sites/default/files/USTR-2010-0003-0245.1.pdf.

[10] Studies developed from the industry-funded Tufts Center for the Study of Drug Development have consistently given the highest estimates—based on self-reported data that is not subject to independent evaluation. *E.g.*, Joseph A. DiMasi & Henry G. Grabowski, *The Cost of Biopharmaceutical R&D: Is Biotech Different?*, 28 MANAGERIAL & DECISION ECON. 469, 475 (2007) (reporting an average capitalized cost of $1.24 billion, with $559 million out-of-pocket costs per approved drug); Joseph A. DiMasi et al, *The Price of Innovation: New Estimates of Drug Development Costs*, 22 J. HEALTH ECON. 151, 165, 166 (2003) (reporting the capitalized cost of a new drug is $802 million). However, critics assert these numbers are inflated. *E.g.*, ANGELL, *supra* note 6, at 37–51; Donald W. Light & Rebecca Warburton, *Extraordinary Claims Require Extraordinary Evidence*, 24 J. HEALTH ECON. 1030, 1030-32 (2005).

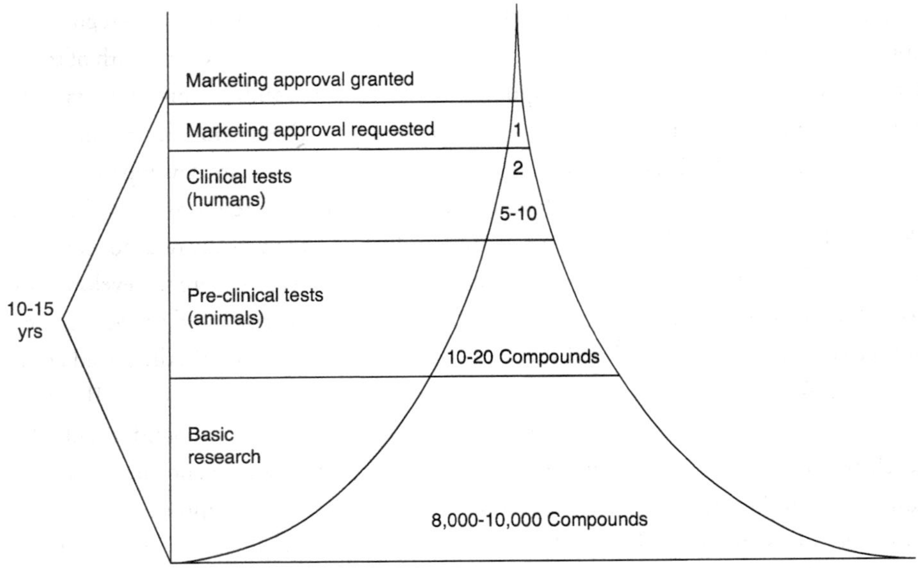

FIGURE 1 Drug Development Process

evaluated, undoubtedly the cost is substantial. However, although companies that develop new drugs are indeed responsible for bearing some of this cost, they are not the only ones shouldering it; governments often play a major role in underwriting the cost of drug discovery, either through direct funding of research[11] or indirect means, such as tax benefits[12] as well as patent protection for government-funded research.[13] In fact, some suggest that if such indirect subsidization were better captured, governments—rather than pharmaceutical

[11] Governments can play an especially important role in funding basic scientific research that is less focused on specific drugs, but may have wide application to a number of areas. The National Institute of Health plays a major role in funding scientific research in the United States.

[12] *E.g.*, Nina Crim, *Tax Proposal to Promote Pharmacologic Research, to Encourage Conventional Prescription Drug Innovation and Improvement, and to Reduce Product Liability Claims*, 29 WAKE FOREST L. REV. 1007, 1039–72 (1994).

[13] A common practice is to grant patent rights to nationally funded research to private companies. The United States first began doing so under the 1980 Bayh-Dole Act, which has been replicated or proposed for replication in other countries. 35 U.S.C. §§ 200–12 (2006); The Protection and Utilization of Public Funded Intellectual Property Bill, 2008, Bill No. 66 of 2008 (India); David C. Mowery & Bhaven N. Sampat, *The Bayh-Dole Act of* 1980 *and University-Industry Technology Transfer: A Model for Other OECD Governments*?, 30 J. TECH. TRANSFER 115, 115–25 (2005); Bhaven N. Sampat, Int'l Ctr. for Trade & Sustainable Dev., *The Bayh-Dole Model in Developing Countries: Reflections on the Indian Bill on Publicly Funded Intellectual Property* 5 UNITED NATIONS CONFERENCE ON TRADE & DEV. 1-6 (2009); Anthony D. So et al., *Is Bayh-Dole Good for Developing Countries? Lessons from the US Experience*, 6 PLoS BIOLOGY 2078, 2080–82 (2008).

companies—would actually contribute more to overall research and development.[14] In addition, consumers help pay for research by paying a premium on new drugs.

a. Long Development (NME)

The development of a NME is broken down into several stages. The first step is scientific discovery, which primarily involves laboratory work. Successful compounds identified in this initial stage then enter into clinical testing. The few compounds that successfully emerge from clinical testing then move toward marketing approval. This section provides a basic overview of scientific discovery and clinical testing.

The first critical step in developing an NCE is drug discovery. The goal of drug discovery is to identify a chemical compound for treating a disease or condition; this step occurs primarily in the laboratory.[15] A compound that appears promising in the laboratory will undergo further testing beyond the lab. However, only a small percentage of laboratory-investigated compounds investigated are considered promising enough for continued testing. Those compounds also may be modified to enhance effectiveness, ease of manufacture, or minimize side effects.[16]

These compounds are then subjected to clinical tests. This is necessary not only to assess whether they can be commercially developed, but to comply with marketing approval requirements if the tests are successful. Testing begins with animals. If results warrant it, further investigation is done in humans. Initially, healthy volunteers are tested to find out basic information about how the drug is absorbed as well as its effects on organs. If the drug is deemed safe, it is then tested on increasingly larger groups of people (including those with the target conditions) to assess the appropriate dose for efficacy and to determine any risks, including side effects.

Many initially promising compounds fail during clinical tests. Sometimes compounds that display promise in a lab setting, fail to do so in animals or humans. A compound may also be rejected because it is not safe, or has intolerable side effects.

Some successful drugs may have an unexpected path to commercialization: sometimes a compound that is initially thought to be useful for one condition proves to be ineffective against it, but highly effective against another. For example, the well-known drug for erectile dysfunction first sold as Viagra was initially investigated for cardiac problems. The drug was not particularly effective for that use. However, when researchers discov-

---

[14] Donald W. Light, *Basic Research Funds to Discover New Drugs: Who Contributes How Much?*, *in* MONITORING FINANCIAL FLOWS FOR HEALTH RESEARCH 2005: BEHIND THE GLOBAL NUMBERS, GLOBAL FORUM FOR HEALTH RESEARCH 26, 32 (Mary Anne Burke & Andrés de Francisco eds., 2005), *available at* http://www.globalforumhealth.org/Media-Publications/Publications/Monitoring-Financial-Flows-for-Health-Research-2005-Behind-the-Global-Numbers.

[15] Drug companies refer to the first step as a *screening* process for biologically active compounds, otherwise known as *targets*.

[16] This step is known as *optimization*.

ered patients were reluctant to stop taking it, further inquiry revealed its use for what is now a whole new class of compounds.

b. Abbreviated Development

An alternate type of drug that is "new" to the consumer has a shorter development trajectory than just described. Such variations may include new dosages or delivery systems of previously approved compounds, combinations of previously approved compounds, new indications for a previously approved compound, or a slight variation of a previously approved compound. This section briefly explains why the development path is shorter in each case; however, precise information on the exact time or cost involved cannot be provided as drug companies generally do not share this proprietary information. In all cases, the abbreviated development is considered a useful type of "life cycle management" for drug companies, which is criticized by some for "evergreening" patent exclusivity, as discussed later in this chapter.

NEW DOSAGES AND NEW DELIVERIES

One way a company can develop a newly marketed drug is to create a new dosage or delivery system for an existing and already commercially successful drug. Starting with a known desired compound immediately shortcuts the multi-year process of finding a potentially promising compound, with scientific development focusing on less challenging issues such as adjusting the known dose to an extended release dose that would continue to be effective. Alternatively, development could focus on a new delivery system that would be more desirable for consumers, such as by pill instead of injection. In both instances, standard methods of achieving these results can be modified to suit the compound in question, reducing unpredictability as well as development time.[17] In addition, some stages of clinical development can be shortened as the compound is already known to be safe.

A number of popular drugs on the marketplace provide new dosages of previously developed compounds. For example, the first treatment for osteoporosis (sold as Fosomax) was a NCE that required daily administration, but the subsequent once-a-week treatment used the same active ingredient (sold as Fosomax Once Weekly). Similarly, the first antidepressant drug in the class of selective serotonin reuptake inhibitors (commonly known as SSRIs) was a NCE sold under the brand Prozac. However, the same active ingredient in Prozac was later approved and sold as a once-a-week treatment (Prozac Weekly).

[17] A new delivery system is reported to take a mere three or four years to develop at a cost of twenty to fifty million. Biswajit Dhar & K. M. Gopakumar, *The Case of the Generic Industry in India*, *in* INTELLECTUAL PROPERTY AND SUSTAINABLE DEVELOPMENT: DEVELOPMENT AGENDAS IN A CHANGING WORLD 97, 120 (Ricardo Melendez-Ortiz & Pedro Roffe, eds., 2009).

NEW COMBINATIONS

A "new" drug (to a consumer) may also be a combination of previously known and approved drugs. This can similarly shorten the development path if each drug included in the combination is already known and has expected properties. For example, the allergy medication sold under the brand Allegra was an NCE, but a later version that included a known decongestant (sold as Allegra-D) was not an NCE. However, although combinations may reduce the development time, the amount of time is likely a function of whether there is knowledge concerning how the respective components interact with each other. For example, given that other allergy drugs had successfully added a known decongestant without a problem, adding the same decongestant to the active ingredient in Allegra likely posed minimal technical issues. Conversely, developing a new HIV treatment that involves several active ingredients may be more complex as the combination of those active ingredients may be less predictable.

NEW USES

Similarly, the development of a new use of an existing compound may have a faster path. Developing a new use begins with a known compound, which is subjected to a battery of tests to assess whether it will be a useful treatment for other conditions. The other conditions may not be known in advance, but starting with a known compound is likely to result in a much shorter path to discovery of a new use. For example, an important and well-known new use is the antiviral agent AZT to treat HIV; the compound was discovered long before HIV-AIDS became an established disease. Because the compound was already known, its path to market was dramatically shortened. In particular, a mere two years after AZT was found effective to treat HIV in the lab, regulatory approval was granted.[18] Another well-known new use involves the use of a compound sold as Retin-A for the treatment of wrinkles, rather than its original use in treating acne. Still another new use involves a drug originally sold as an antidepressant (Prozac), but later approved to treat premenstrual dysphoric disorder.[19]

NEW VARIATIONS

Shortcuts in development may also be achieved when a company develops a new variation of an existing drug—one that is already known to be safe and effective.[20] Often the new

---

[18] ANGELL, *supra* note 6, at 26.

[19] This type of activity has been referred to disparagingly as "disease mongering" for the expansion of treatable illnesses to develop new products. *See, e.g.*, Ray Moynihan et al., *Selling Sickness: The Pharmaceutical Industry and Disease Mongering*, 324 BRIT. MED. J. 886 (2002).

[20] Companies may aim to create a variation by isolating a single enantiomer of a racemic mixture, directly using a drug metabolite, or using a different crystal form or polymorphs. While these terms may all sound exotic, they involve minor changes to those well versed in the field. For additional information concerning how companies

variation is a slight chemical variant of an existing drug. Although some clinical testing may be required, the amount may be limited because the new variation is only slightly different than a known drug. Accordingly, many of the usual early stages of testing for safety can be eliminated; in some cases, the variation may be similar enough to be considered bioequivalent and thereby have very limited stages of testing.

To help put these variations in context, a few examples of popular drug variations may be helpful. For example, the active ingredient in an antidepressant sold as Lexapro is the "purer" version of an older (and related) antidepressant sold as Celexa and alleged to make patients less drowsy.[21] A different example is the newer allergy drug sold as Allegra, which provides similar allergy relief as drug sold as Seldane, but without the negative cardiac effects associated with Seldane. Although ingestion of the drug sold as Seldane should lead to the active ingredient in Allegra (after the drug interacts with normal biological processes), the Allegra drug provides the ultimate desired chemical combination without the side effects.[22]

2. Generic Drug Development

The path to creation of a generic drug is much shorter and simpler than for either a new chemical entity or any of the variations described above. The development of a generic begins with an existing drug the generic manufacturer wants to duplicate—there is no need to find new compounds because another company has already done so. Although the characteristics of the desired drug are known from the publicly available patent, a generic manufacturer must develop an efficient commercial manufacturing process as that information would not necessarily be disclosed in the patent. Yet this process is less difficult and time-intensive than discovery of new drugs.

Generic drug development skips the drug discovery phase entirely and instead focuses on a very limited type of clinical tests. Because the generic company knows the drug it is

---

develop such variations, see, for example, Rebecca S. Yoshitani & Ellen S. Cooper, *Pharmaceutical Reformulation: The Growth of Life Cycle Management*, 7 HOUS. J. HEALTH L. & POLY 379 (2007); A.J. Hutt & J. Valentova, THE CHIRAL SWITCH: THE DEVELOPMENT OF SINGLE ENANTIOMER DRUGS FROM RACEMATES, ACTA FACULTATIS PHARMACEUTICAE UNIVERSITATIS COMENIANAE (2003); Dean Fanelli, *Claiming Novel Polymorphic Forms Of Drug Products*, 88 J. PAT. & TRADEMARK OFF. SOC'Y 1053 (2006).

[21] The official terminology for the process of developing the purer drug is the development of a single enantiomer from a "racemic" mixture. This involves choosing one half of a three dimensional chemical structure, each of which is a mirror image of the other. The single enantiomer is considered "isolated" from the racemix mixture and often has the same therapeutic effect as the racemic mixture, but with fewer side effects. For example, the active ingredient in the allergy drug sold as Xyzal is the isolated enantiomer from the racemic mix sold as Zyrtec, and is similarly effective as Zyrtec for controlling allergies, but with lower incidence of drowsiness.

[22] Officially, the active ingredient in Allegra is the "metabolite" of the active ingredient in the "pro-drug" Seldane. A metabolite is a compound that is formed after ingestion as a result of the body's metabolic processes; the post-ingestion compound is what has therapeutic effects.

duplicating has already been deemed safe and effective, to obtain regulatory approval it only needs to do limited testing to show its proposed generic version is bioequivalent. Essentially, bioequivalence means the products have the same active ingredient at the same dosage such that they are expected to act the same in patients.[23] This showing can be done fairly quickly and inexpensively compared to the multiple stages of clinical testing necessary with a new drug. However, as explained in the last section of this chapter, legal hurdles may prevent a potential generic manufacture from developing a bioequivalent immediately after the original drug is marketed.

B. MARKETING APPROVAL

Most countries require a domestic agency to certify that a new drug is safe[24] and effective[25] before it can be marketed to a consumer; in addition, usually the manufacturing methods must meet certain standards to ensure drug quality and consistency. As the domestic agency regulates what drugs get approved for marketing to consumers, it is often referred to as a regulatory agency.[26] The laws that regulatory agencies apply to decide whether a new drug can be marketed are often referred to as marketing approval.

There are roughly two different ways in which countries approve a drug for sale. In wealthy countries such as the United States, EU Member States, and Canada, the domestic regulatory agency evaluates extensive clinical data submitted by the applicant for a new drug. However, some developing countries do not have the resources to perform such reviews. Accordingly, they may approve drugs based on prior approval by another country. Alternatively, they may rely in part on the World Health Organization (WHO)'s list of drugs that are "prequalified" as safe and effective. (However, the WHO list only addresses treatments for HIV, malaria, and tuberculosis.) Any company may seek WHO approval as a step toward obtaining marketing approval in a country that grants such approval based on WHO approval.[27]

---

[23] Technically, bioequivalence refers to the fact that both drugs have the same blood concentrations for the active ingredient.

[24] Although it would be optimal for all drugs to be completely safe, in most cases safety is assessed relative to risks and benefits.

[25] Although regulatory agencies are sometimes criticized for approving too many drugs—especially when they must be later withdrawn from the market—testing for efficacy is actually a recent development. Originally, such agencies tested compounds only for safety. In other words, they evaluated only whether drugs were likely to harm, and not whether they would likely be effective for their intended purposes.

[26] For example, in the United States, the Food and Drug administration is the regulatory agency. The situation in Europe is more complex, involving the central European Medicines Agency in conjunction with national offices.

[27] A company that seeks prequalification for its product(s) must also be willing to have its manufacturing site inspected to ensure compliance with the WHO Good Manufacturing Practices, or alternatively to have the site inspected by national regulatory bodies.

This section provides an overview of the more extensive approval process in countries with the relevant infrastructure. In particular, regulatory approval for new versus generic drugs will be considered. For the purposes of regulatory approval, a new drug is one that contains at least one active ingredient that has not been previously approved (i.e., a new chemical entity). In contrast, a generic drug is intended to be equivalent (usually bioequivalent) to something previously approved.

1. New Drugs: Regulatory Approval

A company that wishes to commercialize a new drug must get marketing approval in every nation where it desires to sell it. Although most countries have similar standards, approval from one country only authorizes sales in that country. Accordingly, a multinational company that wants to sell a new drug in the United States, Canada, and Japan would need to file separate marketing approval applications in each respective country.

In general, the marketing approval process for an NME requires an applicant to submit data establishing the proposed new drug is safe and effective for its designated use.[28] A large portion of the application is comprised of the extensive clinical data developed by a company that establishes the drug's safety and efficacy. However, as the standard is just safety and efficacy—not *improved* safety or efficacy—a drug may be approved for sale even though it does not provide a significant therapeutic advantage over existing drugs. Indeed, clinical studies that support approval of a drug's safety and efficacy may compare the drug to a placebo. In addition, the party seeking approval typically does the studies; in other words, they are done by a party with an interest in obtaining approval. Although scientific protocols help to minimize investigator bias, some suggest company-sponsored clinical trials are more likely to find treatments to be effective.[29]

Marketing approval may sometimes be granted based on more limited clinical data in conjunction with reliance on published literature or prior approval of another drug.[30] This might be possible, for example, when there is a change in the dosage (from twice a day to once a day). Alternatively, this route may also work where the proposed drug is

---

[28] *E.g.*, Federal Food, Drug, and Cosmetic Act 21 U.S.C. § 355(b)(1) (2006) (US standard); Food and Drug Regulations, C.R.C., ch. 870, § C.08.0002(2) (2010) (Canadian standard).

[29] *E.g.*, Nicholas Bakalar, *Review Finds Drug Makers Issue More Positive Studies,* N.Y. TIMES, Feb. 27, 2007, at 7, *available at* http://www.nytimes.com/2007/02/27/health/27canc.html (finding clinical trials of breast cancer treatments are more likely than independently financed studies to find that treatments are effective); In addition, negative clinical data may not be disclosed—a problem that has led to calls for legal requirements to disclose all data. *E.g.*, Jerry Avorn, *Dangerous Deception – Hiding the Evidence of Adverse Drug Effects*, 355 NEW ENG. J. MED. 2169 (2006); Gardiner Harris, *Diabetes Drug Maker Hid Test Data, Files Indicate,* N.Y. TIMES, July 13, 2010, at A1, *available at* http://www.nytimes.com/2010/07/13/health/policy/13avandia.html. However, even mandatory disclosure would not eliminate the fundamental problem that drug companies have a vested interest in the outcome of their trials.

[30] *E.g.* 21 U.S.C. § 505(b)(2) (2006); Council Directive 2001/83/EC, art. 10, 2001 O.J. (L 311) 67.

comprised of two previously approved drugs such that information exists on the safety and efficacy of the individual components. This abbreviated pathway is also relevant for a new indication of a drug previously considered safe and effective for another purpose. However, the abbreviated approval may be contingent on other issues, such as whether it is possible to use the data submitted with the previously approved drug or if there are any patent barriers. These issues will be addressed in the section regarding legal protection of drugs.

2. Generic Drugs: Regulatory Approval

Before a generic drug can be sold, it must be approved for sale by the same regulatory agency that approves new drugs. The agency will only give this approval if the drug is safe and effective. The generic must also be essentially identical to its counterpart brand product with respect to active ingredient, dosage, route of administration, and intended use such that the generic is considered therapeutically equivalent to the original.[31] An application to market a generic drug does not need to provide direct clinical studies regarding safety and efficacy of the proposed generic. Rather, the application typically refers to and relies on the clinical data of a previously approved drug. In other words, a generic drug application contains a limited amount of data showing only that the proposed drug is "bioequivalent" to the previously approved drug.[32] If the agency finds the proposed generic to be bioequivalent, it infers the proposed generic is just as safe and efficacious as the drug to which the generic is bioequivalent. The generic applicant never has access to the previously submitted clinical data and does not need to: it simply relies on its existence to establish the proposed generic is equally safe and effective.

One important issue is what the regulatory agency considers a previously approved drug.[33] In countries with extensive regulatory approval processes (such as the United States, EU Member States, and Canada), the previously approved drug whose data are relied upon would be a drug approved by the same agency. In other words, an application to sell a generic version of a drug in Canada would attempt to establish bioequivalence with a previously approved drug there. However, some developing countries do not have the resources to evaluate extensive clinical data involved with new drug applications. In those cases, a generic drug application often refers to a previously approved drug from *anywhere in the world*, which is usually an application previously filed in a country with extensive regulatory infrastructure.

Another important issue is *when* a generic drug application may be filed and approved. Often a generic application cannot be filed immediately after the original drug (to which

[31] *E.g.*, 21 C.F.R. § 314.92(a)(1) (2010). Bioequivalence refers to the rate and extent of the drug's absorption being roughly equivalent to the previously approved drug.

[32] *E.g.*, 21 U.S.C. § 355(j)(4)(F) (2006).

[33] This may be referred to as the "reference listed drug."

the proposed generic would be bioequivalent) is approved. Barriers to generic drug applications include both patent and regulatory laws, as will be further discussed in the section on the intersection of patent and regulatory approval.

III. Achieving and Maintaining Commercial Success

Achieving and maintaining commercial success over a new drug often involves much more than scientific development and marketing approval. Most companies have development and marketing strategies they use in tandem with a variety of legal tools that can help to protect a drug from competition. This section begins with an overview of relevant legal tools. Then, the intersection between patent and regulatory approval is revisited to give a sense of the differing time periods and issues involved. Finally, the section ends with a discussion of how development, legal, and marketing strategies all work together to promote and sustain commercial success for drugs.

A. LEGAL TOOLBOX OF PROTECTION

Several types of legal protections may protect the commercial viability of a drug. Patents are most commonly associated with protecting new drugs and often considered the most valuable. Accordingly, they are discussed first. Another common tool is trademarks, otherwise known as "brands" to a consumer. Finally, this section discusses the legal protection provided by the regulatory laws of some countries.

1. Patents

A patent is a legal document granted by a nation that gives its owner certain legal rights. It is analogous to a deed for real property in that it is a document with information establishing legal rights. However, unlike real property, patent rights are limited in time. The basic patent term is generally less than twenty years, with the exact term a function of how long it takes for the application to be reviewed.[34]

A patent provides its owner with a commercial advantage in the marketplace – the patent owner may exclude others from a variety of activities relating to the patented invention within the nation that has granted the patent. In particular, the patent owner can prevent others from making, using, selling, offering to sell, or importing the invention during the patent term. For example, the owner of a US patent on the cholesterol

[34] While patent terms are often referred to as twenty year terms, that is generally not accurate. A typical patent term is calculated as 20 years from the filing of the patent application minus the time necessary to evaluate the patent; there are generally no patent rights before the patent is granted. Accordingly, if the application review took 3 years, the patent term would be 17 years. However, as discussed in Chapter 8, some countries permit this basic patent term to be extended where the patent application or regulatory review process was lengthy.

treatment drug marketed as Lipitor can prevent another competitor from selling the same compound in the United States.

In a free market, a patent owner can charge a premium for its patented product because no one else is legally permitted to make or sell it. However, the owner of a drug patent may still be subject to competition from others who make drugs within the same *class* of treatment. For example, multiple companies own patents and sell drugs relating to acid reflux, depression, and cancer. A patent right covers a specific invention, which in the case of drugs is usually a very specific chemical compound. Often multiple chemical compounds treat the same condition—each of which could be subject to its own patent. The owner of a patent on Lipitor cannot prevent all others from making cholesterol treatment drugs—only those that make the identical chemical compound disclosed in the patent. Nonetheless, a patent still provides a commercial advantage. Generally, when there are several different patented drugs that treat the same condition, they can each be priced at a premium in comparison to an unpatented drug, or a drug whose patent has expired.

An important question is what can be patented. In general, a patent can only exist for inventions that constitute patentable subject matter and meet certain additional requirements. Although the requirements are sometimes stated with slightly different wording, there are essentially three additional requirements. In particular, the invention must be (1) capable of industrial application (or useful), (2) "new," and (3) have an "inventive step" (or be "nonobvious").[35] Although these terms may seem straightforward, they have a special meaning under patent laws that will be discussed here.

Before discussing the meaning of each term, it may be helpful to first consider the process of assessing patentability. The process can be envisioned as a series of stages where each step is a tougher requirement to satisfy. The first element—patentable subject matter—is the easiest one to meet. The second element—useful—is still fairly simple to satisfy. The elements that primarily serve to winnow the scope of what is patentable are that an invention be both new and have an inventive step. This is shown graphically in Figure 2.

As noted above, the most fundamental requirement of patentability is that an invention be patentable subject matter. Historically, some nations excluded certain inventions from patentability for policy reasons. Sometimes governments decided it was bad policy to patent some inventions (such as drugs) as that might lead to higher drug costs and less access to medicine. Today however, drugs are generally considered patentable subject matter; this is largely the result of an international agreement referred to as TRIPS (described more in Chapter 3).

The next requirement is that an invention be either useful or be capable of industrial application. This requirement is an easy one for most new drugs or the active ingredient of a drug. For example, an active ingredient that is shown to reduce tumor growth would be deemed to be useful (or, alternatively, capable of application in the pharmaceutical

[35] In addition, to be granted a patent, the application must explain how to make and use the invention.

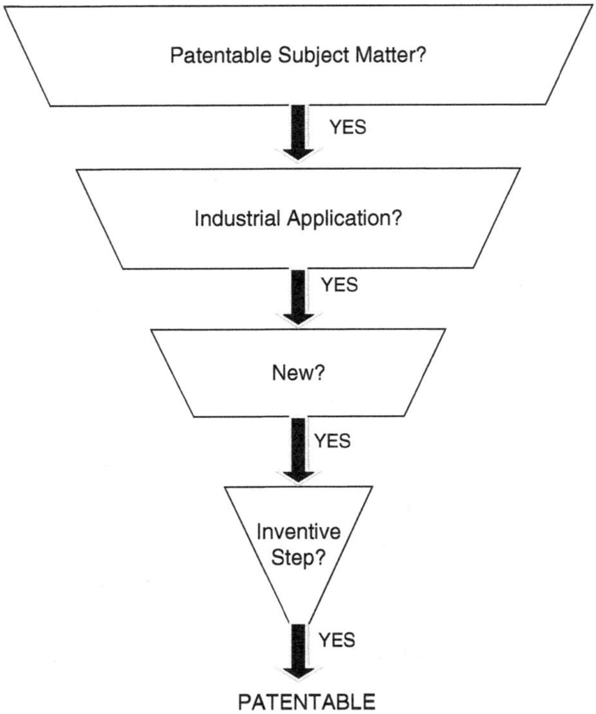

FIGURE 2 Patentability Analysis

industry). Importantly, this requirement of usefulness is a much lower threshold than the regulatory requirement that a drug be safe and effective.[36] For example, an active ingredient may be useful, but not necessarily safe for human use.

The toughest patent requirements to establish are that the invention is new and has an inventive step. These requirements must be assessed with respect to what is known before the date of the patent application—what is referred to as *prior art*. Countries have different definitions of what counts as prior art, but it generally refers to what is publicly known before the date of the application. The prior art may include a patent or publication by someone else, or even the inventor herself. An invention is considered "new" if there is nothing in the relevant prior art that is identical to the invention claimed in the patent application. However, whether an invention has an "inventive step" (or is non-obvious) requires a consideration of whether someone with similar skill to the inventor would have easily created the invention.

[36] *See In re* Brana, 51 F.3d 1560, 1567–68 (Fed. Cir. 1995) (noting the usefulness requirement does not envision that a compound meets regulatory approval and in fact, that the stage at which a patentable compound "becomes useful is well before it is ready to be administered to humans").

One example may help clarify the distinction between new and versus nonobvious. Imagine a world with only wooden doorknobs. In this world, the first ceramic doorknob would be new from a patentability standard if it had never previously existed. However, the ceramic doorknob be obvious. For example, if the process of making ceramics is well-known, and shaping a ceramic doorknob is also well-known, then a ceramic doorknob would be a new item—but one with expected properties, such that it would be obvious and thus unpatentable. But if the ceramic doorknob evidenced surprising properties not common to ceramic items, it might be nonobvious. For example, if the ceramic doorknob was magnetic—a property not typical of ceramics—this could be nonobvious. Although doorknobs are not the topic of this text and we live in a world populated by many doorknobs besides wooden ones, this example hopefully illustrates the elements of what is *new* and what is *nonobvious*. To understand these terms in the field of pharmaceuticals requires an understanding of chemistry. However, the same basic principles apply and are sufficient for an understanding of the discussion in this text.

Even if an invention satisfies all the patentability criteria, a patent does not exist unless and until a patent application is filed and a patent is granted by a national patent office. A patent office generally does an independent examination of patentability. In addition, in some countries the patent office is aided in its evaluation by third parties (such as competing firms) that may suggest an invention fails to satisfy one or more of the patentability requirements.

A patent owner can enforce its patent through private litigation. For example, if Abbott believes its competitor is selling a product that infringes Abbott's patent, Abbott can sue and seek monetary damages as well as an injunction. A patent owner can often also seek a preliminary injunction in advance of the trial on the merits, although there is no guarantee a court will grant this request. A preliminary injunction would be desirable to the patent owner to prevent erosion of its market share and brand name. However, courts usually balance the equities and may consider that money damages later would sufficiently compensate the patent owner, in which case a preliminary injunction is not appropriate.

One important caveat to patent rights is that they only exist in the country that granted the patent. For example, the owner of the US patent on the active ingredient in the drug marketed as Lipitor cannot prevent another company from selling the drug in Australia based on the US patent. Rather, to obtain exclusivity in Australia, an Australian patent must be obtained. Multinational companies generally seek parallel patents in every commercially valuable market.[37] However, they may forgo the time and expense of seeking patents in poor markets that are unlikely to have resources to pay for a premium-priced drug.

[37] Companies can streamline the process of seeking global patents by filing a single "PCT" application that will eventually be reviewed under patent laws of individual countries. Patent Cooperation Treaty, June 19, 1970, 28.7 U.S.T. 7645, 1160 U.N.T.S. 232. This process is only applicable to countries that are members of the agreement, although developing countries are increasingly pressured to become members, as discussed in Chapter 8.

2. Trademarks

A trademark is a word, phrase, logo, or other device used in commerce that serves to identify a good or service.[38] Most commercial drugs use a trademark as the brand name; for example, trademarks for patented (as well as unpatented) items include Lipitor, Oxycontin, Plavix, and Benadryl. A trademark is not required to sell drugs, but, like patent protection, trademark protection can assist commercial sales. Typically, the manufacturer of a new drug will select a name for a drug as a trademark and use trademark laws to prevent another competitor from using a similar name on a similar product. In other words, the manufacture of Tylenol painkillers can use trademark laws to prevent another firm from selling a similar compound as "Tylenol." Sometimes trademark protection can also extend to the exterior package bearing the drug, or even the color or shape of the pill itself.[39]

Trademarks can be most helpful *after* a patent expires when a drug must compete with generic equivalents.[40] In contrast to a patent, which can prevent someone from making a patented drug, a trademark does not prevent its manufacture. However, when a brand name drug becomes unpatented and competes with a generic equivalent, the trademark prevents the generic from being sold with the same brand name. In other words, even after a patent expires on a branded drug, it will likely continue to be sold with its trademark brand. A generic company is barred from using the brand name without violating trademark laws. For example, whereas "Flonase" is the trademark used to market a nasal decongestant spray, the name Fluticasone propionate is the nonproprietary name used by manufacturers of the generic version. The name accurately reflects the active ingredient and also signals to the consumer that this is not the original branded drug. Because a trademark can last indefinitely (unlike patent protection), it can preclude a generic company from ever selling its product as Flonase.

Consumers familiar with the trademark of the patented drug may continue to purchase that drug because of familiarity and trust in the trademark. For example, consumers commonly prefer a brand versus generic products, such as Coca-Cola versus the store cola. Similarly, consumers may prefer the branded drug. In fact, the preference for branded drugs may be stronger than for other consumer goods because branded drugs are often the only option known to consumers for years before a generic equivalent is available.[41]

---

[38] *E.g.*, TRIPS: Agreement on Trade-Related Aspects of Intellectual Property Rights, Apr. 15, 1994, Marrakesh Agreement Establishing the World Trade Organization, Annex 1C, art. 15(1), The Legal Texts: The Results of the Uruguay Round of Multilateral Trade Negotiations 320 (1999), 1869 U.N.T.S. 299 [hereinafter TRIPS].

[39] Technically, colors and shapes are considered to be "trade dress," rather than a "trademark," but the applicable laws are the same.

[40] In addition, trademarks can also be used to suggest an only incrementally different product is "new" and thus deserving of a higher price.

[41] *E.g.*, William M. Landes & Richard A. Posner, The Economic Structure of Intellectual Property Law 189–90 (2003).

3. Regulatory Protection

In some countries, a drug can be protected from competition by not only patents, but also regulatory laws. There are two ways that regulatory agencies can protect a drug from generic competitors. First, a regulatory agency could deny approval of any generic drug that is "linked" to an existing patent such that the proposed generic would infringe the patent if it were sold—a practice first adopted by the United States and now recognized by a handful of countries.[42] Second and more commonly, a regulatory agency can limit when a generic applicant may seek approval by restricting the generic applicant's ability to rely on clinical data submitted by the originator of the initial drug. As discussed earlier, generic drugs are usually approved based on a streamlined application that refers to clinical data of the comparable drug. However, for a certain period of years after approval of the initial drug, many countries do not permit generic manufacturers to refer to the earlier data; this is referred to as *data exclusivity*. Although the second company could undertake its own clinical studies, that is expensive and not cost-effective. Usually the generic company waits until the data exclusivity expires before applying for regulatory approval. Both of these types of regulatory protection are discussed in more detail in Chapter Nine.

B. THE INTERSECTION OF PATENT AND REGULATORY APPROVAL

To understand all the issues involved in bringing a drug to market, it is important to bear in mind not only the existing different types of legal protection, but also how different legal regimes intersect. This section outlines the intersection of patent and regulatory approval processes for both new and generic drugs.

1. New Drugs: Timing and Effective Term of Protection

A drug company seeking to market a new drug would seek patent and regulatory approval at different times, which may result in a shorter effective patent term for drugs than many other patents. This section explains the differing time periods that impact the effective patent term.

a. Timing of Patent versus Regulatory Application

Typically, a patent application for a drug is filed long before an application for marketing approval is sought. The difference in timing is partially a function of differing

[42] As discussed in Chapter 8, patent linkage is one type of protection the United States has been requiring of other states in US-initiated free trade agreements.

requirements as well as when information becomes available.[43] The government agency granting marketing approval typically examines clinical data that establish a drug is safe and effective—data that takes many years to collect. In contrast, a patent may be granted on a chemical without any need to establish its safety or efficacy; the requirement that an invention be "useful" is a much more minimal threshold. For example, a patent may be granted on a chemical compound that shows promise for treating heart disease even if there is no clinical proof of efficacy. Even if the compound is later discovered to have toxic side effects (such that it will not be approved as a drug), the patent is still valid as the patent requirements have been satisfied. Because less information needs to be known regarding a patent application, it can be filed earlier.

In addition, certain patent requirements *mandate* a patent application be filed promptly or else all patent rights may be lost. To meet the requirement that an invention is "new," a patent application must be promptly filed. For example, if a promising new drug compound is described in a scientific publication before a patent application seeking protection of the compound is filed, the publication may constitute "prior art" that bars a patent.[44] A chart comparing the differing requirements of the patent and regulatory system (as well as what is evaluated) appears in Table 1.

b. Effective Patent Term

The different times when patent and regulatory approvals are granted have a direct impact on the effective (as opposed to actual) patent term. The actual patent term refers to the length of patent protection provided under patent laws with the average being seventeen years.[45] However, a drug patent is not *effective* from a commercial standpoint

---

[43] *See In re* Brana, 51 F.3d 1560, 1568 (Fed. Cir. 1995) (noting drugs are eligible for patent protection before they satisfy regulatory approval standards).

[44] Some countries deny patent rights if the invention in a patent application is publicly disclosed at any time before the application. *See, e.g.*, Convention on the Grant of European Patents arts. 54–55, Oct. 5, 1973, 13 I.L.M. 270 (as amended by Revision Act of Nov. 29, 2000), *available at* http://www.epo.org/patents/law/legal-texts/epc.html (very limited exceptions to general rule considering public disclosures to bar patentability). Other countries permit a patent to issue if a patent application is sought within six or twelve months after the public disclosure without any additional limitations. *See, e.g.*, 35 U.S.C. § 102(b) (2006) (one year); Tokkyo Hô [Japan Patent Act], Law No. 121 of 1959, art. 30, *available at* http://www.cas.go.jp/jp/seisaku/hourei/data/PA.pdf (six months); Canada Patent Act, R.S.C., ch. P-4, § 28.2, *available at* http://laws.justice.gc.ca/PDF/Statute/P/P-4.pdf (one year); Patentnyi Zakon Rossiiskoi Federatsii [Patent Law of the Russian Federation] No. 3517-1, Sept. 23, 1992, art. 4 § 1 (six months); Zhuan li fa [Patent Law] (promulgated by the Standing Comm. Nat'l People's Cong. Gaz., Mar. 12, 1984, effective Apr. 1, 1985) (amended 2000), 2000 STANDING COMM. NAT'L PEOPLE'S CONG. GAZ. 491, art. 24 (P.R.C.), *available at* http://www.sipo.gov.cn/sipo_English/laws/lawsregulations/200804/t20080416_380327.html (six months).

[45] The patent term begins the day the patent issues, but ends 20 years from when the patent application was filed. 35 U.S.C. § 154 (a)(2) (2006). Accordingly, the patent term is 20 years less the amount of time to evaluate the patent application. The average amount of time to evaluate a patent is currently 34.6 months; therefore, the average patent term is slightly less than 17 years. U.S. Patents and Trademark Office, Patent Pendency Statistics–FY09, http://www.uspto.gov/patents/stats/patentpendency.jsp (last visited Sep. 3, 2010).

TABLE 1

	Patent vs. Marketing Approval Requirements	
	Patent	Marketing Approval
Basic Requirements	Patentable Subject Matter Useful New & Nonobvious	Safe Effective
What is evaluated	Invention in patent application is compared to "prior art"	Application that includes extensive clinical tests that establish safety and efficacy
Timing	Very early in development (before commercial viability is known)	Close to commercialization

unless and until the patented drug is also granted marketing approval. In other words, the effective patent term is the length of the patent term remaining after marketing approval is granted. The *effective patent term* is usually shorter than the technical patent term because marketing approval usually comes after patent approval, with the average being about eleven years;[46] In some countries, companies have an alternative method of exclusivity provided through the regulatory laws (as discussed later). A visual depiction of how the patent and regulatory approval timing impacts the effect patent term is shown in Figure 3.

Some countries expressly recognize the shorter effective patent term for pharmaceuticals and attempt to provide compensation, whether through direct extension of the patent term or some other type of market exclusivity. The United States provides an actual extension of the patent term, although an application is required during the original patent term and the extension is subject to a complicated formula with a maximum patent term of fourteen years after marketing approval is granted.[47] In the EU, patent-like protection is provided via the Supplementary Protection Certificate (SPC) after the original patent term expires.[48] The basic term of the SPC is five years, but similar to the United States,

[46] *E.g.*, Henry G. Grabowski & John M.Vernon, *Effective Patent Life in Pharmaceuticals*, 19 INT'L J. TECH. MGMT. 98 (2000) (studying drugs approved between 1990 and 1995).

[47] 35 U.S.C. §§ 156, 156(c) (2006); *see also* Australia Patent Act Section 1990, Sect. 70; Japan Patent Act, art. 67, no. 2.

[48] Council Regulation 1768/92 of 18 June 1992, 2001 O.J. (L 182) 2.7.1992, now codified as Regulation No. 469/2009 Of The European Parliament And Of The Council Of 6 May 2009 Concerning The Supplementary Protection Certificate For Medicinal Products, Regulation No. 469/2009 of 6 May 2009, O.J. (L 152) 1 (2009). The SPC must be sought in individual EU Member States.

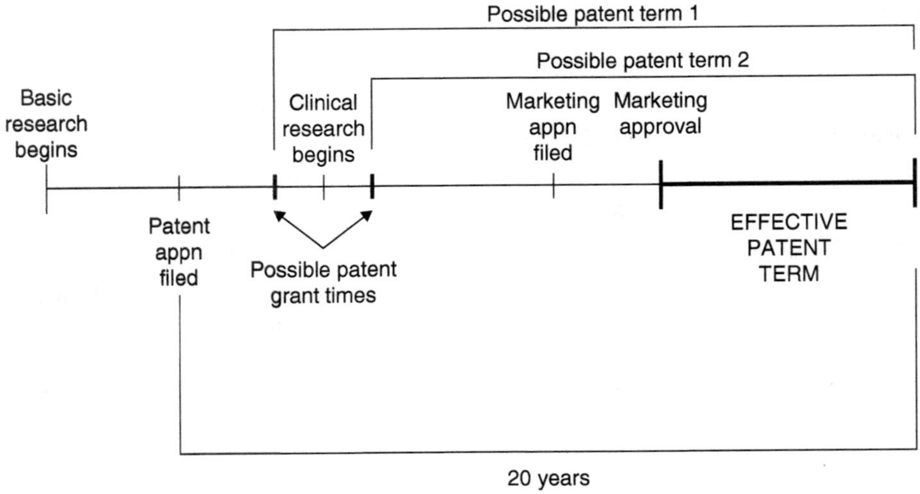

FIGURE 3 Patent versus Marketing Timelines

the total period of exclusivity for the patent and SPC cannot exceed fifteen years. However, drug patents are not always given extended terms, even in developed countries. For example, Canada does not provide for an extension.[49]

2. Legal Barriers to Generic Drugs

Both patent and regulatory laws may present problems for the introduction of generic drugs. Although not all countries presently preclude generic drugs through both patent and regulatory laws, this section will explain how each may impact the introduction of generics.

a. Patent Barriers

To understand how patents impact the creation and approval of generic drugs, it is important to recall what a patent provides, as well as what is required to approve a generic drug for sale. A generic drug will not be approved for sale unless its bioequivalence with the patented product has been established. To do so, the generic company must *make* the patented product. However, a patent permits the owner to exclude others from making the

[49] *E.g.*, Margaret Smith, Government of Canada Parliamentary Research Branch, MR-144E, Patent Term Extensions for Pharmaceutical Products (1997), *available at* http://dsp-psd.pwgsc.gc.ca/Collection-R/LoPBdP/MR/mr144-e.htm; *see also* Ministry of Economic Development, Review of the Patents Act 1953: The Pharmaceutical Patent Term in New Zealand: A Discussion Paper, App. 1 (2005), *available at* http://www.med.govt.nz/templates/MultipageDocumentPage____1727.aspx (noting Canada does not provide patent term extensions; this is also true of Argentina, Brazil, China, Colombia, Ecuador, Hungary, India, Malaysia, Peru, South Africa, and Venezuela).

patented invention. Thus, without some exception to the usual patent laws, a generic manufacturer could not legally take steps to gain regulatory approval during the patent term.

If the generic company cannot take steps to seek regulatory approval during the patent term, the patent owner will benefit from a de facto extension of its patent term due to the several years required to conduct bioequivalence testing and seek regulatory approval. In other words, if the patent owner can completely prohibit creation of the patented drug for bioequivalence testing during the patent term, the patent owner will benefit from an extended period of exclusivity even after the patent expires. Many nations consider this to give the patent owner an undue windfall after the patent term and thereby provide a exception to patent infringement to address this problem. In particular, many nations permit generic companies to make patented drugs during the patent term for the purpose of seeking regulatory approval to sell the drug once the patent expires. This is sometimes referred to as a regulatory approval exception from patent liability. However, this regulatory approval exception may not be enough to allow a generic manufacturer to seek approval during the patent term. Although the exception provides immunity from a patent suit, some countries have other laws that bar a potential generic competitor from seeking approval.

b. Regulatory Barriers

There are two possible barriers to entry under some regulatory laws. First, some countries preclude generic applicants from relying on the clinical data of the original applicant for a certain number of years based on a type of regulatory law called *data exclusivity*. For countries where this exists, the second applicant is prohibited from relying on the original data for a set number of years (usually five to ten), calculated from the date of approval of the prior drug. If data exclusivity extends past the patent term, it will serve as an additional barrier to generic drug entry. A second company can, of course, submit an application for approval of a new drug along with the full complement of clinical tests. However, a company that aims to sell a generic drug most likely does not have the resources to do this testing so that it usually simply waits until exclusivity ends.

In a few countries, generic drugs face yet another barrier to entry under regulatory laws. In particular, some countries preclude approval of generic applications if there are one or more patents that apply to the drug; this type of law is called *patent linkage*.[50] Notably, patent linkage does not refer to a single patent that covers the active ingredient. Rather, patent linkage can bar approval of a generic drug due to subsequent patents. Imagine the active ingredient for a drug is patented and set to expire in 2012. However, a new patent is granted in 2011 on the formulation of the drug, with a patent to expire in 2018. As there is a new patent "linked" to the marketed drug, no generic may be made until 2018.

[50] The US provides patent linkage. Canada and Australia also provide patent linkage as a result of free trade agreements with the US. EU member states, on the other hand, do not provide patent linkage.

The bottom line is that both patent and regulatory laws may impact when potential manufacturers of generic drugs can enter the market. For countries that provide data exclusivity protection, a generic applicant cannot apply for marketing approval until that data exclusivity period has expired. This is true even if there is an exemption from patent infringement for making the bioequivalent drug. Similarly, if a country provides patent linkage, a generic drug will not be approved while there is at least one unexpired patent linked to the bioequivalent original. In other words, a generic drug may only be approved if it is free of patent liability as well as regulatory barriers.

C. MARKET (AND SALES) STRATEGIES

This section aims to describe the final element of the overall picture of how drugs are sold. In particular, this section focuses on the business strategy of both new and generic companies. This strategy includes not only a consideration of what types of drugs to sell, but where to sell them and how to best sustain corporate profits.

1. New Drug Manufacturers Business Model

a. Business Model Overview

Drug companies make their money by charging a premium for new drugs. The drugs are typically patented so that their legal exclusivity enables a company to charge a premium. However, the same drug is not sold at the identical price globally. Rather, companies seek to maximize their profits by selling drugs at the highest possible price in differing regions. The price may be influenced by laws, regulations, or drug-purchasing policies. For example, some nations impose price caps on drugs or profit ceilings on drug companies. Other countries aim to control drug costs by using "reference pricing" to purchase drugs, which means will pay a similar price for drugs that have a similar effect. Still other countries obtain discounts by negotiating in bulk for an entire country. The one country where none of these constraints predominate is the United States.[51] Not surprisingly, drugs cost more in the United States, and the United States accounts for almost half of global sales.[52]

---

[51] Although individual states or private employers may use one or more of these mechanisms to reduce costs, the federal government cannot use its bulk purchasing authority to reduce costs of Medicare drugs—the predominant program of prescription drugs funded by the government. 42 U.S.C. § 1935w-11(i)(2009); *see also* GRETCHEN JACOBSON, SIDATH PANANGALA & JEAN HEARNE, CONG. RESEARCH SERV. RL33802, PHARMACEUTICAL COSTS: A COMPARISON OF DEPARTMENT OF VETERANS AFFAIRS (VA), MEDICAID, AND MEDICARE POLICIES (2007). Legislation was recently proposed to permit the federal government to negotiate Medicare drugs, but that language did not make it into the final law. *Compare* H.R. 3590, 111th Cong. (2009) *with* Patient Protection and Affordable Care Act, Pub. L. No. 111–148, 124 STAT. 119.

[52] *E.g.*, OECD HEALTH POLICY STUDIES, PHARMACEUTICAL PRICING POLICIES IN A GLOBAL MARKET 11 (2008), http://www.centad.org/seminar/2.%20Price%20regulation/OECD%20Pharma%20pricing%20policies.pdf (noting the United States accounted for 45% of global sales in 2006); Press Release, IMS Health Inc., IMS Health Reports Global Pharmaceutical Sales by Region, 2006 (Mar. 20, 2007), *available at* http://www.

Companies selling new drugs seek to maximize profits by keeping each distinct market separate. In other words, because companies can charge more for the same drug in the United States than in Canada (where price caps exist), companies do not want drugs sold in Canada to be imported into the United States. Ideally, drug companies would like to consider such imports a violation of their patent rights. There is a heated controversy about whether this is appropriate, which is given further elaboration in Chapter Two.

Although companies do engage in price differentiation, the price difference for the same drug in a wealthy versus poor country may not be substantial. In some countries with relatively low per capita income, including middle-income countries such as Brazil and India, brand name companies may sell at prices commensurate with those of wealthier countries—even though this means selling to a very small percentage of the population.[53] If the differential is small, it may reduce incentives to buy drugs solely to attempt to sell them in a higher profit area.

b. Drug Development

The types of new drugs developed are a function of several factors. First, diseases and conditions of citizens in wealthier countries receive greater focus as these markets can yield substantial profits. Second, even among diseases of wealthy markets, there is a focus on drugs that address chronic conditions as well as on developing and maintaining "blockbuster" drugs (defined as drugs that each earns over $1 billion per year,[54] although many blockbusters can far exceed this figure). In fact, in 2006, the top-ten selling drugs produced sales of $60 billion—an average of $6 billion per blockbuster.[55]

New drugs are primarily targeted to the problems of countries and regions that comprise the majority of corporate profit: the United States, Europe, and Japan.[56] To the extent other countries have similar conditions, developed drugs will also treat their needs. For example, all countries need drugs to treat heart disease, hypertension, and cancer.

In addition to focusing on conditions that afflict wealthy countries, there is a business incentive to address chronic diseases that require continued treatment versus those for which a drug provides a complete cure.[57] A chronic condition may be a severe one that

---

imshealth.com/deployedfiles/imshealth/Global/Content/StaticFile/Top_Line_ Data/Global%20Pharmaceutical%CC20Sales%CC20by%CC20Region%202006.pdf.

[53] *See, e.g.*, Peter Hammer, *Differential Pricing of Essential AIDS Drugs: Markets, Politics and Public Health*, 5 J Int'l Econ. L. 883, 888 (2002) (noting that only the wealthiest 10% of citizens in developing countries typically purchase drugs—and that these drugs are often sold at higher prices than in developed countries).

[54] *E.g.*, OECD Health Policy Studies, *supra* note 52, at 60–61; Joshua Cohen & Kenneth Kaitin, *Follow-On Drugs and Indications: The Importance of Incremental Innovation to Medical Practice*, 15 Am. J. Therapeutics 89, 91 (2008).

[55] OECD Health Policy Studies, *supra* note 52, at 61.

[56] *Id.* at 58 (noting sales in the United States, Europe, and Japan account for 80% of sales).

[57] Countries recognize these business realities and sometimes aim to work with the profit incentive to promote innovation into "orphan" diseases. *See, e.g.*, Orphan Drug Act, Pub. Law No. 97-414, 96 Stat. 2049 (1982)

may never be cured, such as HIV or multiple sclerosis. However, it could also be one that is not life-threatening, but for which there is a perceived consumer demand and interest (such as depression, asthma, or even male pattern baldness).

Another important business focus of drug companies is on developing and maintaining "blockbuster" drugs. An example of a blockbuster is the cholesterol treatment drug Lipitor, marketed by Pfizer, with annual sales of $12 billion.[58] Corporate profits tend to rely substantially on sales of one or more blockbuster drugs as other drugs are reported to not cover their own development costs, let alone those of failed products that never reach the marketplace.[59] Corporate profits often plummet when a blockbuster drug loses patent protection and is subject to generic competition.[60] For example, the blockbuster antidepressant Prozac provided a quarter of Lilly's revenue during patent protection, but within a year of generic competition, sales of Prozac plunged nearly 70 percent.[61] A com-

(codified as amended at 21 U.S.C. § 360aa–ee, 26 U.S.C. § 45(C), 42 U.S.C. § 236); Regulation No. 141/2000 of 16 Dec. 1999 on Orphan Medicinal Products, 2000 O.J. (L 18) 1. However, such legislation is controversial—critics note it not only results in high prices, but that companies have manipulated the terms of the statute to gain protection for drugs that apply to a broad population. *See, e.g.*, *Anticompetitive Abuse of the Orphan Drug Act: Invitation to High Prices: Hearings Before the Subcomm. on Antitrust, Monopolies & Business Rights of the S. Comm. On the Judiciary*, 102d Cong., 2d Sess. 198 (1992); Jess G. Thoene, *Curing the Orphan Drug Act*, 251 SCIENCE 1158, 1158 (1991); Marlene E. Haffner, *Orphan Products—Ten Years Later and Then Some*, 49 FOOD & DRUG L.J. 593, 596–98 (1994).

[58] Shannon Pettypiece et al., *Pfizer's Offer for Wyeth is Being Weighed by Boards (Update 2)*, BLOOMBERG, Jan. 26, 2009, http://www.bloomberg.com/apps/news?pid=newsarchive&sid=aafnhn_9.58M; *Annual Review 2008: Key Medicines and Their Performance*, PFIZER (Oct. 14, 2010) http://www.pfizer.com/investors/financial_reports/annual_reports/key_medicines.jsp.. Another highly profitable blockbuster drug is Plavix, which has annual sales of about $5 million. *See* Jennifer Corbett Dooren: *Study: Other Drugs May Limit Plavix's Effectiveness*, WALL ST. J., Mar. 3, 2009, http://online.wsj.com/article/SB123609179058219021.html; Peter Loftus & Shirley Wang, *Bristol Posts a Profit as Drugs Make Gains*, WALL ST. J., Jan. 28, 2009, at B3. *See also* Pharmaceutical Sector Inquiry Preliminary Report 34 tbl.2 (European Commission DG Competition, Working Paper, 2008), *available at* http://ec.europa.eu/competition/sectors/pharmaceuticals/inquiry/preliminary_report.pdf (listing blockbusters); Michael Bartholow, *Top 200 Prescription Drugs of 2009*, PHARM. TIMES, May 11, 2010, http://www.pharmacytimes.com/issue/pharmacy/2010/May2010/RxFocusTopDrugs-0510.

[59] *E.g.*, NATIONAL INSTITUTE FOR HEALTH CARE MANAGEMENT, CHANGING PATTERNS OF PHARMACEUTICAL INNOVATION 4 (2002); PHARMACEUTICAL RESEARCH AND MANUFACTURERS OF AMERICA, DELIVERING ON THE PROMISE OF PHARMACEUTICAL INNOVATION: THE NEED TO MAINTAIN STRONG AND PREDICTABLE INTELLECTUAL PROPERTY RIGHTS 9 (2002), *available at* http://www.ftc.gov/os/comments/intelpropertycomments/phrma020422.pdf.

[60] *E.g.*, Barbara J. Williams, *A Prescription for Anxiety: An Analysis of Three Brand-Name Drug Companies and Delayed Generic Drug Market Entry*, 40 NEW ENG. L. REV. 1, 1 (2005) (noting that generic drug entry means prices that are 25 to 50% lower; patent expiration and generic entry signal that the "days of recouping expenses and reaping profits are over"); *see also* Grabowski & Vernon, *supra* note 46 (noting that prices drop to thirty-seven percent of original level two years after the entry of generic manufacturers).

[61] *E.g.*, Benjamin G. Druss et al., *Listening to Generic Prozac: Winners, Losers, and Sideliners*, 23 HEALTH AFFAIRS 210, 211, 215 (2004) (noting that during the five months following the introduction of a generic form of Prozac (fluoxetine), 65.8% of new fluoxetine prescriptions were for generics, and 73.8% of patients on Prozac switched to the generic form, while Eli Lilly's net company profits dropped 22%).

pany with a successful blockbuster drug may also try to sustain blockbuster revenues by developing a close chemical variant. Last but not least, potential blockbuster drugs may be acquired when multinational companies buy other companies with promising portfolios of drugs in development.

Although blockbuster drugs provide a high return on development costs, developing a blockbuster does not need to start from the beginning stages of drug discovery. In particular, rather than develop NMEs from scratch, companies may develop a new dosage or new formulation of an *existing* blockbuster to retain blockbuster profits. For example, as the antidepressant blockbuster Prozac was about to expire, a once-weekly formula was patented and approved. Similarly, a once-weekly dosage of the osteoporosis blockbuster Fosomax was developed after the original daily version. Another popular technique to expedite the development process and maximize profits is to create minor variations of existing drugs that take a substantially shorter time to develop than the benchmark of ten to twelve years. For example, optical isomers (mirror images) of previously approved compounds have been successfully developed and marketed as "new" drugs. As discussed earlier, the newer allergy drug marketed as Xyzal is a mirror image of the patent-expired drug sold as Zyrtec. Similarly, whereas the antidepressant sold as Celexa contained a mixture of two mirror halves, one of which did not contribute to its antidepressant properties, the newer drug that contains only the active half is now sold as Lexapro.

The focus on blockbusters coincides with criticism that companies are not developing enough truly innovative drugs. Only about a third of recently approved drugs are new chemical compounds.[62] The majority of drugs are variations of previously approved drugs, which represents incremental rather than monumental innovation. Critics charge pharmaceutical companies are engaging in "evergreening" of patents by patenting small variations that seek to perpetuate patent profits.[63] In addition, some studies indicate incremental innovation may provide little therapeutic benefit to patients.[64] However, the incremental approach to innovation may also be viewed as a logical approach to maintain corporate profits and shareholder approval in a field that is otherwise unpredictable. Companies defend their practice as a legitimate "life cycle management" of patent terms that is both legal and socially beneficial as even slight chemical variations may have differ-

---

[62] *See, e.g,* Congressional Budget Office, Pub. No. 2589, Research and Development in the Pharmaceutical Industry 7, 14-15 (Oct. 2006).

[63] *E.g.,* John R. Thomas, Cong. Research Serv., RL 31340, Patent "Evergreening": Issues in Innovation and Competition 7-10 (Nov. 13, 2009). In addition, even among NMEs, many recently approved drugs do not provide major advances in treatment. *See, e.g.,* Molly Redfield Wart, KEI Research Note 2010:3, Notes on the 2004 to 2009 United States Food and Drug Administration Approval of New Molecular Entities, (2010), http://keionline.org/misc-docs/research_notes/kei_rn_2010_3.pdf.

[64] *E.g.,* Silvio Garattini & Vittorio Bertele, *Efficacy, Safety and Cost of New Anticancer Drugs*, 325 Brit. Med. J. 269, 270-71 (2002); J.A. Staessen et al, *Cardiovascular Protection and Blood Pressure Reduction: A Meta-Analysis*, 358 Lancet 1305, 1312-13 (2001).

ent side effect profiles that are beneficial to patients.[65] This may be particularly true for some types of medications where response rates tend to be low, such that patients may respond very differently to drugs within the same therapeutic class. Indeed, some supporters of this practice note that an original breakthrough drug is often replaced with an incremental improvement; indeed, a number of drugs on the WHO's list of essential drugs are reported to be such incremental improvements.[66] In addition, they note patients can elect whether to buy the newest patented drug or the drug whose patent has expired—although companies may simultaneously attempt to influence patient decisions through promotional efforts, as discussed in the following section.

c. Market Exclusivity

In addition to selecting what drugs are developed, companies can increase commercial success by successfully utilizing their legal tools of patent and regulatory protection. In addition, strategic marketing and promotion are used in conjunction with legal tools.

Patents on multiple aspects of a marketed drug help protect corporate profits. There are typically multiple patents related to a single commercial pill. The most valuable patent is the one on the active (chemical) ingredient. However, there may be other patents on the formulation, such as a coated tablet. Additional possible patents include the process of making the active ingredient or formulation. As noted earlier, multiple patents can be critical to forestalling generic competition in a country that recognizes patent linkage—the situation where generics are not approved until all related patents have expired. In particular, companies often engage in a practice of sequential filing of patents to ensure they have the longest possible time before a generic drug will be approved.

In addition, in some countries that recognize data exclusivity, new uses of existing drugs are entitled to market exclusivity for a certain number of years. This exclusivity will exist regardless of whether the new use is patented or if a patent has expired, as discussed more thoroughly in the chapter focused on data exclusivity.

---

[65] *E.g.*, Martin A. Voet, The Generic Challenge: Understanding Patents, FDA & Pharmaceutical Life-Cycle Management (2d ed. 2008). Some in the pharmaceutical industry argue that while incremental innovation may have some overlap with sequential filing of patents that is referred to as evergreening, the two concepts are distinct.

[66] Albert I. Wertheimer & Thomas A. Santella, Competitive Enterprise Institute, Pharmaceutical Evolution: The Advantages of Incremental Innovation 9 (April 2009), *available at* http://cei.org/cei_files/fm/active/0/Wertheimer%20and%20Santella%20-%20Pharmaceutical %20 Evolution.pdf

d. Marketing

Companies can also successfully market their drugs through a combination of clinical studies and advertising. In fact, drug companies have been criticized for spending more money on marketing expenses than on research.[67]

For example, although a drug only needs to establish safety and efficacy for a given indication, a company may provide clinical tests to become the first drug approved for a given indication that can then be marketed—even if that indication does not necessarily provide a substantial therapeutic benefit. For example, the allergy drug marketed as Zyrtec was the first approved and promoted allergy drug for indoor and outdoor allergies; other allergy drugs actually are effective for this purpose as well, but only the manufacturer of Zyrtec did the clinical tests to obtain this approval.[68] Schering-Plough, the maker of blockbuster allergy drug Claritin, took a similar approach when Claritin was about to go off patent in 2002 and received FDA approval for a variation of Claritin—marketed as Clarinex—that was advertised as an improvement for treating year-round indoor and outdoor allergies.[69]

Sometimes a company conducts clinical tests to show its product is superior to others. This is not usually required to establish a drug is safe and effective as the standard can be satisfied by comparing a drug to a placebo.[70] However, clinical studies demonstrating some advantage in efficacy or decrease in side effects may be beneficial in persuading consumers (as well as others who purchase or subsidize drugs) that the new drug is worthwhile. Knowing the possible risks, companies sometimes compare their drug with a competitor's drug at a lower dosage level.[71] Although careful analysis of the scientific studies would reveal the study is not a fair comparison, this level of detail may not reach actual consumers.

Companies can and do market new drugs to doctors in a variety of ways. Marketing efforts aimed at doctors include not only advertisements in medical journals, but office visits, free sample products, and medical education. Companies may characterize their

---

[67] *See, e.g.,* Marc-Andre Gagnon & Joel Lexchin, *The Cost of Pushing Pills: A New Estimate of Pharmaceutical Promotion Expenditures in the United States,* 5 PLOS MED. 29, 32 (2008). The advertising expenses are particularly large in the United States where direct-to-consumer advertising is permissible.

[68] *See, e.g.,* Press Release, PR Newswire Service, Announcing Availability of Zyrtec for Seasonal and Perennial Allergies (Mar. 14, 1996); Elyse Tanouye, *Allergy Drugs Wage Bitter War of the Noses,* CHI. SUN-TIMES, May 26, 1996, at 40.

[69] *E.g.,* Gardiner Harris' *Schering-Plough Wins New Approval For Allergy Drug,* WALL ST. J., Feb 12, 2002, at B10; Stephen S. Hall, *The Claritin Effect—Prescription for Profit,* N.Y. TIMES MAG., Mar. 11, 2001, at 40.

[70] In some limited situations it is considered unethical to compare a new drug to a placebo, such as with antibiotics. In those cases, the comparison must be to the ordinary treatment. However, even when this is done, the new drug may be approved without being superior; rather, it may simply need to be *no worse* than the standard treatment. CONGRESSIONAL BUDGET OFFICE, *supra* note 62, at 23-24; Garattini & Vittorio, *supra* note 64, at 269–70.

[71] *E.g.* ANGELL, *supra* note 6, at 80–82.

marketing as a helpful tool to inform busy doctors of new treatments and their benefits. But this promotional activity is controversial. It is not an objective presentation of information, but can nonetheless influence doctors—even though they may not acknowledge it.

In addition, although only the United States and New Zealand permit direct advertising to consumers of prescription drugs, as noted earlier, the US market is a very important one to drug companies. Since 1977 when the United States clarified what constitutes permissible advertising to consumers, [72] there has been a substantial increase in ads—along with evidence that these ads promote sales.[73] The ads can prompt patients to request marketed drugs from physicians and thus dovetail with a company's promotions to physicians. In addition, even in countries that do not technically permit direct advertising, companies may have mechanisms to provide information concerning their products to consumers, as further discussed in Chapter 8.

Where advertising to consumers is permissible, companies may engage in heavy and expensive advertising campaigns to persuade consumers to switch to a new (often patented) variant drug just as the originally patented drug is about to become available in generic form. For example, just as Prilosec was about to become generic, its manufacturer Astra Zeneca launched a very similar drug under the new brand Nexium. Nexium was heavily marketed to consumers as an improvement based on studies that actually used higher doses of Nexium as compared to Prilosec.[74] In addition, Astra Zeneca priced Nexium slightly below Prilosec while both were on the market, with additional discounts and free samples provided to hospitals and doctors.[75] Most Prilosec patients switched to Nexium and thus profits continued to flow to Astra Zeneca.[76]

2. Generic Drug Manufacturers Business Model

Generic drug manufacturers have a very different model for making money than the multinational companies that make new drugs. As generic drug manufacturers will not be the only seller of a drug, they cannot price drugs at a premium. However, they may still be profitable. After all, unlike the companies that make new drugs, generic manufacturers have essentially no research costs. Generic companies also have no advertising costs

[72] *E.g.*, "Draft Guidance for Industry on Consumer-Directed Broadcast Advertisements," 62 Fed. Reg. 43,171 (August 12, 1997) (notice by the FDA of standards for direct-to-consumer advertising).

[73] *E.g.*, Dhaval Dave & Henry Saffer, *The Impact of Direct-to-Consumer Advertising on Pharmaceutical Prices and Demand*, (National Bureau of Economic Research, Working Paper No. 15969, 2010).

[74] *E.g.* ANGELL, *supra* note 6, at 77–79; Yoshitani & Cooper, *supra* note 20, at 394.

[75] *E.g.*, ANGELL, *supra* note 6, at 77–78. Similarly, Schering-Plough patented the active metabolite of the drug sold as Claritin, and after receiving marketing approval for Clarinex to treat indoor and outdoor allergies, marketed it as an improvement under the name Clarinex at slightly lower prices. *Id.* at 79–80.

[76] Gardiner Harris, *As a Patent Expires, Drug Firm Lines Up Price Alternative*, WALL ST. J. June 6, 2002, at A1; Gardiner Harris, *Two Fronts in Heartburn Market Battle*, N.Y. TIMES, Aug. 20, 2003, at C12; Neil Swidey, *The Costly Case of the Purple Pill*, BOSTON GLOBE MAG., Nov. 17, 2002, at 11.

because unlike new drugs, consumers are already aware of the utility of generics. Indeed, generic drugs are often long-awaited entrants to a market that need no introduction to consumers. In addition, generic manufacturers have the luxury of focusing only on making and selling drugs that have already been commercially successful in the marketplace.

However, although generic drug companies have relatively minimal costs in bringing a drug to market, they need a high volume of sales to compensate for their not having exclusivity in the marketplace.[77] At a minimum, a generic company will have competition from the original manufacturer of the drug. However, even if there are only two suppliers, the generic manufacturer cannot expect to get half the sales because it is a new entrant to a field that has been long occupied by a single manufacturer. But if there are only two suppliers, the generic drug can be sold at a higher cost; studies indicate that when there is only one generic competitor, the generic drug will offer only a modest discount as compared to the originator drug.[78] Given the small difference, consumers are likely to choose the more expensive brand name with which they are familiar.[79] However, if there are multiple generic manufacturers, the price will further drop and sales volume will be especially important to attain profits. Some studies show prices are most competitive when there are at least five competitors.[80] The most dramatic price reduction as a result of generic competition has been seen with HIV treatment. The initial regiment most people in developing countries today receive costs 99 percent less than it did ten years ago due to generic competition from multiple manufacturers in India, Brazil, and Thailand.[81] Although generic competition will generally not result in such intensive price drops, substantial discounts are common for blockbuster drugs where there tends to be more generic entrants.[82] In general, the more generic entrants, the lower generic companies must price their goods—and the more important it is that they sell in high volumes.

---

[77] Some generic manufacturers have been successful in obtaining their own patents on formulations of existing drugs and thus attained a measure of market exclusivity. However, this is the exception rather than the rule.

[78] *E.g.*, R.G. Frank, *The Ongoing Regulation of Generic Drugs*, NEW ENG. J. MED 1993, 1995 (2007) (indicating the average price drops to 94% of the original with a single generic, but to 52% with two manufacturers); CONG. BUDGET OFFICE, HOW INCREASED COMPETITION FROM GENERIC DRUGS HAS AFFECTED PRICES AND RETURNS IN THE PHARMACEUTICAL INDUSTRY 28 (1998) (suggesting the first generic enters the market at 70–80% of the price of the original drug).

[79] In one recent study of insured patients, although the majority recognized generic drugs were a "better value," that did not translate into a personal preference for generics. William Shrank et al., *Patients' Perceptions of Generic Medications*, 28 HEALTH AFFAIRS 546, 549, 553–54 (2009).

[80] David Reiffen & Michael Ward, *Generic Drug Industry Dynamics* (U.S. Federal Trade Comm'n Bureau of Econ., Working Paper 248, 2002), http://www.ftc.gov/be/workpapers/industrydynamicsreiffenwp.pdf (suggesting eight competitors are necessary for significant price drops); Frank, *supra* note 78 (noting price drops to a third or lower of the original when there are five or more generics).

[81] MÉDECINS SANS FRONTIÈRES, UNTANGLING THE WEB OF ANTIRETROVIRAL PRICE REDUCTIONS 9, 33 (13th ed. 2010), *available at* http://utw.msfaccess.org/downloads.

[82] *E.g.*, Henry Grabowski & Margaret Kyle, *Generic Competition and Market Exclusivity Periods in Pharmaceuticals*, 28 MANAGERIAL & DECISION ECON. 491, 494 (2007) (noting that for drugs that have more than $500 million

IV. Conclusion

The full picture of drug development in marketing is a complex and nuanced one. New drugs do indeed take a substantial time and money to develop to market. However, many drugs that seem new to consumers may have shortcuts in development while still being fully protected by patent and possibly regulatory laws. This discussion should provide a useful overview of drug development so as to better understand subsequent chapters on more detailed aspects of patent and regulatory protection of drugs in the international arena.

in sales, there are likely more than seven competitors after year one, whereas those with sales of only $50–100 million had on average only between two and three generic entrants after the first year).

2 A Limitation on the Patent Right to Exclude: An Introduction to International Exhaustion

This chapter provides an important clarification to the usual patent rights that were first explained in Chapter 1. In particular, it explains that patent owners may sometimes lack the ability to exclude imports of patented drugs they previously sold in another country. Whether the patent owner lacks this ability depends on whether a nation rejects a legal doctrine called *international exhaustion*, to exclude what are referred to as *parallel imports*. This chapter explains the controversy concerning this doctrine, and how it impacts the commercial sale of drugs. Consistent with this book's focus on presenting current realities, this chapter focuses on existing laws that are often driven by patent-owning companies; as will be explained, such companies are opposed to international exhaustion of patent rights. This chapter provides a helpful background to other chapters, as controversy over international exhaustion often lurks behind discussions of other international issues. This chapter answers some key questions concerning international exhaustion, such as:

- What is international exhaustion?
- What are the reasons for and against recognizing international exhaustion?
- Why do patent-owning drug companies generally oppose international exhaustion?
- How do companies effectively market their drugs in a world where some countries recognize international exhaustion?

I. Introduction 36
II. Explaining Exhaustion 37
A. Domestic Exhaustion 37
B. From Domestic to International Exhaustion: An Uneasy Transition 39
C. Deciphering Related Terms 43
III. Principal Arguments 45
A. For International Exhaustion (and Parallel Imports) 45
B. Against International Exhaustion (and Parallel Imports) 46
1. Alleged Benefits of Price Discrimination 47
2. Alleged Health and Safety Concerns 47
IV. Commercial Implications 48
V. Conclusion 51

I. Introduction

This chapter explains how domestic patent laws may impact international trade in patented drugs as well as the profits of drug companies. Although many factors affect free trade, the issue here is whether patent laws permit—or prevent—the free trade of drugs first sold by the patent owner in another country. The relevance of this can be illustrated by considering the following scenarios:

- The owner of a patent on an HIV drug in Rwanda is considering selling the drug there at below-market prices in recognition of the AIDS epidemic and poverty in the country, as well as public pressure to do so. However, the patent owner is hesitant to do so for fear that some of these drugs could be purchased and resold in markets where the patent owner typically sells the same drug for a much higher price.
- A cancer drug is patented in both Canada and the United States. However, the drug costs less in Canada (where drugs are subject to price caps). The patent owner should be able to sell the patented drug for a higher price in the United

States—but only to the extent it can prevent the drugs sold in Canada from being imported into the US market at a lower price.

In both of these scenarios, the patent-owning drug company has a clear interest in ensuring lower-priced drugs in one market do not enter a higher-price market where the patent owner sells the identical drug. Consumers generally prefer access to lower-priced goods, even if originally from another market. Whether they have such access depends on the doctrine of international exhaustion.

One of the basic rights granted to a patent owner is the right to exclude imports of the patented product. It would seem this basic right would protect the patent owner in the above scenarios. However, some countries have an exception to this usual right. In particular, some nations' patent laws bar a patent owner from excluding importation of a patented product that the patent owner first sold in another country.[1] This is referred to as *international exhaustion* of patent rights because the first unrestricted global sale of the patented product is considered to have exhausted the patent owner's rights to that product. Some suggest exhausting the patent owner's rights is fair not only because has the patent owner already received a profit, but also because consumers are entitled to the lowest possible global price. On the other hand, some suggest that exhaustion may reduce prices in the short term, but ultimately increase overall prices because companies would uniformly price goods higher. These arguments will be explained later in more detail after first laying the foundation.

II. Explaining Exhaustion

Because international exhaustion can be confusing in addition to controversial, this part begins with a brief review of patent rights along with the related and relevant concept of domestic exhaustion. International exhaustion will then be explained, followed by a discussion of terms that are frequently used in conjunction with discussions of international exhaustion.

A. DOMESTIC EXHAUSTION

A patent owner's usual rights include the right to exclude others from making or distributing the patented product within the patent-granting nation. The right to control distribution covers a variety of activities, including selling, offering for sale, and importing. However, when a patented product is first sold with the patent owner's authorization—

[1] As discussed in Chapter 1, a patent is granted by a nation and only has effect within that nation. Accordingly, in this scenario, the reference to a "patented product" is to a product that has been successfully patented in two separate countries.

either from the patent owner himself or a licensee with no contractual limitation—the patent owner generally loses the ability to further control that product.

Figure 4 provides a visual depiction of two different scenarios of domestic exhaustion that occur in country A. In the first case, the patent owner ("P") sells the drug ("d") to a wholesale company ("W"). At this point, the patent owner's rights in the drug are exhausted. Accordingly, when the wholesale company sells that drug to the consumer (denoted as a stick figure), there is no patent infringement. Although sale of a patented drug usually violates the patent owner's right, there is no such violation if the patent owner's rights have been exhausted, as in this case. In the second scenario, the patent owner P does not sell the drug directly, but instead licenses a company ("L") to make and sell the drug. L is thus a company authorized to sell the patented drug. When L sells the drug to the wholesale company W, the patent rights are similarly exhausted before reaching the consumer. Accordingly, the end result is the same whether the patent owner originally makes and sells the patented good, or the patent owner authorizes a licensee to do the same—after the first authorized sale, the patent owner has no ability to control further distribution of the product.

The reason a patent owner's right to domestically control the patented item is exhausted after the first legitimate and unrestricted sale is that the patent owner is presumed to have obtained an adequate economic return with the initial sale, such that further control of the patented item is not appropriate. The purchaser of the patented item is then free to use the patented item or sell it to others—despite the fact that using and selling a patented invention typically gives rise to liability. This principle is enshrined in domestic laws and referred to as *domestic exhaustion*; alternatively, it is referred to as *national exhaustion*.[2] After all, the rights of the patent owner are exhausted only with respect to goods that are sold domestically in the patent-granting nation.

Before proceeding further, let us consider an example of how domestic exhaustion works. Assume that patent owner Novexis sells a box of a patented drug for attention-deficit disorder to wholesaler X without any restrictions. Once the sale is completed, patent owner Novexis has exhausted its patent rights in that box and cannot control what X does with it. X is free to use the drug or resell it.[3] The patent owner's usual right to exclude others from using or selling the patented invention does not apply to X because Novexis's rights in that box have been exhausted.

However, Novexis *can* prevent X from *making* a duplicate of the drugs that X bought. Exhaustion of rights only applies to those specific goods sold by the patent owner. The right to exclude others from making goods is not impacted by the doctrine of exhaustion. Accordingly, the patent owner still retains the right to exclude X from making a new version of the patented good. If X makes the patented product without authorization,

---

[2] Countries may provide an explicit exception in the same statute that provides patent rights, or recognize domestic exhaustion through case law.

[3] Of course, there may be other laws that regulate whether it is permissible for a buyer of a patented drug to give the patented drug to another.

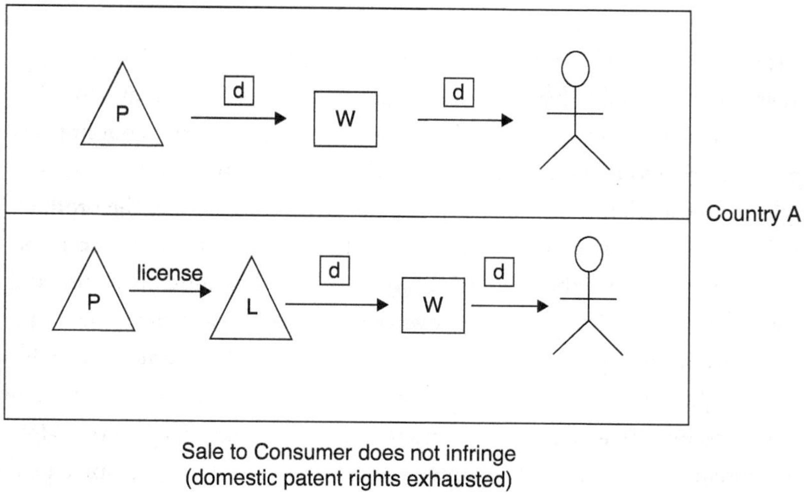

FIGURE 4 Domestic Exhaustion

X would violate Novexis's patent rights. In addition, the patent owner Novexis can and likely will make more of the patented drug—the exhaustion of its rights only applies with respect to goods the patent owner sold. Exhaustion has no impact on the patent owner's ability to make and sell additional copies of the patented good. However, because patent owners know and expect they will not be able to control further distribution of their patented drug, they charge a high price for its first sale.

Before moving on to international exhaustion, the following questions should help recap the principles of domestic exhaustion:

Q: Does domestic exhaustion of patent rights bar the patent owner from suing others who make the patented drug?

A: No. Domestic exhaustion of patent rights only applies to sales of products from the patent owner itself (or those authorized by the patent owner).

Q: Does domestic exhaustion of patent rights prevent the patent owner from making more products like the one that is exhausted?

A: No. The exhaustion principle refers to rights in a *single product* (for example, a single box of Novexis's attention deficit disorder drug) that are exhausted when the patent owner sells it. The patent owner can make unlimited additional products—and sell them subject to the same domestic exhaustion principle.

B. FROM DOMESTIC TO INTERNATIONAL EXHAUSTION: AN UNEASY TRANSITION

Domestic exhaustion is the basis for the related, but more controversial doctrine of *international exhaustion*. Just as the first unrestricted sale of a patented product exhausts the patent owner's right to control the product domestically, *international* exhaustion limits the patent owner's right to control a product that has been sold without restriction anywhere in the world (i.e., the first *global sale*). The first global sale is considered to exhaust the patent owner's right to continue to control the product it sold because the patent owner is presumed to have received an adequate reward with that first sale.

International exhaustion has a similar effect as domestic exhaustion, but impacts a different patent right. Domestic exhaustion provides an exception from the patent owner's usual right to restrict domestic use or sale of patented products when the products were first sold by the patent owner. In contrast, international exhaustion is an exception from the patent owner's usual right to exclude *imports* (as well as subsequent use and sale) when the first global sale is by the patent owner. However, because the initial import is a prerequisite to any subsequent use or sale, the focus here (and for patent owners) is on the importation right. The importation right is important because, from the patent owner's perspective, a country that adopts international exhaustion limits its preferred global distribution system. As discussed in Chapter 1, companies prefer to maximize profits by selling the same drug at the maximum price in each market, which results in differing global prices. However, companies may not be able to exclude lower-priced imports that it first sold without restriction in another country if the importing country recognizes international exhaustion.

Importantly, domestic and international exhaustion only limit some of the usual patent rights: domestic exhaustion limits only the patent owner's right to control *some* sales, and international exhaustion only impacts the patent owner's right to control *some* imports. For both types of exhaustion, the patent owner rights are limited only with respect to products that the patent owner itself sold or authorized to be sold. A patent owner always retains the right to exclude patented products made or sold by *others*. In the domestic context, the patent owner can exclude others from making the patented product. Similarly, a patent owner can always bar imports of a patented product made by another.

Although domestic and international exhaustion have many similarities, one major distinction is that domestic exhaustion is universally recognized, but international exhaustion is not. A patent owner can not control a patented product it sold domestically–absent an agreement to the contrary . However, a patent owner's ability to control a patented product in global trade depends on whether the *importing* country recognizes international exhaustion. If so, the patent owner cannot bar imports of a patented product it first sold elsewhere. Conversely, if the importing country *rejects* the principle of international exhaustion, the patent owner will have the right to bar importation of its patented product first sold in another country.

Figure 5 emphasizes that in evaluating international exhaustion, the focus is on the laws of the importing country, which is noted as "Country B." As shown here, the patent owner first sells the patented drug "d" in Country A. That drug is then imported to Country B. Whether the importation in Country B constitutes patent infringement depends on whether Country B recognizes or rejects international exhaustion. The border of Country B is bolded to emphasize that only the importing country's laws (not the exporting country) are relevant to this issue. If Country B recognizes international

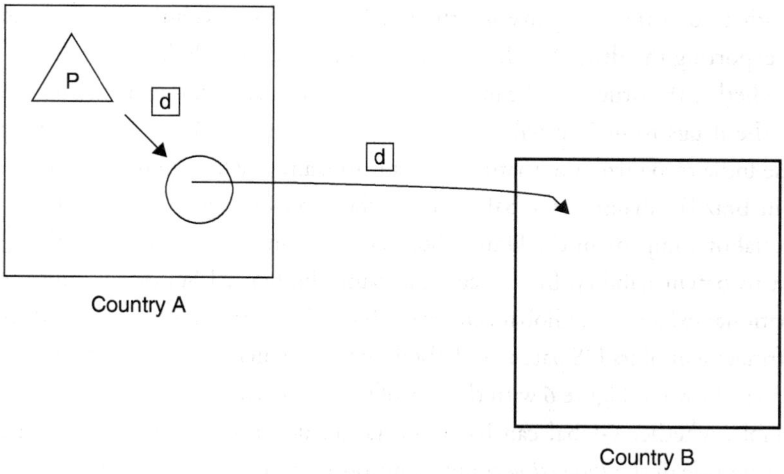

FIGURE 5 International Exhaustion—Look to Importing Country (B's) Law

exhaustion, the importation does not violate any of the patent owner's rights in Country B. If, however, Country B rejects international exhaustion, the importation would constitute an infringement of the patent owner's rights in Country B.

The following questions recap some basic international exhaustion.

Q: What do domestic and international exhaustion have in common besides using the word exhaustion?

A: They are both exceptions to the patent owner's usual right to exclude all others from a host of activities relating to the invention. In addition, the policy reason is similar—in both cases, the patent owner is presumed to have received the full benefit of the patent with the first sale (whether that took place domestically or globally).

Q: Do all countries recognize international exhaustion?

A: No! This is a highly contentious issue on which countries have been unable to agree.

Q: Does international exhaustion only exist if recognized by both the exporting and importing country?

A: No. International exhaustion refers to the *importation right*; accordingly, only the law of the importing country is relevant.

A hypothetical may be helpful to further reinforce how the doctrine of international exhaustion determines whether an imported product violates national patent laws. Assume multinational drug company Global has patents over a drug in all relevant countries. Global sells a bottle of patented drugs in Brazil to businessman Bob. Whether those drugs may be exported to another country without patent liability depends on whether those other countries recognize international exhaustion. Technically, the issue is not about exporting the drugs (as there is no patent right to exclude exports), but, rather, about whether the drugs may be imported. An individual or company in India that purchases the drugs from Bob will be able to import them without any patent problem because India recognizes international exhaustion, and the first global sale of the pills to Bob (in Brazil) exhausted Global's right in those pills under Indian law. However, an individual or company in the United States that attempts to do the same thing will be subject to patent liability. In particular, because the United States does not recognize international exhaustion, Global can legally block importation of those goods by suing for infringement of its US patent and obtaining an injunction. A visual depiction of this scenario is shown in Figure 6 with the box of patented drugs labeled as "d."

Notably, whether Global can bar importation depends solely on the international exhaustion rule of the *importing* country. Accordingly, Figure 6 bolds the borders of the importing countries to underscore it is the laws of those countries that are critical to this issue. It is of no relevance whether the exporting country (Brazil) recognizes international exhaustion, as there is never a right to bar exports. Global's patent rights can only control import, but not export of goods. There is never a patent right to exclude others from exporting. So, Global's ability to control distribution of its goods will always focus on the importing country. Global can use its patent to bar importation of goods that it previously sold in another country if the importing country does *not* recognize international exhaustion.

Although Global may not be able to bar importation of drugs at the border of every country, it can always bar a variety of other activities that do not involve goods first made by the patent owner. In particular, Global may bar others from making, selling, or offering

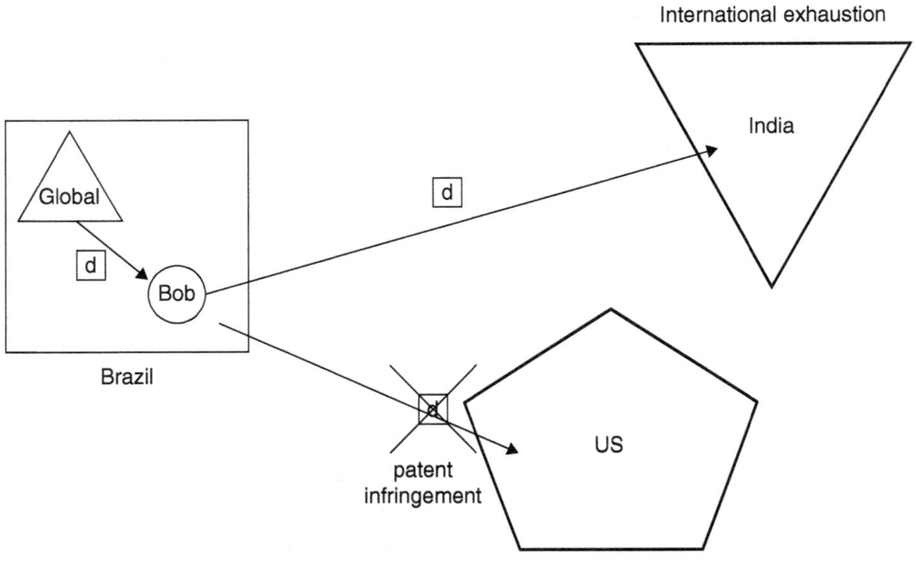

FIGURE 6 "Global" Example

to sell a duplicate copy of the patented pills. So, for example, even though international exhaustion exists in India, no one in India can make duplicate copies of the pills (or sell them) without permission from Global.

C. DECIPHERING RELATED TERMS

Now that the implications of international exhaustion have been explained, this section will introduce related terms that are often used in conjunction or in lieu of the term *international exhaustion*.[4] In particular, when goods first sold globally by a patent owner are imported into a second country without permission of the patent owner, those goods are referred to as *parallel imports*. The label *parallel* refers to the fact these imported goods will be sold in parallel with those of the patent owner in that second country. Alternatively, parallel imports may be referred to as *gray market goods*. The word *gray* can be considered to refer to what is a gray legal area with respect to whether the goods are considered to infringe a patent right. This is in contrast to the term *black market*, which refers to illegal goods.

One important note in considering related terms is that they are not limited in application to patents. Rather, international exhaustion, as well as the related terms of parallel imports and gray market goods also apply to other types of intellectual property, such as trademark and copyright.

Labeling a good a parallel import or gray market good does not indicate whether the good infringes a patent, as this is a function of whether the import occurs in a country that recognizes international exhaustion. If it does, the parallel import does not infringe the patent right. Conversely, in a nation that does not recognize international exhaustion, the parallel import would infringe and the patent owner would be entitled to its usual panoply of remedies, including the ability to enjoin the importation and seek monetary damages.

Patented goods imported without the patent owner's authority—whether labeled parallel imports or gray market goods—are not fake or counterfeit goods. In other words, although these goods may potentially infringe a patent, they are legitimate goods as these goods were first sold by the patent owner. They are considered unwelcome imports to the patent owner not because they are inferior products, but because the patent owner prefers to segregate markets to attain the highest selling price in each.

Table 2 provides a summary of terms that often arise in connection with discussions of international exhaustion.

---

[4] While the focus of this book is on patents, international exhaustion and all related terms discussed here also apply to other types of intellectual property, such as trademark and copyright. The specifics of when parallel imports of goods protected by these other rights differ, but the concept is the same in all cases—whether importation is permissible depends on the importing nation's rules on exhaustion.

TABLE 2

International Exhaustion—Definitions of Related Terms

Term	Definition
National Exhaustion	The general principle that a patent owner's rights in a patented good end after the first legitimate and unrestricted sale in the patent-granting nation. In other words, if patent owner Abbott sells a drug to a distributor, Abbott cannot claim infringement against the distributor for selling the drug to a retailer or consumer. This is a universally recognized principle.
International Exhaustion	The first authorized and unrestricted sale of a patented product *anywhere in the world* exhausts the patent owner's usual right to prevent import of the patented product.
Parallel Import (Good)	Importation of an authentic (non-counterfeit) product first legitimately sold in another country. The imported good is then sold "in parallel" with products sold directly from the patent owner in the importing country. Parallel imports are imported without permission of the patent owner.
Gray Market Good	Parallel Import (see above definition).

The terms in Table 2 are often used in different combinations to refer to the same activity. International exhaustion refers to the legal principle regarding when a patent owner's right to import is limited. However, there are alternative ways of stating the same thing. To help keep the terms straight, Table 3 provides the different terminology used to refer to whether imported goods are exempt from patent infringement or not by the importing nation.

TABLE 3

Are Imported Goods Permissible? —Summary of Related Terms

Imports of patented goods previously sold by patent owner in another country are NOT INFRINGING	Imports of patented goods previously sold by patent owner in another country are INFRINGING
International exhaustion recognized	International exhaustion *not* recognized
Parallel imports permitted	Parallel imports *not* permitted
Gray market goods permitted	Gray market goods *not* permitted

Now that a full panoply of terms related to international exhaustion has been introduced, a few questions may help to reinforce key points:

Q: Are parallel imports counterfeit goods?
A: No. The word *counterfeit* means that goods are fake and illegitimate. However, parallel imports are legitimate goods from the original manufacturer.
Q: Do gray market goods violate patent rights?
A: Not necessarily. The label *gray market good* only indicates the good was not authorized for importation by the patent owner. However, a patent owner might have the authority to bar parallel imports as an infringing import in a country that does not recognize international exhaustion.
Q: What does it mean that parallel imports are barred?
A: It means the patent owner can block imports of patented goods that the patent owner previously sold in another country. Stated differently, parallel imports are barred from a nation that rejects international exhaustion.

III. Principal Arguments

This section evaluates the arguments for and against recognizing a principle of international exhaustion for patents. Some of these arguments relate to international exhaustion of all types of intellectual property rights—not only patents, but also trademarks and copyrights. However, exhaustion of patent rights in drugs raise unique concerns because unlike other consumer goods, there are arguably safety issues at stake. In addition, patents are considered much more critical than other types of intellectual property rights to promoting an incentive for new research and development.

A. FOR INTERNATIONAL EXHAUSTION (AND PARALLEL IMPORTS)

The most often-cited argument for international exhaustion is that it can be a means to lower drug costs. This is a major issue because pharmaceutical companies often sell the same drug at different prices in different markets. Given the global environment of differing prices, permitting parallel imports should enable consumers' access to the lowest-priced drug. This can make drugs more accessible to consumers with less income and also reduce strain on national and local governments that would otherwise assist low-income consumers in purchasing them.

Some argue that universal international exhaustion is an optimal policy from the standpoint of efficiency. In particular, if all countries recognize international exhaustion, goods will flow freely in the global market. This would permit drugs to be produced in low cost regions (such as developing countries) and arguably result in an efficient allocation of resources. However, this is solely a theoretical argument that is unlikely to be tested in reality; some companies and even countries are strongly opposed to international exhaustion.

Arguments for international exhaustion have become more important in recent years as more countries have been required to adopt patent laws pursuant to TRIPS. For a nation that has previously not provided patents and drugs and is philosophically opposed to doing so because of the impact on access to medicine, permitting parallel imports is a reasonable mechanism to comply with global laws permitting patent protection. Many have suggested developing countries embrace international exhaustion of patent rights as one mechanism for promoting access to medicine in the wake of international obligations mandating patent rights.[5] In addition, some suggest that even if international exhaustion were uniformly recognized, it would provide an adequate return to the patent owner who received profits with the first global sale.

B. AGAINST INTERNATIONAL EXHAUSTION (AND PARALLEL IMPORTS)

Patent-owning drug companies are strongly opposed to parallel imports; they would like all nations to universally reject international exhaustion in their patent laws. In essence, the doctrine of international exhaustion of patent rights presents a threat to their current business model of obtaining the highest price in each separate market. If a multinational drug company can block imports, it can maintain its desired business model of segregating different geographic markets. This market segregation is considered critical to maximizing patent profits as many countries have laws or regulations that restrict the price of drugs. A rational patent owner thus seeks to sell the same drug at the maximum possible price in each market. If a patent owner has an absolute right to exclude all imports—including imports of drugs it previously sold elsewhere in the world—that owner can effectively ensure it has total control over the price in each market.

There are two primary reasons that (innovator) drug companies raise in opposing parallel imports of their patented products. First, they assert geographic price discrimination actually benefits consumers by increasing availability of locally lower-priced drugs and providing the necessary revenue to fund development of new drugs.[6] Second,

---

[5] *E.g.*, WORLD HEALTH ORG., PUBLIC HEALTH, INNOVATION AND INTELLECTUAL PROPERTY RIGHTS: REPORT OF THE COMMISSION ON INTELLECTUAL PROPERTY RIGHTS, INNOVATION AND PUBLIC HEALTH ¶ 4.19 (2006); UK COMMISSION ON INTELLECTUAL PROPERTY RIGHTS, INTEGRATING INTELLECTUAL PROPERTY RIGHTS AND DEVELOPMENT POLICY 42 (2002); UN Gen. Assembly, Human Rights Council, *Report of the Special Rapporteur on the right of everyone to the enjoyment of the highest attainable standard of physical and mental health*, A/HRC/11/12, at 11 (Mar. 31, 2009); *see also* Carlos Correa, *Integrating Public Health Concerns into Patent Legislation in Developing Countries*, South Centre (2000) (providing model options for different versions of international exhaustion).

[6] Although this argument is couched in terms of consumer benefit, it can also be framed to focus more specifically on drug companies. In particular, companies sometimes assert that parallel imports have an unfair impact on their ability to maximize profits—especially to the extent that some countries have price controls.

they suggest parallel imports threaten public health and consumer safety. Each of these arguments will be addressed and critiqued.

1. Allege Benefits of Price Discrimination

Drug companies assert the increased profits gained from price discrimination ultimately provide a social gain to consumers worldwide. In particular, they contend price discrimination enables them to price drugs according to consumer ability to pay, with higher countries paying higher costs for the same drug. They suggest that without effective price discrimination, they will be forced to charge uniform prices globally, which would make drugs less available for lower-income regions. In addition, they state the increased profits they gain from price discrimination are necessary to subsidize research and development into new drugs that ultimately benefit all consumers.

Although each of the above arguments sounds plausible, a closer examination reveals some flaws. First, although companies may sometimes provide low-cost drugs to the poorest countries, this is not always the case, and is seldom so for the so-called "middle income" developing countries that have low per capita income. Many consumers in poorer nations already cannot afford drugs that are marketed in their countries not only because they often lack insurance, but because they are paying prices equivalent to those set in wealthier countries despite having less disposable income. Although drug companies are generally not transparent about pricing, it would be in their economic interest to price drugs in middle-income countries so that only the wealthiest could purchase them because the poorest consumers who live on a few dollars a day are unlikely to pay even a reduced price. In addition, drug companies may be resistant to charging less because some large purchasers of medicine in rich countries may use existing prices in other countries to bargain down the purchase price.

The second argument concerns whether parallel imports undermine innovation. Although drug companies do spend a substantial amount on research, they actually spend more money on advertising and promotion of drugs. Moreover, many people have questioned whether the research the drug companies focus on provides much of a social benefit. Most companies today are more interested in developing new and often minor variations of existing drugs that are not likely to provide substantial new health benefits to consumers. In addition, companies often focus more on "lifestyle" conditions (such as depression and erectile dysfunction) that may yield a substantial profit in wealthy markets rather than life-threatening diseases and conditions that predominantly affect poorer countries. Although companies tend to present the issue in stark terms, there are more nuanced approaches. For example, regulatory laws could require licensed firms engaging in parallel trade to pay a fee that would go toward research and development.

2. Alleged Health and Safety Concerns

Drug manufacturers suggest international exhaustion is actually dangerous for consumers. In making this claim, companies emphasize that parallel imports are not sold by them but by an unauthorized distributor, such that the company cannot guarantee safety or efficacy. Companies suggest that although the imports are not fake, they may be of an inferior quality compared to goods sold directly by the company. First, the drug manufactures contend that because they are not in control of the goods, they cannot guarantee the drugs will remain safe; they indicate safety could be compromised from tampering, or because the product is fragile and needs special storage and handling. In addition they suggest there may be a safety issue if consumers cannot read instructions written in a different language.

Although drug companies frequently suggest that parallel goods are not safe or of poorer quality, there is no substantial evidence for this assertion. Counterfeit goods may indeed be unsafe. However, parallel imports are legitimate and not countefeit goods. Unfortunately, parallel goods are sometimes discussed together with counterfeits such that there may be an improper suggestion that parallel goods are unsafe. However, parallel goods are legitimate goods made and sold by the patent owner—even if the patent owner objects to their subsequent distribution, that does not necessarily mean they are unsafe.[7] Drug companies may protest they cannot vouch for distribution of parallel imports, but in the modern economy, drugs are manufactured worldwide. In addition, if safety is the issue, that can be addressed through regulatory laws, such as regulating either the parallel trade products or those that promote such trade.

IV. Commercial Implications

This section addresses how drug companies currently operate in an environment where some, but not all countries recognize international exhaustion. It aims to provide a picture of the current landscape of international exhaustion, including the extent to which international exhaustion currently impacts corporate profits. In addition, this section explains how other laws and strategies beyond patents help to aid in the goal of market segmentation.

An important initial question is which countries currently recognize or reject international exhaustion of patent rights. Importantly, if the markets of major profitability reject this doctrine, drug companies will still be able to mostly maximize their profits. Basically, the current global landscape is favorable to patent-owning drug companies. In particular, the market of highest profits—the United States—rejects international exhaustion.[8] Accordingly, drug companies can utilize patent laws to keep out parallel imports of lower-priced drugs from other countries. The EU also rejects the principle of international

[7] To the contrary, parallel goods may in fact be safe. *E.g.*, Cary Coglianese et al., Import Safety: Regulatory Governance in the Global Economy 116 (2009).

[8] *Jazz Photo Corp. v. U.S. Int'l Trade Comm'n*, 264 F.3d 1094, 1105 (Fed. Cir. 2001); *see also Fujifilm v. Benun*, 605 F.3d 1366 (Fed. Cir. 2010)(affirming that the US continues to reject a rule of international exhaustion).

exhaustion.[9] Thus, drug companies can use patent laws to preclude low-cost drugs from a country such as Rwanda from entering the EU. Although rejection of international exhaustion is technically a legal issue, it has the result of protecting pharmaceutical industries in these nations.

Not surprisingly, countries that presently recognize international exhaustion include a number of the developing ones. For example, India, a number of Latin American countries, and South Africa all recognize international exhaustion. However, they are not the only ones, as New Zealand, a developed country, also does. Notably for drug companies, New Zealand is a much smaller market than either the United States or the EU.

Table 4 provides a snapshot of countries that currently recognize international exhaustion, as well as those who do not.

TABLE 4

International Exhaustion—Sample Countries

International Exhaustion Recognized (Parallel Imports Permitted)	No International Exhaustion (Parallel Imports Barred)
India	United States
Latin American countries (Andean Commission Decision 85)	EU (but, regional exhaustion okay)
South Africa	Australia (if contractual restriction)
New Zealand	
Japan (unless contractual restriction)	

The current legal global landscape concerning international exhaustion seems to largely preserve drug company interests in preventing infiltration of lower-priced imports into the wealthiest markets. Whether due to favorable laws in wealthy jurisdictions or other factors, there seems to be a relatively minimal amount of parallel importation as a percent of world trade.[10] This is perhaps not surprising since drugs sold to LDC countries account

[9] However, the EU recognizes *regional exhaustion*, which means that once a patented product has properly entered any EU Member State, the patent owner has no control over the subsequent sale or distribution of that product to another EU Member State. Parallel trade accounted for about two percent of the pharmaceutical market in 2003 but varies tremendously by EU Member State; parallel trade may be negligible or up to twenty percent. *E.g.*, OECD, OECD HEALTH POLICY STUDIES, PHARMACEUTICAL PRICING POLICIES IN A GLOBAL MARKET 41 (2008). Although some have studied parallel trade in the EU and suggested that cost savings do not get passed along to the consumer, and that the parallel trade in the EU is very different than potential parallel trade for developing countries as most EU citizens have insurance. *E.g.*, Robert C. Bird & Peggy E. Chaudhry, *Pharmaceuticals and the European Union Managing Gray Markets in an Uncertain Legal Environment*, 50 VA. J. INT'L L. 719 (2010)

[10] *E.g.*, Frederick M. Abbott, *First Report (Final) to the Committee on International Trade Law of the International Law Association on the Subject of Parallel Importation*, 1 J. INT'L ECON. L. 607, 613 (1998) ("[E]mpirical studies

for a minute fraction of the global market. Nonetheless, pharmaceutical companies are vigilant in taking a variety of actions to protect their interests.

One way they aim to minimize trade in low-cost drugs is to use contracts in conjunction with in patent laws. As noted earlier, international exhaustion generally only applies with respect to the first *unrestricted* global sale. Accordingly, drug companies place contractual restrictions on further sales. In addition, drug companies (through their national-government trade representatives) may try to ensure other countries agree that contractual restrictions will in fact be respected. For example, as discussed in Chapter 8, the United States has included provisions in some free trade agreements that require nations to bar imports of goods sold under contractual restriction. One important market the United States has imposed this obligation on is Australia.[11]

Even in countries that already have desirable patent laws from the drug companies' point of view (i.e., ones that reject international exhaustion), the companies may still seek protection of their goals through additional laws. As noted in Chapter 1, drug companies try to protect drugs from direct competition not only through patent laws but regulatory ones. For example, in the most profitable market—the United States—regulatory laws also enable drug companies to exclude unwanted parallel imports from other countries such as Canada.[12] Despite an exception for limited supplies for personal use, this regulatory rule prevents large-scale imports.[13] Although pharmaceutical countries have been very concerned about cross-border trade from Canada to the United States, the amount of such trade is minimal compared to total sales in the United States.[14]

Another method companies use to supplement effective enforcement of desired patent and regulatory laws is through the use of trademarks. A trademark is a word, symbol, etc., that is used to identify and distinguish its source. Drug companies can sell the same product with different names or different packaging. These differences make it easier for drug companies to help identify products that should be seized by customs officials as infringing. In addition, drug companies may use the differing trademarks to help them identify unauthorized products that they can then try to block through private litigation.

---

on parallel importation in the international context are sparse, and in many cases quasi-anecdotal. . . . [but] parallel import goods may not reflect a significant proportion of world trade.").

[11] United States-Australia Free Trade Agreement art. 17.9(1)(4), May 18, 2004, T.I.A.S. No. 6422, *available at* http://www.ustr.gov/trade-agreements/free-trade-agreements/australian-fta/final-text.

[12] 21 U.S.C. § 331(t) (2006). Technically, the Medicare Modernization Act of 2003 permitted an increase in imports from Canada *if* the Secretary of Health and Human Service certified that such imports would not pose an additional risk to public health and would result in significant cost savings. 21 U.S.C. § 384(b) (2006); *see also* RICHARD ABOODD, PHARMACY PRACTICE AND THE LAW 146 (6th ed. 2010). However, the Secretary of Health and Human Services has yet to make such a certification. [Page 120 STAT. 1355] Public Law 109-295 109th Congress

[13] FDA REGULATORY PROCEDURES MANUAL, 9–12 (Mar. 2010); 21 U.S.C. § 956 (2006); Pub. L. No 109-295, § 535, 120 Stat. 1355 (2006).

[14] In 2005, the total sales volume of goods represented 0.5% of US sales. OECD, *supra* note 9, at 41–42.

Another approach patent-owning companies take is to limit drugs sold. A company may decide not to sell drugs in certain areas where expected sales (and profits) are low. The example provided in the introduction to this chapter illustrates this concept. As noted earlier, a drug company may be reluctant to sell its patented HIV drug in Rwanda for fear that the drug might be imported into wealthier markets—even though there are patent and possibly regulatory laws that should minimize this. Although drug companies have provided some patented drugs at low cost to poor markets, such a decision is likely not made lightly, as it must be balanced against the risk of undesirable imports. Even in countries with desirable laws that bar parallel imports, the patent owner may prefer to avoid taking legal action. The ideal situation for the drug company is to bar all possible imports, with the effect being enhanced if there are few drugs available to import in the first instance.

A drug company may also limit the sale of goods that have a high probability of becoming imports in a wealthier market, without regard to whether the initial market has low sales. For example, the same drug is often sold at a much lower price in Canada than the United States because Canada imposes price caps on drugs, but the United States does not. Although there are already US patent and regulatory laws that favor the drug company interests, the drug company may still need to take legal action to enforce the laws. Moreover, even legal action will not bar all imports. After all, a consumer who travels to Canada and imports a drug for personal use will not violate regulatory laws, given the exception for personal use. Technically, the consumer who imports a small sample of drugs would violate the patent right of importation. However, drug companies are unlikely to sue individual consumers: the value of the drugs will be relatively low and the public relations impact would be very negative. Given all these factors, a drug company would want to limit sales of drugs in Canada as another tool to limit imports. Indeed, after cross-border sales briefly increased after the development of Canadian Internet pharmacies, patent-owning companies limited supply to Canada to minimize this problem.[15]

V. Conclusion

Although most countries today must provide patent rights, differing prices of patented drugs will continue to provide an incentive for some to buy a cheaper version of the drug in one market to sell in a second market. This incentive does not exist in markets that bar parallel imports (by rejecting the doctrine of international exhaustion). Accordingly, patent-owning drug companies want to bar parallel imports in all markets to maximize profits. There is currently no uniform role on whether parallel imports are permissible; in

---

[15] *E.g.*, OECD, *supra* note 9, at 42.

other words, countries take differing positions on whether to recognize international exhaustion of patent rights. Although the law in many wealthy markets already favors patent owners, they continue to lobby for a more uniform approach, as will be discussed in Chapter 8 on TRIPS-Plus. Moreover, fear that parallel trades hinder desired market segregation may also impact patent owners' approach to other patent concepts (such as compulsory licensing), as will be later explored.

II The Current Framework

3 An Introduction to TRIPS

This chapter provides a brief overview of TRIPS, with emphasis on provisions (some of which are reprinted in the Appendix) that impact patented and generic drugs. In particular, this chapter discusses what TRIPS requires in terms of what must be patented, the scope of patent rights, and how patent rights are enforced for WTO member countries. In addition, this chapter discusses the TRIPS requirement to protect some clinical data. This chapter answers some key questions, such as the following:

- Does TRIPS establish uniform patent standards?
- Do all WTO member countries need to provide patents on drugs?
- Does TRIPS prevent generic companies from relying on data of other companies in seeking regulatory approval?
- What happens if a WTO member country violates TRIPS?

I. Background 56
A. Introduction 56
1. The Importance of TRIPS 57
2. The Negotiation of TRIPS 58
B. Interpretation of TRIPS 60
II. Patent Requirements 62
A. What is Patentable 62
1. "Industrial Application" 63
2. "New" 64
3. "Inventive Step" 64
B. Patent Process 65
1. Patent Examination Process 65
2. Challenges to Patents and Patent Applications 66
III. Patent Rights & Remedies 66
A. Standard Patent Rights 67
1. Exclusionary Rights 67
2. Term of Protection 68
B. Exceptions to Patent Rights 70
1. Compulsory Licenses 70
2. "Limited" Exceptions 71
C. Enforcement 74
1. Regular Enforcement 74
2. Provisional Measures 75
3. Border Measures 75
IV. Data Protection 76
A. "Unfair Commercial Use" 76
B. When Data Protection is Required 81
C. Term of Protection 82
V. TRIPS Compliance & Transitional Protection 83
A. Overview 83
B. Transitional Provisions 84
1. Mailbox Applications 85
2. Exclusive Marketing Rights (EMR) 85
VI. Conclusion 87

I. Background

A. INTRODUCTION

The Agreement on Trade Related Aspects of Intellectual Property Rights, known more commonly as TRIPS, sets global intellectual property requirements for all members of the World Trade Organization (WTO),[1] which is comprised of over 150 countries at all levels of economic development.[2] The intellectual property requirements mandated by TRIPS include patents.

TRIPS requirements fall into three categories. First, TRIPS requires all member states to comply with several preexisting international agreements on intellectual property. Second, TRIPS imposes fairness requirements, such as ensuring equal treatment to citizens of all members.[3] Third, TRIPS requires members to provide minimum standards of intellectual property rights. This chapter focuses on these minimum standards primarily with respect to patents, addressing what must be patented, the scope of patent rights, and their enforcement under TRIPS.

Before discussing the details, it is important to understand the general concept of minimum standards of intellectual property protection under TRIPS. Although member states must meet these minimum standards, they are permitted to exceed them.[4] There may also be great diversity in laws because the minimums are often undefined, leaving room for variation. Thus, TRIPS does not contemplate or result in uniform laws. For example, an identical invention could be patentable in one country, but not another. How could this happen? Consider this analogy: there is a minimum building standard that a house have walls, but no definition as to what constitutes a wall. So, the house could be a traditional one, a tepee, or an igloo. Just as the wall requirement may permit different variations, so too the TRIPS requirements permit different standards of patentability. In explaining the minimum—but largely undefined—patent requirements under TRIPS, this chapter will aim to show how the requirements can be defined to promote access to medicine.

1. The Importance of TRIPS

International agreements prior to TRIPS primarily focused on procedures to streamline patent applications worldwide (in countries that provided patents) and on ensuring that

---

[1] TRIPS: Agreement on Trade-Related Aspects of Intellectual Property Rights, Apr. 15, 1994, Marrakesh Agreement Establishing the World Trade Organization, Annex 1C, THE LEGAL TEXTS: THE RESULTS OF THE URUGUAY ROUND OF MULTILATERAL TRADE NEGOTIATIONS 320 (1999), 1869 U.N.T.S. 299 [hereinafter TRIPS].

[2] *See* WTO, Members and Observers, http://www.wto.org/english/thewto_e/whatis_e/tif_e/org6_e.htm (providing a list of current member states).

[3] TRIPS imposes a "national treatment" requirement, which means that each member state must provide the same type of protection to all members; no member can give preferential treatment to its own citizens. TRIPS art. 3. In addition, TRIPS has a "most favored nation" (MFN) requirement that requires a member state to provide the same advantage to all other WTO members if it grants an advantage in an agreement beyond the WTO. TRIPS art. 4.

[4] However, members may not contravene other provisions of TRIPS in going beyond the minimum standards. TRIPS art.1(1); *see also* Annette Kur & Henning Grosse Ruse-Khan, *Enough is Enough—The Notion of Binding Ceilings in International Intellectual Property Protection* (Max Planck Papers on Intellectual Property, Competition & Tax Law, Research Paper No. 09-01), *available at* http://ssrn.com/abstract=1326429.

foreign applicants would not be treated less favorably than domestic ones.[5] However, TRIPS changed all that: TRIPS is the first international agreement that *requires* countries to provide patent rights.

TRIPS is also important because it has built-in enforcement mechanisms. Prior international agreements could not effectively sanction members who failed to comply, but TRIPS provisions are enforceable pursuant to the Dispute Settlement Understanding (DSU) that applies to all WTO agreements.[6] The DSU provides WTO member countries with a quasi-judicial process for addressing alleged violations when voluntary negotiations fail.[7] A three-person WTO panel decides the dispute, with the possibility for review by the Appellate Body.[8] If the WTO panel determines that a country is not in compliance with TRIPS, the country will be given a short period to bring its laws into compliance.[9] Failure to do so may result in monetary sanctions or suspension of benefits under a WTO agreement.[10]

2. The Negotiation of TRIPS

TRIPS was negotiated as part of the broader WTO, framework including other related agreements on goods and services.[11] Developing countries who wanted access to markets

[5] Paris Convention for the Protection of Industrial Property, art 3, Mar. 20, 1883, revised July 14, 1967, 21 U.S.T. 1583, 828 U.N.T.S. 305 (requiring equal treatment for foreign applicants); Patent Cooperation Treaty, June 19, 1970, 28.7 U.S.T. 7645, 1160 U.N.T.S. 232 (providing a streamlined process for submitting patent applications in multiple jurisdictions).

[6] Understanding on Rules and Procedures Governing the Settlement of Disputes, art 3.2, Apr. 15, Marrakesh Agreement Establishing the World Trade Organization, Annex 2, Legal Instruments—Results of the Uruguay Round, 33 I.L.M, 1125 (1994) [hereinafter DSU]. Currently, a country can only be found in violation of actual TRIPS provisions. There is the possibility of potential liability for "non-violation" complaints based upon what member countries expected to achieve under TRIPS, even if this is not explicitly in any TRIPS requirement; however, there is presently a moratorium on such complaints. TRIPS art. 64(2); WTO, TRIPS "Non-Violation" Complaints (Article 64.2), Background and the Current Situation (Dec. 2009), http://www.wto.org/english/tratop_e/trips_e/nonviolation_background_e.htm (noting the moratorium has been consistently extended since TRIPS was concluded).

[7] Where possible, member states are encouraged to resolve their disputes amicably without the need for an official interpretation. DSU art. 3(7) (noting that members "shall exercise its judgment as to whether action under these procedures would be fruitful").

[8] Each panel is comprised of different members, but the Appellate Body has a constant number of seven who are each appointed to serve (staggered) four-year terms. *See* DSU art. 8 (composition of panel); *Id.* art. 17 (Appellate Body composition).

[9] DSU art. 21(3).

[10] *Id.* art. 22. Usually sanctions are imposed under the agreement violated; however, there is authority to impose sanctions in different sectors when sanctions within the same agreement would be impracticable or more harmful for the prevailing country, which is referred to as "cross-retaliation." *Id.* art. 22.3(b)–(c).

[11] Other agreements include the Agreement on Agriculture (AoA) (providing rules on market access and export subsidies), the General Agreement on Trade in Services (GATS) (providing rules on trade in services parallel to WTO rules on physical goods), the Sanitary and Phyto-Sanitary Agreement (SPS) (limiting member

of wealthy countries for their products had an incentive to agree to TRIPS. However, it would be inaccurate to say that developing countries willingly negotiated away traditional sovereign rights concerning the existence and scope of a patent system in return for broader market access. Although that is certainly part of the story, the conclusion of TRIPS was bitterly contested and never a foregone conclusion over its multi-year negotiating period.

It is also important to note that TRIPS is only one point in the history of intellectual property rights: it is neither the beginning nor ending. Prior to TRIPS, both developed and developing countries wanted to change the international landscape, but in different directions. Developed countries wanted to increase global patents (and other intellectual property rights), while developing countries wanted more flexibility. For example, to ease then-existing global restrictions on compulsory licensing of patents. Neither side was able to come to agreement on amending the earlier and more limited international agreement on patents, called the Paris Convention. Moreover, intellectual property rights were becoming increasingly more important to wealthy countries whose developing countries wanted exports often relied on intellectual property such as trademarks and copyrights.

A small group of powerful corporate interests successfully lobbied their governments to include intellectual property (IP) norms within the agenda for the negotiations surrounding the General Agreement on Tariffs and Trade (GATT), the agreement on goods that preceded the WTO. IP norms had never been part of the GATT framework. To the contrary, GATT recognized that IP rights could pose a barrier to trade.[12] However, these rights were initially proposed as being relevant to free trade under GATT because wealthy countries might be reluctant to trade goods unless and until counterfeit trademarked goods were addressed as a problem. Notably, patent rights were not mentioned in the initial discussions of incorporating IP in the GATT framework. Nonetheless, the GATT negotiating document forming the seed that blossomed into full-fledged patent and other intellectual property standards under TRIPS referred to "intellectual property rights" rather than solely trademarks.[13]

---

policies relating to food safety and animal and plant health), and the Agreement on Technical Barriers to Trade (TBT) (aiming to ensure technical standards and certifications do not unduly burden trade).

[12] GATT art. XX(d) permitted members to adopt measures that would otherwise not be consistent with the GATT if such measures were necessary to protect intellectual property—so long as the measures would not result in unjustifiable discrimination between countries or be a disguised restriction on international trade. General Agreement on Tariffs and Trade art. XX(d), Oct. 30, 1947, 61 Stat. A-11, 55 U.N.T.S. 194 [hereinafter GATT]. The GATT regime continues within the framework of the WTO, as established by the Uruguay Round of trade negotiations. Marrakesh Agreement Establishing the World Trade Organization art. II(4), Apr. 15, 1994, 1867 U.N.T.S. 154, 156, 33 I.L.M. 1144, 1145.

[13] TRIPS grew out of the 1986 Punta Del Este Declaration, as part of the same GATT multilateral trade negotiations that led to the establishment of the WTO. GATT, Ministerial Declaration of Punta Del Este, of Sept. 20, 1986, *reprinted in* LAW AND PRACTICE UNDER GATT III.A.3 (K. Simmonds & B. Hill eds., 1989).

Throughout the negotiating process that led to TRIPS, developing countries resisted the incorporation of IP norms within the broader trade framework. Some countries suggested IP should be included only where a trade distortion could be proven, and with more favorable treatment being given for developing countries—proposals that are consistent with the earlier GATT framework. The final TRIPS agreement imposes new IP norms, largely as desired by wealthy nations such as the United States, EU Member States, and Japan. However, some key terms are undefined, or vaguely phrased, thus permitting both developed and developing countries to initially claim that the agreement reflected their interests. All of these issues form an important backdrop to subsequent controversies concerning TRIPS.

B. INTERPRETATION OF TRIPS

Before discussing what TRIPS requires, it is important to explain how TRIPS should be interpreted. Sometimes parties tend to claim that TRIPS requires what they seek. However, what countries seek does not govern what TRIPS actually requires. Rather, TRIPS is to be interpreted according to the Dispute Settlement Understanding (DSU), which provides rules for interpreting all WTO agreements, including TRIPS.[14]

According to the DSU, all terms are interpreted according to customary rules of international law, meaning the final text generally controls. Disputed terms are interpreted in light of the treaty's broader context, including its preamble and its sections titled *objectives* and *principles*.[15] This context must also be interpreted in conjunction with subsequent agreements between the parties, which includes the 2001 Doha Declaration on Public Health (as will be later discussed).[16] Secondary material, such as negotiating history, is only considered to confirm an interpretation achieved by the usual interpretation of disputed terms in light of the broader context, or if the usual interpretation leads to an interpretation that is ambiguous or reaches an absurd result.[17]

Although the interpretive framework is easy to state, predicting how WTO panels will apply it can be more challenging. Typically, WTO panels rely extensively on dictionary definitions for undefined terms. However, interpreting TRIPS' objectives and principles

[14] DSU art. 1(1).

[15] Vienna Convention on the Law of Treaties, art. 31(1), May 23, 1969, 1155 U.N.T.S. 331, 8 I.L.M. 679 (stating that a treaty shall be interpreted in good faith in accordance with the ordinary meaning to be given to the terms of the treaty in their context and in the light of its object and purpose) [hereinafter Vienna Convention].

[16] *See* Vienna Convention art. 32(3).

[17] Vienna Convention art. 32. Recourse may be had to supplementary means of interpretation, including the preparatory work of the treaty and the circumstances of its conclusion, in order to confirm the meaning resulting from the application of article 31 of the Vienna Convention, or to determine the meaning when the interpretation according to article 31: (a) leaves the meaning ambiguous or obscure; or (b) leads to a result which is manifestly absurd or unreasonable. *Id.*

has proven more difficult, perhaps because these provisions contain competing goals. For example, the objectives state that IP rights should contribute to innovation and also benefit users "in a manner conducive to social and economic welfare, and to a balance of rights and obligations."[18] Although this statement may appear clear, *how* this balance can be achieved is much less certain—especially as patent rights by definition will limit access to patented inventions, such as drugs. WTO panels sometimes recognize that these provisions are pertinent, but thus far they have not seemed to play an explicit role in the panels' final decisions. However, according to a proper interpretation of the DSU (as well as the opinion of most scholars), these provisions should be relevant—even if thus far they have played only a minimal role in panel decisions.[19]

Although the interpretive tools may not yield great predictability, WTO panel decisions on specific cases of alleged violations of TRIPS do provide some guidance. Technically WTO panels are prohibited from creating new law pursuant to the DSU. However, given that TRIPS is a skeletal agreement with undefined key terms, every panel decision provides important clarification. Also, although WTO cases are only binding on the parties before the particular WTO panel, subsequent panels traditionally give substantial deference to prior decisions, as do WTO member states.

So, what is the ultimate impact of TRIPS interpretation on member countries? Essentially, member states get to decide how to interpret and implement TRIPS rules, unless and until their decisions are formally challenged through the WTO dispute settlement process. And formal disputes are relatively rare—for all TRIPS issues (including not only patents, but also trademarks and copyrights), there have been a total of only twenty-five notifications of disputes, with eight panel decisions,[20] only four of which

---

[18] TRIPS art. 7.

[19] *See* Appellate Body Report, *Canada—Term of Patent Protection*, ¶ 101, WT/DS170/AB/R (Sept. 18, 2000); Denis Borges Barbosa et al., *Slouching Towards Development in International Intellectual Property*, 2007 MICH. ST. L REV. 71, 101–02 (2007); Susy Frankel, *WTO Application of "The Customary Rules of Interpretation of Public International Law" to Intellectual Property*, 46 VA. J. INT'L L. 365, 385 (2005–2006); Robert Howse, *The Canadian Generic Medicines Panel: A Dangerous Precedent in Dangerous Times*, 3 J. WORLD INTELL. PROP. 493, 496–501 (2000). However, there is some sign that in a future dispute, a WTO panel may give greater consideration to contextual provisions based on a ruling from the WTO Appellate Body—albeit in a case that does not involve the tougher contextual clauses under TRIPS. *See* Appellate Body Report, *United Sates-Continued Existence of and Application of Zeroing Methodology*, ¶¶ 268–73, WT/DS350/AB/R (Feb. 4, 2009) (suggesting a holistic approach to interpretation).

[20] Panel Report, *India—Patent Protection for Pharmaceutical and Agricultural Chemical Products*, WT/DS79/R (Aug. 24, 1998) (EU–India mailbox); Panel Report, *India—Patent Protection for Pharmaceutical and Agricultural Chemical Products*, WT/DS50/R (Sept. 5, 1997) (US–India mailbox); Panel Report, *Canada—Patent Protection of Pharmaceutical Products*, WT/DS114/R (Mar. 17, 2000); Panel Report, *Canada—Term of Patent Protection*, WT/DS170/R (May 5, 2000); Panel Report, *European Communities—Protection of Trademarks and Geographical Indications for Agricultural Products and Foodstuffs*, WT/DS290/R; WT/DS174/R (Mar. 15, 2005); Panel Report, *United States—Section 211 Omnibus Appropriations Act of 1998*, WT/DS176/R (Aug. 6, 2001), *aff'd* by Appellate Body Report, WT/DS176/AB/R (Jan 2, 2002); Panel Report,

specifically relate to protection of drugs under patent or patent-like rights.[21] Countries seem to be cautious in bringing disputes for fear of receiving an undesirable interpretation of TRIPS.

II. Patent Requirements

The basic rule is that a member state must grant a patent for an *invention* in a *field of technology* if the invention satisfies three requirements:[22] (1) the invention must be capable of industrial application, (2) the invention must be new, and (3) the invention must have an inventive step.[23] Each of these requirements will be explained below after first discussing what constitutes an invention in a field of technology.

A. WHAT IS PATENTABLE

The first question in determining what is patentable is defining what is meant by an *invention* and a *field of technology*. These terms are important because members do not

China—Measures Affecting the Protection and Enforcement of Intellectual Property, WT/DS362/R (Jan. 26, 2009); Panel Report, *United States—Section 110(5) of the Copyright Act*, WT/DS160/R (June 15, 2000).

[21] Panel Report, *India—Patent Protection for Pharmaceutical and Agricultural Chemical Products*, WT/DS79/R (Aug. 24, 1998) (EU–India mailbox) & Panel Report, *India—Patent Protection for Pharmaceutical and Agricultural Chemical Products*, WT/DS50/R (Sept. 5, 1997) (US–India mailbox), *as modified by* Appellate Body Report, *India—Patent Protection for Pharmaceutical and Agricultural Chemical Products*, WT/DS50/AB/R (Dec. 19, 1997); Panel Report, *Canada—Term of Patent Protection*, WT/DS170/R (May 5, 2000), *aff'd by* Appellate Body Report, *Canada—Term of Patent Protection*, WT/DS170/AB/R (Sep. 18, 2000); Panel Report, *Canada—Patent Protection for Pharmaceutical and Agricultural Chemical Products*, WT/DS114/R (Mar. 17, 2000).

[22] There are two exceptions that permit member states to bar inventions in certain fields of technology from patentability, but neither is directly relevant to the granting or denying patents on drugs. TRIPS permits a categorical exclusion for methods of treating humans and animals. TRIPS also permits nations to decline to protect genetically engineered plants and animals, although nations must protect "plant varieties" through some effective system. TRIPS art. 27(3). In addition, TRIPS provides an exception from patentability for *individual* inventions (as opposed to an entire class of inventions) when such exclusion is necessary because commercial exploitation would harm *ordre public* or morality, including human life or health. *Id.* art. 27.(2). In other words, not all drugs could be precluded from patentability based on this exception as it is to be applied to individual inventions. In addition, because the term *necessary* has been interpreted narrowly in WTO jurisprudence to be an objective standard, one country's preference to not patent a drug would not be likely to satisfy this standard.

[23] Although not usually an issue with respect to access to medicine, TRIPS does require that a patent application provide a sufficiently clear description of the invention to enable someone with similar technical skill as the applicant/inventor to make the invention. *Id.* at art. 29(1). This requirement is common among patent laws. *See, e.g.*, 35 U.S.C. 112 (2006); Convention on the Grant of European Patents, art. 83, Oct. 5, 1973, 13 I.L.M. 270 (*amended by* the Revision Act of Nov. 29, 2000) [hereinafter EPC], *available at* http://www.epo.org/patents/law/legal-texts/epc.html; Tokkyo Hô [Japan Patent Act], Law No. 121 of 1959, art. 36(4), *available at* http://www.cas.go.jp/jp/seisaku/hourei/data/PA.pdf.

need to provide patents on things that do not constitute an invention or that are not inventions in a field of technology.

Although TRIPS does not define either of these important terms, it is difficult to argue that a new drug compound would either not be an invention or not be in a field of technology. However, TRIPS does clarify that inventions include both products as well as processes and that members cannot "discriminate" against inventions in a specific field of technology (subject to a temporary exception for least developed countries); in other words, members must provide patents to inventions in all fields of technology.[24] This means members cannot deny patents to inventions in the field of medicine and can also not decline to provide patents on products such as chemical compositions that are active ingredients of drugs. The nondiscrimination requirement, as well as the mandate to protect products, means countries must now provide patents on new drugs as well as new active ingredients for drugs. This is a major change to the laws of many countries; prior to TRIPS, about fifty countries did not provide patents on drug compounds, as they wanted to promote greater access to medicine.[25] TRIPS does not permit such a policy justification to completely exempt drugs from patentability.

Although TRIPS now requires drugs and drug compounds to be generally considered patentable, the undefined term *invention* still leaves some room for flexibility. One area of possible flexibility is to consider slight molecular variations of previously existing drugs or chemical compounds to not be adequately inventive.[26] Other possible ways to limit the scope of patentability of drugs—and to thereby increase access to medicine—lie in the definition and application of the other TRIPS requirements for patentability, as will be described below.

1. "Industrial Application"

An invention can only be patentable according to TRIPS if it is capable of *industrial application*.[27] TRIPS does not specify what industrial application means. Nonetheless, a compound that is useful for treatment of a disease or condition would seem to be capable of industrial application in the healthcare industry.

---

[24] TRIPS art. 27(1) (noting the nondiscrimination requirement is "[s]ubject to paragraph 4 of Article 65, paragraph 8 of Article 70 and paragraph 3 of this Article").

[25] As noted in Chapter 1, a patent on a product (such as a drug) excludes all others from making the same drug—even if they do so using a different method.

[26] Even if such a variation were considered an invention, it may still be potentially be barred from patentability in a country that narrowly defines what constitutes an "inventive step."

[27] The text of TRIPS uses the phrase "industrial applicability," but a footnote states that this is synonymous with "useful." TRIPS art. 27, note a. The United States uses the term "useful" in its patent laws whereas many other countries use the phrase "industrial applicability," which is generally more narrowly defined. *E.g.*, 35 U.S.C. § 101 (2006); EPC art. 101 (industrial applicability); Japan Patent Act art. 29.1; *see also* John R. Thomas, *The Post-Industrial Patent System*, 10 FORDHAM INTELL. PROP. MEDIA & ENT. L.J. 3 (1999).

Importantly, the requirement is only that the invention be *capable* of use in an industry. This is a wholly different inquiry than the usual regulatory requirement that a drug actually work as intended. Accordingly, most chemical compounds are likely to satisfy this requirement—regardless of whether the compound will eventually be commercially marketable.

2. "New"

TRIPS also requires that a patentable invention be *new*. Most countries consider an invention new if it did not previously exist (in the *prior art*, which is a patent term for what existed before the invention). However, countries have taken different views in assessing whether an invention previously existed. The strictest definition considers an invention new only if it was not previously known anywhere in the world, without regard to whether that knowledge was documented in writing. But some countries have more relaxed definitions that aim to balance other goals; for example, they provide an inventor a slight grace period to file a patent application after initially disclosing the invention to the public through a published paper or presentation rather than have the inventor lose all rights. A country that wants to promote access to medicine would want the strictest definition of new because whether an invention is new is one of the gatekeepers to patentability.

Another important issue for drugs in particular is whether new *uses* of known compounds are considered patentable. As noted earlier, TRIPS requires that patents be granted on products and processes. However, whether a newly discovered use of an existing compound should be considered new is not clarified in TRIPS and thus becomes another area for variance. Some countries, such as the United States, clearly provide patents on new uses. However, TRIPS does not require all countries to follow suit as it does not define *new*. Therefore, under TRIPS, countries can define the word to exclude newly discovered uses of an existing compound.

3. "Inventive Step"

The last requirement for a patentable invention under TRIPS is that the invention have an *inventive step*. This is often considered a critical criterion for patentability as it is meant to be the ultimate gatekeeper to ensuring the exclusivity inherent in a patent is justified by a deserving invention. Whereas an invention may be "new" because there is nothing identical to it previously, an invention may still not deserve a patent if it is not a significant advance over the prior art (i.e., if it lacks an inventive step).

Evaluating whether inventions satisfy this requirement can be notoriously difficult. Whereas assessing whether something is "new" usually only takes into consideration whether it is identical to what existed before, assessing an inventive step requires subjectivity. Some countries require a determination of whether the invention could or would have been discovered. Others require a showing that the invention provides an actual technical advance.

Once again, the lack of definition of this term under TRIPS presents an opportunity for countries to decide how best to tailor patent laws to their interests. Accordingly, with respect to access to medicine, there is not necessarily a uniform view on the best way to define inventive step. On the one hand, a strict definition would arguably help to ensure that more products were freely available. On the other hand, in countries with the technical capacity for making incremental developments (such as India), it could deprive local companies of patent rights. Although some Indian companies have argued that they should have patent rights for their incremental advances, every patent—even by a local company—inherently limits access to medicine because it enables its owner to charge higher prices. Granted, a patent owner could set a more modest price, but the cost is still likely to be higher than for an unpatented product. However, some argue that even incremental advances might result in helpful products to consumers and thus still deserve patents to provide an incentive.

Before considering how TRIPS impacts the way patents are obtained, the following questions provide a quick review of what is patentable TRIPS:

Q: Does TRIPS require member countries to have the same patent standards?

A: No! TRIPS requires only "minimum" standards of protection, such that countries are free to provide more than the minimum. In addition, although TRIPS sets forth these minimum standards of patent protection, it does not define some of these standards, which gives nations flexibility to adopt differing standards.

Q: Does TRIPS require that the same invention be entitled to patent protection in all WTO members?

A: *No!* For example, although one country may consider a new use of a known compound to be adequately new for patentability, a different country can consider that same use to not be new under TRIPS.

B. PATENT PROCESS

TRIPS sets forth a few procedural parameters concerning how patent applications are evaluated, as well as how both applications and issued patents may be challenged.

1. Patent Examination Process

TRIPS requires that patent applications be examined within a "reasonable" time, with a focus on ensuring the term of protection is not unduly limited.[28] The duration of the examination process is important because TRIPS calculates patent terms from when patent applications are filed. In other words, the length of the patent term will be a func-

[28] TRIPS art. 62(2) (procedures for granting rights, such as patent rights must be done within a "reasonable period of time" to prevent "unwarranted curtailment of the period of protection").

tion of the examination period: the longer the examination time, the shorter the patent term. This TRIPS requirement regarding patent examination aims to ensure that the process itself does not unduly curtail patent terms.

Notably, the focus here is on the general process for examination of patent applications in a given nation and not on individual patents. Accordingly, as long as the *process* is reasonable, there would not be a TRIPS violation. TRIPS does not define reasonable, and no suggestions are given about any presumptive length of examination of patents. Given the lack of definition, as well as the inherent subjectivity in interpretations of the term *reasonable*, member states should easily comply with this requirement.

2. Challenges to Patents and Patent Applications

TRIPS contemplates that nations may provide procedures to enable third parties to challenge patent applications as well as issued patents. Although TRIPS does not require such procedures, it does set procedural guidelines for nations that do so. In particular, TRIPS requires such procedures be (1) fair, (2) equitable, (3) not unnecessarily complicated or costly, and (4) without undue delay.[29] TRIPS is silent, however, on what the grounds should be for the bases of such challenges, as well as who would be a proper party to raise one. Accordingly, member states are free to decide not only whether to include such procedures, but who may challenge patents and patent applications and the grounds on which a challenge may be made. In addition, members may determine how many opportunities third parties should have to challenge applications and patents—limited only by the TRIPS requirement that such procedures occur without undue delay.

III. Patent Rights & Remedies

Under TRIPS, each patent owner has the usual right to exclude unauthorized others from certain activities regarding the patented invention for a particular term. Patent rights are to be provided without discrimination as to the field of technology. In other words, inventions in all fields should be provided the same rights. Beyond the nondiscrimination requirement, standard patent rights include exclusionary rights that last for a certain period of time. Those standard rights are subject to two exceptions that will be later discussed.

[29] *Id.* art. 62(4).

A. STANDARD PATENT RIGHTS

1. Exclusionary Rights

The general rule is that the owner of a patent on a product (such as a drug compound) is permitted to exclude all others from a variety of activities regarding that patented invention—including making, using, selling, offering to sell, or importing (subject to an important caveat) it.[30] The patent owner has the right and authority to prohibit *each* of these activities. In addition, there is no requirement the exclusion be limited to commercial activities.

However, the usual right to exclude imports of the patented product (into the patent-granting nation) is not absolute. TRIPS contemplates that nations may limit the patent owner's importation right in one instance—if the patent owner attempts to exclude the importation of a patented product the owner first sold in another country. For example, if Ford sold a patented engine in Germany at a premium due to its patent rights, Ford might be barred from preventing a third party from importing the patented engine into Bolivia on the assumption that Ford already profited from its patent in Germany. This situation generally only arises when the same good is priced differently in different nations. However, that is the very business model of drug companies that sell patented drugs. Accordingly, whether nations permit patent owners to exclude imports of their own products (which would likely be sold at a lower cost) is important to patent owners. TRIPS essentially permits each member to decide whether to permit or bar such imports by precluding nations from raising formal disputes concerning this issue.[31] (This exception to imports, which is often referred to as *international exhaustion*, is further explained in Chapter 2.) What is important to understand with respect to TRIPS is that countries *may*—but need not—provide an exception to the importation right. In particular, if the patented product was first sold by the patent owner (or under her authority) abroad, a nation may consider the patent owner's right to exclude imports to be exhausted by the first international sale. If a nation provides such an exception, it must be applied uniformly to all consistent with the TRIPS requirement of nondiscrimination.[32] In other words, a country cannot create an exception for some WTO member countries but not others.

---

[30] TRIPS art. 28. TRIPS also provides rights to owners of patents on methods, although these are much less valuable to the pharmaceutical industry. *Id.* art. 28.1(b).

[31] TRIPS art. 6 ("For the purposes of dispute settlement . . . subject to the provisions of Articles 3 and 4, nothing in this Agreement shall be used to address the issue of exhaustion of intellectual property rights.") *See also* World Trade Organization, Ministerial Declaration on the TRIPS Agreement and Public Health of November 14, 2001, ¶ 5(d), WT/MIN(01)/DEC/2, 41 I.L.M. 755 (2002) [hereinafter Doha Public Health Declaration] ("The effect of the provisions in the TRIPS Agreement that are relevant to the exhaustion of intellectual property rights is to leave each Member free to establish its own regime for such exhaustion without challenge, subject to the MFN and national treatment provisions of Articles 3 and 4.").

[32] TRIPS arts. 3–4.

2. Term of Protection

A patent owner can legally enforce the right to exclude others from the patented invention during the patent term. Accordingly, the length of that term is just as important as what activities may be excluded.

TRIPS requires that there be a minimal term of protection.[33] However, TRIPS does not limit protection to issued patents. Rather, TRIPS permits (but does not require) members to also provide rights to holders of pending patent applications.

Regardless of when member states desire to begin protection (as this date is not specified under TRIPS), the protection may not end before twenty years from the patent filing date. This time frame is somewhat confusing because although it refers to the patent application filing date, TRIPS does not actually require rights to begin with the filing date. Although nations have the option to begin rights then, that is not the only possibility, as a nation could alternatively not provide any rights until the patent is granted. In particular, a nation could provide that the patent term begins with issuance of the patent and ends twenty years from filing, so in essence, the patent term is twenty years from filing minus the period of patent examination. So, if the patent examination process took three years, the actual patent term would be seventeen years from the date the patent issued.

Importantly, TRIPS does not require a uniform term for all patents. In addition, it does not require or even permit a uniform patent term calculated from the date of the patent grant. A WTO panel has clarified that it is improper to set a uniform term for all patents that begins from the date of the patent grant as that would not necessarily always comply with the requirement that the term *end* at least twenty years from the original filing date. In particular, Canada's law that provided a uniform patent term of seventeen years from the patent grant was deemed improper because there were some cases where the patent term would end before twenty years from filing.[34] Even though most patent terms would end no earlier than twenty years from filing, the WTO panel found this term impermissible, and the Appellate Body affirmed.

Members are free to provide more protection in two ways. First, members can extend the patent term beyond twenty years. Some members do in fact provide an extension of the patent term where the examination process is particularly lengthy, or if the effective term of a drug patent is considered curtailed because of a lengthy regulatory review process. Second, members can (but need not) provide rights to holders of patent applications. Although extending the patent term is a straightforward concept, providing rights to a holder of a pending application requires a bit more explanation—especially as not all patent applications result in patents.

---

[33] *Id.* art. 33. (The term of protection available shall not end before the expiration of a period of twenty years counted from the filing date").

[34] Panel Report, *Canada—Term of Patent Protection*, WT/DS170/R (May 5, 2000), *aff'd by* Appellate Body Report, *Canada—Term of Patent Protection*, WT/DS170/AB/R (Sept. 18, 2000).

Nations that provide some rights before a patent is issued generally only permit enforcement of those rights *if and when* a patent issues. In other words, some countries will permit a patent owner to sue for monetary damages based on the use or sale of the invention before it became a patent, but only if and when an actual patent is granted. In addition, retroactive damages are only applicable to the time between when the application was published and the final patent issued.[35]

In all countries, the owner of an issued patent must have certain rights. In particular, a patent owner can bring a lawsuit to enjoin the violation of its right to exclude as well as for monetary damages. In fact, nations are required to provide such remedies to patent owners.

A visual depiction of possible patent terms, as well as what rights are provided, is shown in the following figures below that illustrate two different types of laws. In Figure 7 (Scenario 1), the law permits a patent owner to obtain retroactive damages for activity that occurred after the patent application was published, but before the patent was granted. By contrast, Figure 8 (Scenario 2) illustrates what remedies are available to a patent owner if the law only provides remedies for the period after the patent is granted. The phrase "full exclusionary rights" refers to the fact the patent owner can seek both monetary damages and an injunction. In contrast, "retroactive $" refers to the patent owner being able to obtain monetary damages after the patent is issued, but no injunction would be granted during this time.

In addition, in each figure, two time lines are provided to underscore that the length of the patent term is not finite—rather, it is a function of the time it takes the patent office to evaluate and grant the patent. The first line depicts a situation where a patent is granted three years after the application is filed. The minimum TRIPS patent term ends twenty years after filing such that the actual patent term is the difference between twenty years

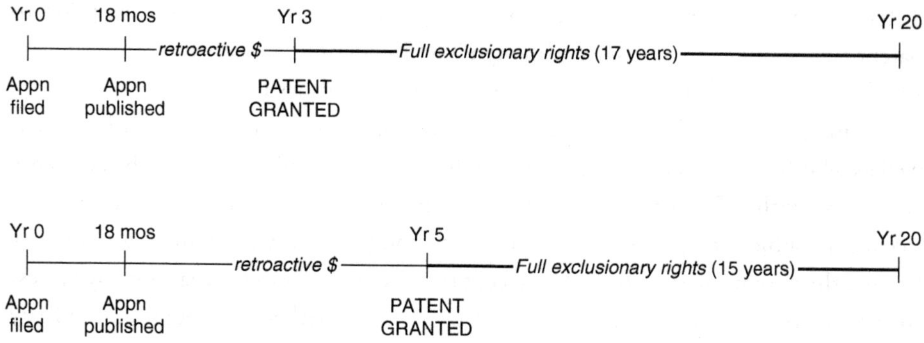

FIGURE 7 TRIPS Patent Term—Scenario 1

[35] *E.g.*, 35 U.S.C. 154(d) (2006) (providing a reasonable royalty for the time from publication until patent issuance in limited situations that all require actual notice of the published patent application); Patents Act, 1977, c. 37, art. 69 (Eng.)(permitting an action or damages for activity that occurred after publication and before the patent is granted, but only if a patent is granted).

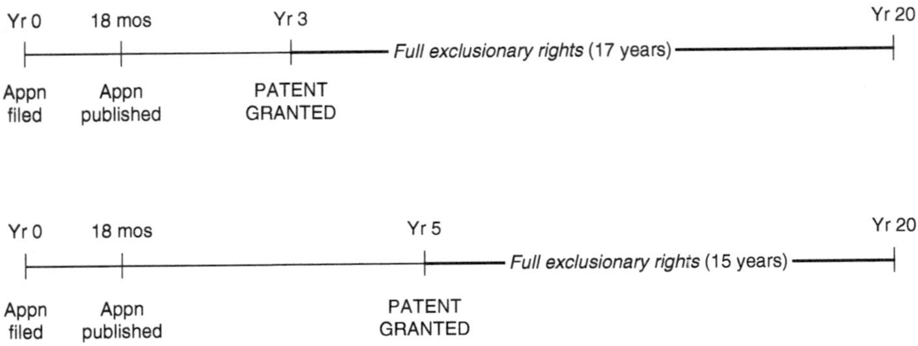

FIGURE 8 TRIPS Patent Term—Scenario 2

and the three years during which the patent application was pending; in other words, if a patent issues three years after the application, the patent term is seventeen years. Similarly, the second line in each figure shows a shorter patent term where there is a longer (five-year) pendency between the filing and patent grant.

B. EXCEPTIONS TO PATENT RIGHTS

TRIPS provides two important exceptions to patent rights. First, it imposes procedural requirements for compulsory licenses of patents that apply only to the extent that a nation grants compulsory licenses. Second, TRIPS permits "limited" exceptions other than compulsory licenses if they comply with several requirements set forth in article 30.

1. Compulsory Licenses

TRIPS permits nations to continue to provide a historically recognized exception to patent rights: *compulsory licenses*. Although TRIPS does not actually use this term, it is widely understood that TRIPS article 31 relates to compulsory licenses.[36]

An important initial issue is to determine what constitutes a compulsory license, as well as whether TRIPS mandates that such licenses be provided. A compulsory license consistent with TRIPS would be the government's unauthorized use of a patented invention, subject to payment of some fee to the patent owner. Unlike a voluntary license, the patent owner is forced to accept this use, as well as a fee that is likely far less than what the owner would like to charge in a free market. Although compulsory licenses have a long history of use in many countries, they also have a history of global controversy. TRIPS does not eliminate their use or the surrounding controversy. TRIPS does not require member countries to provide compulsory licenses; however,

[36] TRIPS art. 31. Most of this provision is reprinted in the Appendix.

to the extent they do, TRIPS imposes restrictions regarding the *form* of compulsory licenses. TRIPS does not dictate what subject matter may be licensed, as this is solely an issue for national determination; however, no patented matter—including drugs—are exempt from possible licenses. As TRIPS clearly permits the use of compulsory licenses, the many procedural requirements for them are given more detailed treatment in Chapter 5. However, the following questions may nonetheless help to reinforce some fundamental issues concerning such licenses under TRIPS:

Q: Does TRIPS make compulsory licenses illegal?
A: No! TRIPS permits nations to grant compulsory licenses and only imposes procedural requirements surrounding how the license is issued.
Q: Can drugs be subject to compulsory licenses under TRIPS?
A: Yes! There is no provision under TRIPS that exempts patented drugs from consideration. Member states are allowed to decide what subject matter is subject to a compulsory license.

2. "Limited" Exceptions

Besides permitting compulsory licenses as an exception to patent rights, TRIPS also permits countries to provide "limited exceptions" to patent rights under article 30. A limited exception exists when all the following requirements exist:

(1) the exception is a "limited" one in that it is a small deviation from the usual patent rights; *and*
(2) the limited use must not "unreasonably conflict with normal exploitation" of the patent; *and*
(3) the exception must not "unreasonably prejudice" the "legitimate interests of the patent owner"—taking into account the "legitimate interests of third parties."[37]

Although many of the important terms are not defined, a WTO panel has helped to provide insight into some of these provisions and also ruled on two situations of direct relevance to access to medicine. In particular, a WTO panel analyzed whether two different provisions of Canadian law intended to expedite the entrance of generic drugs *after patent expiry* qualified as permissible exceptions under article 30.[38] One exception was

---

[37] TRIPS art. 30. The full text is reprinted in the Appendix. Panel Report, *Canada—Patent Protection of Pharmaceutical Products*, WT/DS114/R, ¶¶ 7.20–7.21 (Mar. 17, 2000) [hereinafter WTO Panel Report, *Canada Generics*] (noting that these are cumulative requirements).

[38] *Id.* This panel report has been criticized and its continued application questioned. In particular, this panel report preceded the Doha Public Health Declaration that clarified the importance of using TRIPS articles 7 and 8 to interpret all TRIPS provisions. *E.g.*, Robert Howse, *The Canadian Generic Medicines Panel: A Dangerous*

found appropriately "limited," but the other was not. The exception held to be appropriately limited is common in many countries and is often referred to as either a *regulatory review* exception or a *Bolar* exception (referring to a case that was legislatively overruled to create the exception).[39]

a. Regulatory ("Bolar") Review Exception is Appropriately Limited

First, the panel found that permitting a generic company to make limited quantities of a patented drug during the patent term for the purpose of seeking regulatory approval to sell the drug *after* patent expiry was a permissible exception to the patent owner's usual right to exclude others from making the patented invention. The exception was "limited" because the patented drug was made solely to obtain regulatory approval, with no commercial sales until after the patent term expired.[40] However, the panel found that although a patent owner is generally entitled to exclude others during the patent term, "normal exploitation" of a patented drug does not include the several years of de facto exclusivity that would occur if generic companies were prohibited from seeking regulatory approval until after the patent expired. Similarly, the panel found the patent owner had no "legitimate interest" in an identical patent term as owners of patents not subject to regulatory review. In other words, even though owners of drug patents often have a shorter *effective* patent term than patent owners of other inventions that do not require regulatory review, they have no "legitimate" interest in extending their exclusivity period beyond the actual patent term.[41] An implicit, but not fully articulated issue is that there may be a legitimate interest of consumers or the broader society in ensuring that generic drugs are not unduly delayed after the patent term has expired.

b. Stockpiling Exception—Can It Ever Be Appropriately Limited

Although seeming to acknowledge the interest in lower-cost drugs, the panel did not go so far as to suggest that all activities during the patent term that expedite entry of generic drugs upon patent expiry were permissible. In fact, the panel found a different provision of Canadian law, referred to as the "stockpiling exception," not appropriately limited and thus impermissible.[42] Canada permitted generic manufacturers who were granted regulatory approval to sell generic versions of the patented drug after patent expiry to make and

Precedent in Dangerous Times, 3 J. W. INTELL. PROP. 493, 496–501 (2000); CHRISTOPHER GARRISON, EXCEPTIONS TO PATENT RIGHTS IN DEVELOPING COUNTRIES, 19–42 (UNCTAD/ICSTD 2006), *available at* http://www.iprsonline.org/resources/docs/Garrison%20-%20Patent%20Exceptions%20DC%20-%20Blue%2017.pdf.

[39] *See* Roche v. Bolar, 733 F.2d 858 (1984).

[40] WTO Panel Report, *Canada Generics*, ¶ 7.45.

[41] *Id.* ¶ 7.82. Of course, the panel's finding does not mean that nations are precluded from providing owners of drug patents with longer terms if they elect to provide more than the minimum term required under TRIPS.

[42] *Id.*, ¶ 7.38.

"stockpile" unlimited quantities of the patented drug during the last six months of the patent term. Canada argued that this was helpful for ensuring that generic competition existed on the very day the patent expired; without the exception, consumers would not be able to purchase generics for several months after the patent expired, as that is the time necessary for a firm to gear up for commercial production.

The WTO panel found that Canada's stockpiling provision was not adequately limited. Importantly, the panel did not say that stockpiling of generic versions can never be permissible; however, it seemed important that there was no limitation in the *quantity* that was made in the last six months.[43] Also, although not explicitly stated, it may be relevant that the stockpiling exception only reduced a de facto extension of the patent monopoly for a few months—in contrast to the several years involved with the regulatory approval exception.

c. Other Limited Exceptions?

Although the WTO panel did not clearly define the scope of all possible exceptions to patent rights, it at least clarified that countries can continue to provide an exception for regulatory review of generic applications, which most countries recognize as important for promoting faster and cheaper access to medicine. Another common exception the WTO panel has not ruled on (but is likely permissible) is "experimental use" of a patented invention.[44] In many countries, making a patented invention for experimentation is permissible. The WTO panel seemed to implicitly suggest that countries had a legitimate interest in facilitating use of inventions during the patent term to ensure that the overall goal of patents to advance technical knowledge is not frustrated.[45] No countries have challenged whether experimental use is permissible. Although this is not dispositive of the meaning of article 30, unless and until there is a challenge and a WTO panel holds otherwise, countries should be able to continue to recognize such an exception, which could assist in the creation and development of new drugs.

d. Does a Limited Exception Need to Comply with the Nondiscrimination Requirement?

The same WTO panel asserted that the nondiscrimination requirement with regard to fields of technology applied to this exception. Although the panel's statement on this issue is not well supported in the text, its application did not seem to change the ultimate result.[46] For example, the panel noted that the regulatory review exception did not

---

[43] *Id.* ¶¶ 7.30–7.36.

[44] *Id.* ¶ 7.69 (noting that although the panel purported to assert no opinion on whether experimental use was permissible under TRIPS article 30, it recognizes that scientific experiments constitute a type of legitimate interest).

[45] *Id.*

[46] *Id.* ¶ 7.91; Carlos M. Correa, Trade Related Aspects of Intellectual Property Rights: A Commentary on the TRIPS Agreement 283 (2006).

constitute either explicit or implicit discrimination. The panel held that the text did not explicitly discriminate because it was not limited to drugs. The panel accepted Canada's representation that the law applied to all products subject to marketing approval—despite the fact that a lower court held that it applied only to drugs. In addition, the panel determined that the law was not implicitly discriminatory because the EU failed to show clear evidence of a discriminatory purpose despite evidence that Canada intended to regulate pharmaceuticals. The panel suggested that "preoccupation" with the impact of law to one area does not indicate a discriminatory purpose unless there is evidence that the broader purpose is a "sham."[47]

C. ENFORCEMENT

Although discussion of TRIPS tends to focus primarily on the fact that some member states must for the first time provide patents, an equally important requirement of TRIPS is effective *enforcement* of patent rights. After all, a patent is of no commercial utility unless it is enforceable against competitors.

TRIPS provides comprehensive requirements for enforcement of all intellectual property rights, including patents. TRIPS does not require a separate judicial system for enforcing these rights, but enforcement must comply with certain basic ground rules. Notably, each member must ensure that enforcement procedures are "available" under its law so as to permit "effective action" together with "expeditious remedies." A WTO panel has clarified that it is not enough for a procedure to exist if remedies are essentially unavailable for some works.[48] In addition, all enforcement procedures must be "fair and equitable" and shall "not be unnecessarily complicated or costly or entail unreasonable time limits or unwarranted delays."[49]

1. Regular Enforcement

TRIPS requires that each member's judicial authorities have certain rights. In particular, all WTO members must have the authority to stop infringement, as well as to order that an infringing party pay damages that are "adequate to compensate" for the infringement.[50] TRIPS does not set forth specific requirements on how such damages should be assessed other than that they be "adequate to compensate." As with other terms, this lack of

---

[47] WTO Panel Report, *Canada – Generics*, ¶ 7.104.

[48] Panel Report, *China—Measures Affecting the Protection and Enforcement of Intellectual Property Rights*, WT/DS362/R (Jan. 26, 2009) (adopted Mar. 20, 2009). The case involved Chinese copyright law that refused to permit enforcement of rights for copyrighted works whose publication or dissemination are prohibited by law (as being unconstitutional or immoral). Although these works were not denied copyright protection, they were denied effective enforcement of rights.

[49] TRIPS art. 41(2).

[50] *Id.* art. 44 (authority to issue injunctions); *Id.* art. 45 (authority to order infringer to pay damages).

definition should give nations discretion to decide what is adequate. Similarly, although TRIPS requires that courts be empowered to stop the infringing activity, it does not mandate specific situations where this should happen. In many countries, courts have the authority to stop infringement by issuing an injunction against the infringing party, which they decide to impose based on equitable grounds. TRIPS does nothing to disturb this.

2. Provisional Measures

TRIPS also requires provisional measures to be given to rights-holders before a final judgment of rights. In other words, TRIPS mandates that judicial authorities must have the authority to prevent alleged infringement from occurring before they make a final decision on whether infringement has in fact occurred. One way judicial authorities have been known to do this is through issuance of a preliminary injunction before holding a trial on the merits.

As with other remedies, TRIPS only requires these possible preliminary remedies be within the scope of judicial power, but does not dictate when courts must actually impose them. Traditionally, many countries have permitted courts to order a preliminary injunction before a trial on the merits if the rights-holder is likely to prevail and the equities suggest that an injunction is appropriate. TRIPS does not change this. However, a court could also elect to refrain from stopping allegedly infringing conduct in advance of a final determination of infringement—especially when doing so would result in increased public access to a lower-cost drug.

3. Border Measures

TRIPS also imposes some obligations for the suspension of infringing goods by customs authorities at national borders. The only official obligation is to prevent importation of counterfeit trademark or pirated copyright goods.[51] However, the text explicitly states that members may permit similar suspensions for infringements of other types of intellectual property, which could extend to patents.[52] An important current international issue is whether border measures to police patent infringement might be technically permissible under this provision, yet nonetheless inconsistent with TRIPS because of contravening other provisions of TRIPS.[53] For example, this issue arises in the context of a number of shipments of generic goods from India that have been stopped en route to other countries pursuant to an EU Regulation that permits seizure of suspected

---

[51] *Id.* art. 51.

[52] *Id.* (stating "members may enable such an application to be made in respect of goods which involve other infringements of intellectual property rights. . . .").

[53] One issue is whether permitting customs officials to stop drugs that allegedly infringe a local patent would result in enforcement procedures that involve an "unwarranted delay" in violation of TRIPS article 41.

infringement of patented goods (and includes in-transit goods within its reach). This is an important, yet complicated issue given extended discussion in Chapter 10.

IV. Data Protection

TRIPS also requires protection of clinical data submitted to government authorities in conjunction with obtaining government approval to sell a new drug; governments must protect the data from *unfair commercial use* by others, although as will be explained, exactly what constitutes such use is controversial.[54] This protection is wholly separate and in addition to any potential patent protection that may exist for the related drug. Notably, the protection is for clinical data *associated with* a drug rather than protection of the drug itself.

The scope of this protection is of great importance to access to medicine, as the extent to which clinical data is protected impacts when generic drugs may be approved. As noted in Chapter 1, generic drug applicants are generally not required to submit original studies on the safety and efficacy of their proposed generic equivalent of a previously approved drug. Rather, generic applicants rely on the existence of data submitted by the patent owner of the previously approved drug that it was safe and efficacious. The generic applicant submits only limited data to show its proposed drug is bioequivalent to the prior one and relies on the fact that the equivalent drug was previously deemed safe to justify an inference that its proposed product is safe as well.

There are two primary issues with the TRIPS data protection requirement. The first is the scope of protection: TRIPS only requires data to be protected from "unfair commercial use." The second is when member states have an obligation to protect data.

A. "UNFAIR COMMERCIAL USE"

A critical point of contention is what it means to protect data from "unfair commercial use"—another important, yet undefined phrase in TRIPS. Although there are three separate words in this critical phrase that could each be separately analyzed, the key issue is whether a generic applicant's *reliance* on prior data constitutes not only a "use," but a "commercial use" that is unfair. PhRMA, as well as countries that strongly support PhRMA, consider this to be clearly the case. However, as "unfair commercial use" is not defined, countries retain some flexibility to define it, as further explained below.

[54] TRIPS art. 39(3) ("Members, when requiring, as a condition of approving the marketing of pharmaceutical or of agricultural chemical products which utilize new chemical entities, the submission of undisclosed test or other data, the origination of which involves a considerable effort, shall protect such data against unfair commercial use.").

Some have argued that reliance on data is clearly not use, such that countries can permit generic applicants to rely on prior data. This conclusion is a permissible reading of TRIPS. As "use" is not defined, member states should be free to define it as requiring actual use, but not reliance. Of course, "use" could be alternatively defined to include reliance. The important issue, however, is that *both* definitions are consistent with the undefined TRIPS term. A Canadian court has explained that since regulatory authorities never actually use the data of an originator application, a generic applicant would not—in its view—be relying on the data of the originator.[55]

However, even if reliance were arguably use, whether it would necessarily be *unfair commercial* use is debatable.[56] PhRMA, as well as the USTR, suggests that a generic applicant's reliance on previously submitted data is inherently unfair and impermissible under TRIPS because it inappropriately permits free riding on the work of another.[57] However, this is not the only possible interpretation.

Not only are member states free to define what is "unfair", but there is importantly no generally accepted interpretation of what the word means in this context. In assessing the ordinary meaning of a term pursuant to international interpretation rules, a dictionary definition is a common reference point. The term unfair may be defined as "not equitable or honest or impartial according to the rules."[58] However, what is considered "not equitable" is likely to differ depending on point of view. Indeed, differing views of what is a proper use of clinical data by a generic applicant were reflected in differing laws of countries prior to the conclusion of TRIPS. Although some countries adopted PhRMA's approach in precluding subsequent applicants from relying on original clinical data for a certain period of time, other countries permitted generic applicants to immediately rely on such data in the interest of promoting increased competition and lower-cost drugs.

---

[55] Bayer Inc v. Canada, 1999 1 FC 533 (FCTD), 84 CPR (3d) 129, *aff'd by* 87 CPR (3d) 293 (FCA), leave to appeal refused (259 NR 200). Interestingly, the court came to this conclusion in interpreting a provision of NAFTA that explicitly bars reliance, whereas TRIPS only bars unfair commercial use, and does not use the word "rely" or "reliance." *See* North American Free Trade Agreement, U.S.-Can.-Mex., art. 1711(6), Dec. 17, 1992, 107 Stat 2057, 32 I.L.M. 289 (1993) [hereinafter NAFTA].

[56] There is technically a sub-issue of what constitutes a commercial (as opposed to noncommercial) use. However, the views on this are predictable and do not change the analysis. In particular, although PhRMA suggests any reliance must be a commercial use as it leads to commerciality, others consider reliance as possibly not commercial as it is merely an application for government review with no immediate commercial benefit.

[57] *E.g.*, Request for Consultation by the United States, *Argentina—Patent Protection for Pharmaceuticals and Test Data Protection for Agricultural Chemicals*, WT/DS196/1 (June 6, 2000); *see also* Organisation of Pharmaceutical Producers of India, OPPI Opposition Paper, Regulatory Data Protection—A Building Block for Pharmaceutical R&D (2008) (stating that unfair commercial use was understood by TRIPS negotiators to mean that data will not be used to "support, clear or otherwise review other applications" unless approved by the data originator because any other definition would be "inconsistent with the logic and negotiating history of this provision").

[58] Concise Oxford Dictionary (7th ed. 1989).

As the definition of *unfair* is ambiguous, the next step in proper interpretation is to consider the broader interpretive context. Although protection for drugs is required pursuant to TRIPS article 39(3), article 39(1) provides additional context. In particular, protection of all information under article 39 is stated to be in the context of ensuring effective protection against "unfair competition" as required by the Paris Convention,[59] which defines it as "any act of competition contrary to honest practices in industrial or commercial matters."[60] However, this definition still does not provide a definitive answer on what would be unfair, as honest practices may differ among countries. Indeed, academics have suggested that not only is the term *unfair* flexible, but that *competition* is a term that should be defined by individual countries.[61]

As the ordinary meaning of the phrase "unfair commercial use" cannot be clearly divined from the usual sources, the next interpretive step pursuant to the Vienna Convention rules of interpretation (which govern interpretation of TRIPS) is to consider supplementary means of interpretation, including preparatory work regarding the treaty such as its drafting history. The history of article 39(3) reveals that negotiators did consider language that would have explicitly precluded a generic company from relying on data of another (for a period of time). In particular, a prior proposal stated that "*data may not be relied upon* for the approval of competing products for a reasonable time."[62] This proposal also reflected an existing international agreement that required data exclusivity.[63] TRIPS negotiators were well aware of what language to use to clearly bar generic applicants from relying on data, but ultimately adopted a more ambiguously

---

[59] TRIPS art. 39.1.

[60] Paris Convention for the Protection of Industrial Property, Art. 10*bis*, Mar. 20, 1883, revised July 14, 1967, 21 U.S.T. 1583, 828 U.N.T.S. 305.

[61] *E.g.*, G.H.C. Bodenhausen, Guide to the Application of the Paris Convention for the Protection of Industrial Property as Revised at Stockholm in 1967, at 144 (1991); Carlos Maria Correa, *Protection of Data Submitted for the Registration of Pharmaceuticals: Implementing the Standards of the TRIPS Agreement* (South Centre, Geneva), 25 (2002); *see also* 3 Stephen Ladas, Patents, Trademarks, and Related Rights, National and International Protection 1685 (1975) (noting that there is no clearly objective standards for finding unfair competition).

[62] Brussels Draft, para. 4A, *reprinted in* Daniel Gervais, The TRIPS Agreement: Drafting History and Analysis 421 (3d ed. 2008) (emphasis added) [hereinafter Brussels Draft]; *see also Draft Agreement on the Trade-Related Aspects of Intellectual Property Rights*, Communication from the United States, art. 33, GATT Doc. MTN.GNG/NG11/W/70 (May 11, 1990) ("contracting parties . . . shall not use the trade secrets for the commercial or competitive benefit of . . . any person other than the right holder except with the right-holder's consent, on payment of the reasonable value of the use, or if a reasonable period of exclusive use is given the right holder."); *Draft Agreement on Trade-Related Aspects of Intellectual Property Rights*, Communication from the European Communities, art. 28, GATT Doc. MTN.GNG/NG11/W/68 (Mar. 29, 1990) (explicitly stating that countries should protect clinical data "against *unfair exploitation by competitors*")(emphasis added).

[63] NAFTA art. 1711.5-.6. WIPO apparently also suggested data exclusivity in a model law, although its proposal was not published until after TRIPS came into force. WIPO Provisions on Protection Against Unfair Competition, art. 6(4) (1996) ("any act in the course of industrial or commercial activities, shall be considered an act of unfair competition if it consists or results in an unfair commercial use of secret test or other data, the

worded provision. A proper interpretation of this history is that members considered but rejected an absolute bar on reliance by declining to adopt specific language in favor of the more ambiguous wording. Accordingly, members are not limited to the language that was considered but rejected.[64] Of course, countries can require that generic applicants be precluded from reliance on prior data consistent with TRIPS; however, that is because TRIPS only provides minimum standards and permits members to adopt stronger protection.

This interpretation is also supported by the general interpretive principle of *in dubio mitius,* which has been recognized by the WTO Appellate Body to suggest that countries cannot be assumed to impose more onerous (rather than less burdensome obligations) when the meaning of terms are not clear.[65] Accordingly, given the ambiguity of "unfair commercial use," this term should not be narrowly interpreted to preclude reliance on data of an originator.

This interpretation is also supported by the lack of a uniform subsequent practice. Countries have a variety of different approaches to protecting data, with only some providing for data exclusivity. As noted in a prior WTO panel decision, unless there is nearly universal support for a proposition, it cannot be considered a subsequent practice that would constitute a binding interpretation.

So, the ultimate question is what TRIPS requires when data is "protected" if it does not preclude reliance on data by generic applicants. One possibility is that the protection means a second applicant is precluded from using data that was obtained through dishonest commercial practices. This would generally not be relevant to the usual case of a generic applicant that relies on previously submitted data because access to the data is unnecessary. Although PhRMA might suggest this interpretation is nonsensical in that it would not provide any true obligation, this interpretation does seem to fit with an appropriate interpretation of TRIPS that permits parties to define terms as they wish—especially where the more specific requirements PhRMA desired were rejected during negotiations.

Although data exclusivity could be consistent with TRIPS article 39(3) as providing more than the minimum of TRIPS, there is still a question of the minimum protection

---

origination of which have been submitted to a competent authority for the purposes of obtaining approval of marketing of pharmaceutical or agricultural chemical products which utilize new chemical entities").

[64] As stated by Professor Reichman, "to ignore the clear evolution of the text in favor of quasi-exclusive rights in regulatory data, in a form that was proposed but ultimately excised from the Final Act, would in effect amount to imposing un-bargained for trade concessions . . . that has no legal foundation whatsoever." Jerome Reichman, Undisclosed Clinical Test Data under the TRIPS Agreement and Its Progeny: A Broader Perspective, UNCTAD-ICTSD Dialogue on Moving the Pro-Development IP Agenda Forward: Preserving Public Goods in Health Education and Learning (Bellagio, 29 Nov–3 Dec), at 10 (2004) [hereinafter Reichman, Undisclosed Clinical Test Data].

[65] *See* Appellate Body Report, European Communities—Measures Concerning Meat and Meat Products (Hormones), ¶ 165, WT/DS26/AB/R (Jan. 16, 1998).

that is required. As noted above, there is a reasonable argument that countries need only prevent competitors from unscrupulous use of the data. Some have suggested the minimum requirement should at least provide some compensation to the originator of the data if its data is relied upon to approve a subsequent drug.[66] However, language that would have expressly required a compensation scheme was proposed, but not ultimately adopted.[67]

This section concludes that although data exclusivity, as well as a compensatory liability regime, would be *consistent* with TRIPS, neither are clearly required. Both were discussed, but the final text is sufficiently ambiguous to permit member states discretion to adopt the more minimal requirement of simply prohibiting unscrupulous use of data that does not involve relying on previous clinical data. This conclusion is consistent with the views of many scholars.[68] To help clarify what is permissible under TRIPS, Table 5 lists options that would each be consistent with TRIPS, although some of them provide more than TRIPS requires.

Importantly, although TRIPS does not mandate that generic applicants be precluded from relying on original data, this may be required in agreements subsequent to TRIPS that are more specific in their terms. This is further discussed in Chapter 9, as such agreements expressly provide for data *exclusivity*, which essentially imposes the types of requirements that PhARMA has tried to divine from the ambiguous TRIPS terms.[69]

---

[66] *E.g.*, Aaron Xavier Fellmeth, *Secrecy, Monopoly and Access to Pharmaceuticals in International Trade Law: Protection of Marketing Approval Data under the TRIPS Agreement*, 45 Harv. Int'l L. J. 443, 464 (2004) (suggesting that use is only unfair if not adequately compensated); Shamnad Basheer, *Protection of Regulatory Data Under Article 39.3 of TRIPS: The Indian Context*, 28-29 (2009), *available at* http://papers.ssrn.com/sol3/papers.cfm?abstract_id=934269.

[67] *Draft Agreement on the Trade-Related Aspects of Intellectual Property Rights*, Communication from the United States, art. 33, GATT Doc. MTN.GNG/NG11/W/70 (May 11, 1990) (suggesting one alternative to permitting use of originator data would involve "payment of reasonable value of the use").

[68] *E.g.*, Carlos Correa, Trade Related Aspects of Intellectual Property Rights: A Commentary on the TRIPS Agreement 383–92 (2007); Brook Baker, *Ending Drug Registration Apartheid: Taming Data Exclusivity and Patent/Registration Linkage*, 316 Am. J. L & Med. 303, 315–316 (2008); Reichman, Undisclosed Clinical Test Data, *supra* note 64. *But see* Nuno Pires de Carvalho, The TRIPS Regime of Patent Rights 392–93 (Kluwer Law Int'l 2d ed. 2005); Ingo Meitinger, *Implementation of Test Data Protection According to Article 39.3 TRIPS: The Search for a Fair Interpretation of the Term "Unfair Commercial Use"* 8 J. World Intell. Prop. 123, 133 (2005) (arguing that it is reasonable to consider language in drafts to give "real meaning" to the term "unfair commercial use" that would result in barring generic applicants from relying on prior data).

[69] The pharmaceutical industry has repeatedly suggested that TRIPS already requires data exclusivity. *E.g.*, Pharmaceutical Research and Manufactruers of America special 301 submission 2010, *available at* http://keionline.org/sites/default/files/USTR-2010-0003-0245.1.pdf.

TABLE 5

"Unfair Commercial Use"—Possible Implications

Data Reliance Allowed? (where originator has not authorized use)	Consistent with TRIPS?
Generic applicant may rely on data of originator without any payment	Yes—complies with ***minimum*** protection under TRIPS article 39(3)
Generic applicant permitted to rely on data of prior applicant, but only if generic applicant *pays* for use in some type of compensation scheme	Yes—but exceeds minimum protection
Generic applicant that attempted to rely on data of prior applicant would engage in *unfair commercial use* (US/EU/PhARMA position)	Yes—but exceeds minimum protection

B. WHEN DATA PROTECTION IS REQUIRED

Not all countries have an obligation to protect data under this provision, as it only applies to countries that require submission of data to obtain marketing approval for drugs having new chemical entities. Many developing countries approve the sale of drugs based on their prior approval in another country, such that this TRIPS requirement would have no applicability.

Even for countries that require submission of data to gain approval for drugs having new chemical entities, this requirement will not be applicable in every case. The TRIPS obligation only arises with respect to applications for drugs that use a "new chemical entity" where the submitted clinical data involves "considerable effort" that is not disclosed (other than in the application process). Most clinical data submitted for new drugs likely involves considerable effort along with time and expense[70] and is unlikely to be disclosed because there is a competitive interest in keeping it secret. However, not every application for marketing approval of a drug involves a "new chemical entity"— especially given that this term is not defined.

An important question is thus: when does an application involves a "new chemical entity," such that a country has some obligation to protect the data under TRIPS? As with interpretation of other undefined terms under TRIPS, member states should be permitted to define this term as they see fit. The narrowest definition of "new chemical entity" would be that the chemical entity is new to the entire world; in other words, it has

[70] Theoretically, countries can define the term "considerable effort" narrowly to situations where the data involved a certain amount of time or money. However, this is not necessary as there are other aspects of TRIPS article 39(3) that may be more easily interpreted to avoid any new obligation under this provision.

never been submitted for regulatory approval in any country. However, developed countries do not use this narrow definition; rather, they consider an application to involve a new chemical entity whenever no prior application *in that country* used the entity. However, as with other TRIPS requirements, there is not necessarily any need for developing countries to model their laws after developed countries—especially when they have different interests at stake.

Even if "new chemical entity" is not defined narrowly to mean first in the world, there are other possible limitations. For example, an application to market a new use or new dosage of a previously approved compound should never give rise to any obligation under this provision because the compound is not new in any sense of the word. In other words, the TRIPS requirement to protect clinical data from "unfair commercial use" would never apply to data associated with a new use of an old compound as there is no TRIPS obligation to protect this data; however, that does not mean that generic applicants will always be able to rely on data regarding a new use, as member states may provide *more* protection than TRIPS requires. Nonetheless, with respect to the basic TRIPS requirements that a developing country is attempting to meet (but not surpass), there should be no obligation with regard to protecting data regarding new uses or new dosages of a previously approved compound as that compound is not "new."

To summarize, there are several ways that a member state may not be obliged to provide protection to data under this TRIPS provision. First, it may not be obligated because TRIPS never requires submission of original data. Second, even if TRIPS requires submission of such data, a member state may avoid this requirement in at least some cases by narrowly defining certain terms, such as what constitutes a "new chemical entity." Although countries have some ability to limit protection of data by narrowly defining some of the above terms, the pivotal issue is what protection is actually required by TRIPS. This is the next important question.

C. TERM OF PROTECTION

Another ambiguity with data protection is that no time frame is specified. This would seem to suggest that countries can decide for themselves what duration is appropriate. Although PhRMA has suggested that TRIPS requires a minimum of five to ten years of protection, that suggestion is not grounded in a proper interpretation of TRIPS. In particular, flexibility to define the period of appropriate protection is reinforced by a proper examination of the negotiating history; in particular, countries proposed, but *rejected* language with specific time lines.[71] As none of the more specific language was adopted, it

[71] Prior language suggested "generally no less than five years" or a "reasonable time commensurate with efforts involved." Brussels Draft, ¶ 4A, reprinted in DANIEL GERVAIS, THE TRIPS AGREEMENT: DRAFTING HISTORY AND ANALYSIS 421 (3d ed. 2008).

does not govern. This is especially true as even among states that provide data exclusivity, there is no uniform term—such that there is no subsequent practice that could be used to impose a term upon TRIPS.

The following questions help to revisit some key points about data protection under TRIPS.

Q: Must all countries provide data protection?

A: No. This requirement only applies to countries that require submission of (usually clinical) data as part of a process for assessing whether a drug containing a "new chemical entity" is sufficiently safe and effective to be sold.

Q: Does TRIPS preclude generic companies from relying on prior data of another company in seeking approval of a generic equivalent?

A: Not necessarily. Although pharmaceutical countries have argued otherwise, TRIPS does not explicitly prevent *reliance* on data—it only sets the amorphous standard that data be protected from "unfair commercial use," which is up to member states to define. So, although some countries may consider reliance to be unfair commercial use, that is not the only possible interpretation. Generic companies may be able to rely on the data in countries that define the term to permit reliance (whether with or without payment).

V. TRIPS Compliance and Transitional Protection

A. OVERVIEW

Different time periods were set for when nations were required to bring their laws into compliance with TRIPS in recognition of the fact that TRIPS required a fundamental change to the laws of many countries. Developed countries, such as the United States and Japan, were required to bring their laws into compliance first: within one year of the signing of TRIPS (i.e., by 1996). Developing countries were required to comply immediately with some provisions, but were given five years to provide product protection for drugs if they had not previously done so; during the interim, they had to comply with additional transitional provisions as discussed in the next section.

Least developed countries (LDCs) were given the longest time line for compliance. The original date for LDCs to fully comply with TRIPS was ten years after TRIPS entered into effect. However, the deadline has since been extended another ten years (to 2016), with the potential for another extension.[72] In addition, whereas developing

[72] Council for Trade-Related Aspects of Intellectual Property Rights, *Extension of the Transition Period under Article 66.1 of the TRIPS Agreement for Least-Developed Country Members for Certain Obligations with Respect to Pharmaceutical Products*, IP/C/25 (July 1, 2002) [hereinafter *Extension of the Transition Period*].

countries were required to provide some patent-like protection to drugs during the transition period, no such requirement is currently applied to LDCs. However, whether they may eventually need to provide such protection is an issue subject to periodic review.

As developing countries and LDCs have different obligations, it is important to determine what countries fall into each category. The WTO refers to UN rules for determining a LDC, such that what constitutes one should be consistent with international norms.[73] The WTO is silent on definition of a developing country. This is subject to negotiation and possible dispute settlement.

All of the above dates were for those who were WTO members at the conclusion of TRIPS. However, WTO membership is not static. Since the conclusion of TRIPS, over 150 countries have become members. One of the requirements is that countries must bring their laws into compliance with TRIPS before they are admitted for membership.

Table 6 summarizes the different dates of TRIPS compliance for various countries.

TABLE 6

TRIPS Compliance Dates

Country	Compliance Date
Developed Countries (i.e., US, EU Member States):	1996
Developing Countries (i.e., India):	2000 (basic)—no product patents on drugs until 2005 if not previously granted, but transitional requirements before 2005
Least Developed Countries (i.e., Bangladesh):	2006; 2016 for product patents on drugs (with potential for further extension)
New Members (i.e., China as of 12/01):	Upon Admission to WTO (or as negotiated)

B. TRANSITIONAL PROVISIONS

Although all developing countries must currently be in full compliance with TRIPS, it is helpful to understand what was required during the transition period as it impacts what is currently patented—or not subject to patent—in these countries. As noted above, developing countries that did not provide product protection for drugs at the conclusion of TRIPS were given five years to do so. However, during the interim period, developing countries were required to comply with two other requirements. First, they needed to

[73] Marrakesh Agreement Establishing the World Trade Organization, Apr. 15, 1994, The Legal Texts: The Results of the Uruguay Round of Multilateral Trade Negotiations 4 (1999), Art. XI 1867 U.N.T.S. 154, 33 I.L.M. 1144 (1994).

provide an official means to accept patent applications for drugs and drug compounds to be examined after the transition period.[74] This requirement is often referred to as the "mailbox" provision (with a metaphorical mailbox to receive applications). Second, countries were required to provide some market exclusivity for inventions subject to these pending, yet unexamined applications.

The transitional provisions are best understood in the context of the broader negotiating process. Developed countries with strong pharmaceutical industries (such as the United States and EU Member States) wanted developing countries to comply with all TRIPS provisions—including patents on drug compounds—as soon as possible. Conversely, developing countries wanted to delay implementing this requirement as it was a major change to their patent systems. The compromise reached was to allow a five-year transition period, but also provide some means to protect the interests of developed countries. The transitional provisions foster the interests of developed countries as well as pharmaceutical companies interested in protecting their drug compounds.

1. Mailbox Applications

The mailbox provision is important for ensuring that inventions created before patents are examined (and properly filed pursuant to the mailbox provision) are likely to become issued patents once patents on products are permitted. It is important to remember the patent office evaluates whether an invention is new and has an inventive step as compared to the prior art before the date of the invention, which is usually the date of the patent application. If patent applications were not permitted until the date product patents were granted, inventions created before that time would likely be denied a patent because new developments (including even the inventor's own activities) would make the invention no longer new. The mailbox provision eliminates this problem by ensuring that when a nation begins to grant patents, the filing date of the mailbox application (as opposed to the date that it is examined) is used to assess whether the invention deserves a patent.[75]

2. Exclusive Marketing Rights (EMR)

A complementary requirement to the mailbox requirement is that developing countries provide *exclusive marketing rights* (EMRs) to some inventions filed as mailbox applications. EMRs are a new creation that emerged as a compromise during

---

[74] India was found to be in noncompliance when it had administrative instructions to permit mailbox applications, but those administrative rules conflicted with the actual law. Appellate Body Report, *India—Patent Protection for Pharmaceutical and Agricultural Chemical Products*, WT/DS50/AB/R (Dec. 19, 1997).

[75] Mailbox applications may result in a patent term that is shorter than patents filed and granted after product patents exist because the patent term for mailbox applications is calculated from the date of the original mailbox application.

TRIPS negotiations. An EMR offers some patent-like protection, but not the full scope of rights that accompany patents.

An important issue is what an EMR provides. As with many key TRIPS terms, there is no definition of an EMR. However, the term exclusive *marketing* right seems to focus on providing its owner with rights that only relate to marketing; in other words, rights only exist for *commercial* sales. There is no EMR against someone who makes and uses the invention without a commercial sale. This focus on commercial use of an invention distinguishes an EMR from a traditional patent right that permits an owner to exclude others from a variety of activities without regard to whether those activities happen to be commercial.

Not every invention is entitled to an EMR. The invention must be claimed in a mailbox application filed in the country that is granting the EMR *and* the invention must have previously been granted both a patent and marketing approval in another WTO member country. The only drugs that could have qualified for an EMR in a developing country such as India are those that were already patented and approved for sale in at least one other WTO country *and* subject to a pending mailbox patent application. The requirement that the invention be subject to a mailbox application substantially limits the scope of qualifying drugs, as a mailbox application would only be appropriate for a drug that satisfied patent requirements as of the date the application was filed. As noted earlier, patents are only appropriate for *new* inventions. So, a drug that has been long patented in another country would likely be inappropriate for a mailbox application (as it would probably not be "new").[76] These old inventions can continue to be made in developing countries because they were not patented and cannot be patented in the future because they are not new. However, an EMR could issue for a drug that was *recently* patented and marketed in at least one prior WTO country.

An EMR is limited in time. The EMR does not begin until marketing approval is granted in the developing country. Once granted, an EMR lasts for five years *or* until the mailbox application is either granted or rejected—whichever period is shorter. Importantly, until a company is granted marketing approval for an EMR, other companies can make, use, and even sell in the developing country at issue the invention that is the subject of the application.

[76] Although countries have discretion to define terms such as what is "new," no country thus far has defined "new" to include something that was publicly known for more than one year.

LDCs are not currently obligated to provide either product protection on drugs or EMRs pursuant to the Doha Public Health Declaration.[77] However, the waiver of the obligation to provide EMRs is to be periodically reviewed.[78]

The following questions help to revisit some key points about data protection under TRIPS.

Q: Does an EMR provide patent rights under a different name?
A: No. Although an EMR provides market exclusivity, a patent is broader in that it allows exclusion of use even beyond the commercial market.
Q: Is every application filed under the mailbox rule entitled to an EMR?
A: No. EMRs are only available if the invention subject to the application has been previously granted both a patent and marketing approval in another WTO country.
Q: Is an EMR available for any invention previously patented and marketed in another country?
A: No. It is true that an EMR cannot be granted to a drug unless it was previously patented and granted regulatory approval. However, to be awarded an EMR, an applicant must also establish that the invention is subject to a mailbox application, which would mean the invention was "new" at the time the mailbox application was filed. This may be possible for recently patented drugs for which the mailbox application was filed at the same time as in other countries such that the eventually issued patent (or published patent application) would not be prior art to the mailbox application.
Q: Does an EMR last twenty years?
A: No. An EMR only lasts for five years from issuance or until the related mailbox patent application is granted or rejected—whichever period is shorter.

VI. Conclusion

TRIPS introduces many new requirements to the global landscape. Here is an overview of some of the main requirements:

- all countries must grant patents, but only on "inventions"

[77] Doha Public Health Declaration ¶ 7; *Extension of the Transition Period*, *supra* note 72. Such countries have no obligation to provide patent protection until January 1, 2016—subject to the possibility of an additional extension.

[78] Doha Public Health Declaration ¶ 7.

- TRIPS provides patentability requirements (new, inventive step, industrial application), but does not define these terms, thus giving member states the flexibility to define them
- all countries must provide patent rights, but TRIPS permits countries to grant a compulsory license on any ground deemed appropriate by a country, subject only to its following procedural requirements under TRIPS
- a clear exception to patent rights is that member states may permit generic manufacturers to make limited quantities of a patented drug for purpose of obtaining regulatory review to sell a generic equivalent after the patent expires
- patent rights must be enforceable in a judicial system, but the system does not have to be dedicated to adjudicating only IP rights, and courts retain discretion to decide what constitutes appropriate monetary compensation and whether to prevent infringement via injunctions
- countries must in certain cases protect clinical data submitted to regulatory authorities from "unfair commercial competition," but that protection does not preclude generic applicants from relying on that data—despite the fact PhARMA may feel differently

What's Left?

With the above description as a backdrop, the subsequent chapters aim to provide more detail on how various provisions of TRIPS have been interpreted or contested, as well as how countries are endeavoring to negotiate subsequent agreements that go beyond TRIPS in one or more areas, including patentability, patent rights, and data protection, as well as enforcement/examination of patents.

4 Freedom under TRIPS: India as an Example

This chapter illustrates how individual nations may promote access to medicine while still complying with TRIPS, using India's patent laws as an example.[1] As noted earlier, TRIPS requirements are not only skeletal, but often undefined. Accordingly, under TRIPS, nations have freedom and flexibility to define important terms and requirements.

This chapter discusses various aspects of India's patent laws that are intended to promote access to medicine in a post-TRIPS world where patents must be granted. Although TRIPS flexibility may sometimes be constrained by international agreements or pressures (as discussed in Chapter 8), this chapter explores what flexibilities countries should have under TRIPS. Understanding existing flexibilities under TRIPS is important because it provides a foundation to better understand how subsequent obligations may limit these flexibilities.

[1] As this book goes to press, India is contemplating entering into a free trade agreement with the EU that might require India to modify some of its existing and innovative provisions. Nonetheless, the examples in this chapter should be of continuing relevance to illustrate how domestic rules can accommodate access to medicine while simultaneously complying with TRIPS.

By the conclusion of this chapter, the reader should be able to answer the following questions:

- How can a nation define patentability or patent rights to enhance access to medicine consistent with TRIPS?
- How can procedures for examining patent applications help promote access to medicine?
- How can limited remedies for patent infringement promote access to medicine?

I. Introduction 91
II. Patentability 91
  A. Invention Exclusion 92
    1. New Uses of Known Inventions 92
    2. New Forms of Known Substances 93
    3. TRIPS Compliance 94
  B. Inventive Step 96
  C. Implications of Patentability Rules for Innovation and Access 97
III. Reinforcing Patentability Requirements 98
  A. Patent Application Procedure 99
  B. Indian Law 100
    1. Pre-Grant Opposition 100
    2. Post-Grant Opposition 102
    3. No Presumption of Validity for Issued Patents 103
IV. Patent Rights and Remedies 103
  A. Patent Rights 104
    1. Infringement 104
    2. Patent Term 104
  B. Limitations on Patent Rights and Remedies 105
    1. Regulatory Review 105
    2. Compulsory Licensing 106
    3. International Exhaustion 107
    4. Revocation 109
    5. Limited Remedies—Injunctions 110
    6. Limited Rights and Remedies—Mailbox Application Patents 112
V. Regulatory Protection of Drugs 118
  A. Overview 118
  B. Data Protection (Not Data Exclusivity) 118
  C. Patent Linkage 120

VI. Concluding Thoughts 121
A. Will Generics Dry Up in India? 121
B. Market for India's Generic Drugs? 122
C. Will India Have Any Followers? 123
D. Other Unintended Consequences 123

I. Introduction

This chapter aims to the emphasize countries freedom and flexibility under TRIPS to promote access to medicine in two primary ways. First, this chapter evaluates the extent to which India's current laws enable it to remain a supplier of low-cost generic drugs globally. Second, this chapter considers whether India's laws (or some version of them) provide a useful model for other developing countries desirous of improving access to medicine. Importantly, even if some of India's laws may promote low-cost drugs in India, they may not have the same effect in other countries; India's success in making and supplying low-cost, yet high quality drugs is a function not only of its laws, but also of an existing infrastructure that has had long-term success with manufacturing generic drugs.

India's current patent laws illustrate several approaches to promoting access to medicine while still allowing patents. Access to medicine can be promoted by restricting the scope of what is patented; what is unpatentable may then be sold without waiting for another's patent term to expire. Restricting the scope of patentability can be aided by providing mechanisms to permit parties outside the patent office to challenge patent applications and patents. In addition, even as to patented drugs, access to medicine can be promoted by increasing exceptions to patent rights. Last but not least, courts may also play a role in promoting access to medicine.

The order of topics in this chapter roughly mirrors those in Chapter 3 (An Introduction to TRIPS). The chapter begins with highlights of India's patentability rules, followed by patent examination procedures, patent rights and exceptions, enforcement of patents, and finally, regulatory protection of drugs. Each section begins with a brief recap of the pertinent TRIPS requirements followed by a description of the relevant Indian law and an analysis of TRIPS compliance. The chapter concludes with a discussion of which of India's laws are most important in enabling India to continue to be a major supplier of generic drugs, as well as which laws should be considered for adoption by other countries who want to promote access to medicine.

II. Patentability

TRIPS requires all member states to patent inventions. In particular, TRIPS requires member states to grant a patent on any invention that satisfies three requirements: (1) it is capable of industrial application, (2) it is new, and (3) it has an inventive step. However, TRIPS does not define what constitutes an *invention* or the other requirements of patentability. Member states are thus permitted to define these terms so long as they do not violate any other obligation of TRIPS, such as the prohibition on discriminating against a field of technology. How each of these terms is defined can have an important impact on access to medicine. After all, if a drug is considered unpatentable, it may be made by anyone and will be available at a low cost because there will be no patent premium.

A. INVENTION EXCLUSION

India's patent laws provide an example of an attempt to define what constitutes an invention in a way that promotes access to medicine. India provides two important exclusions from the scope of inventions that have significant implications for the pharmaceutical industry. As noted Chapter 1, the pharmaceutical industry often focuses on finding new uses and developing new variations of existing blockbuster drugs; these strategies help to minimize development costs while maximizing possible profits. Although the industry is likely to continue this model, companies may have difficulty obtaining patents for such developments in India. In particular, Indian patent law section 3(d) declares the following to be unpatentable inventions:

(1) new uses of known substances, and
(2) new "forms" of known substances—unless there is a significantly different "efficacy"[2]

1. New Uses of Known Inventions

The first exclusion is a new use of a previously known compound. For example, the discovery that a topical skin cream originally intended to treat acne also is useful for treating wrinkles is a new use. Although some countries (such as the United States) permit both the original compound and the new use to be patentable, India's law would not permit the method of treating wrinkles to be patentable. The impact of this is that companies cannot bar others from using a previously discovered compound for new uses—even if the original company did in fact discover these uses.

Excluding new uses from the scope of what is an invention seems permissible under TRIPS. There is no definition of what constitutes an "invention" under TRIPS, so nations are free to define it themselves. In general, countries can define terms as they see

---

[2] The Patents Act, 1970, No. 39, § 87, Acts of Parliament, 1970 (India) [hereinafter India Patents Act] (*as amended by* The Patents (Amendment) Act, 2005, No. 15, § 3(d), Acts of Parliament 2005 (India)).

fit—especially if there are already a variety of global approaches such that there is no clear view. Views definitely diverge on whether new uses should be patentable. Indeed, scholars widely agree that countries need not consider new uses patentable, with one even referring to it as "based on a legal fiction of novelty."[3]

2. New Forms of Known Substances

The most controversial provision of Indian patent law is its exclusion of new forms of known substances from patentability unless there is proof of significantly different efficacy. It is important to understand what is a new *form* as well as what constitutes *efficacy*. The Indian law provides a detailed list of substances that would be considered a new form of the same substance. Essentially, the list reflects slight variations on a microscopic level.[4] The terms may sound to a lay reader as though they represent different compounds; however, to a scientist trained in drug discovery, these represent only minor variations.

Developing variations is a known strategy of drug companies that has been criticized as an attempt to "evergreen" their patents with new patents over minor variations. In the United States and EU, pharmaceutical companies typically first obtain a patent on an active ingredient, then successive patents on minor chemical variations to retain market exclusivity as each new patent will provide a new patent term of exclusivity, such that the patent term is "ever green." The practice is criticized by health advocates as unduly permitting companies to retain exclusivity and block the entry of lower-cost generics based on patents that provide minimal innovation. As noted in Chapter 1, many patented and marketed compounds fall into this category, such as Xyzal (a variant of allergy drug Zyrtec), Nexium (a variant of Prilosec), and Clarinex (a variant of allergy drug Claritin).

However, Indian law does not completely preclude variations of old compounds from patentability, as they may still be patented if they show significantly improved "efficacy." Accordingly, a critical question is what *efficacious* means. This important term is not defined in the Indian patent laws or under present Indian patent office guidelines. A plain English definition of efficacy requires only that something have a desired effect. In one closely watched decision, the High Court of Madras dismissed a challenge to the Indian patent office determination that Novartis's new variation of an anti-cancer drug was inadequately efficacious even though it had 30 percent more bioavailability, meaning

---

[3] Carlos Correa, *Patent Rights*, *in* INTELLECTUAL PROPERTY AND INTERNATIONAL TRADE: THE TRIPS AGREEMENT 189, 201 (Carlos Correa & Abdulqawi Yusuf eds., 1998); *see also* CARLOS M. CORREA, TRADE RELATED ASPECTS OF INTELLECTUAL PROPERTY RIGHTS: A COMMENTARY ON THE TRIPS AGREEMENT 274 (2007) [hereinafter CORREA, TRADE RELATED ASPECTS] (noting TRIPS does not require protection of a new use of a pharmaceutical product).

[4] Section 3(d) states that compounds that are considered the same known substance include "salts, esters, ethers, polymorphs, metabolites, particle size, isomers, mixtures of isomers, complexes, combinations, and other derivatives of known substance—unless they differ significantly in properties with regard to efficacy." India Patents Act § 3d.

it was more easily absorbed into the bloodstream,[5] while also declining to consider Novartis's challenge to the law as either inconsistent with the Indian Constitution or TRIPS.[6] The case is presently on appeal before the Indian Supreme Court.[7] However, regardless of the ultimate result in this case for India, this matter provides an example of how the term efficacious could be defined explicitly in national patent laws or guidelines so as to limit patentability.

The incorporation of an efficacy requirement into patentability is a novel approach that poses some serious challenges for the patent examination process. In general, evaluations of efficacy have been the province of regulatory agencies that assess whether drugs are sufficiently safe and efficacious to be sold to the public (as discussed in Chapter 1). An efficacy determination is made based on substantial clinical data developed over many years. Although similar clinical data could theoretically be provided to the patent office, these offices are not trained to examine such data. Moreover, patent applications on potential drug compounds are filed very early in the overall commercial development process—long before data exists regarding possible efficacy. Even more importantly, patent applications cannot realistically be delayed to obtain such information, because a delay would result in the invention failing to satisfy the requirement of being "new."[8]

3. TRIPS Compliance

An important issue is whether section 3(d) complies with TRIPS. The United States has noted this provision is "of concern," but has not actually suggested it violates TRIPS.[9]

---

[5] Novartis AG v. Union of India, (2007) 4 M.L.J 1153, *available at* http://www.scribd.com/doc/456550/High-Court-order-Novartis-Union-of-India.

[6] The court declined to find the term *efficacy* to be unconstitutionally vague. *Id.* ¶ 4 (asserting "everyone in the pharmaceutical field understands what is meant by enhancement in the known efficacy" such that the requirement in the Patent Act is not unconstitutionally vague or arbitrary). In addition, the court asserted that it lacked competence to rule on the TRIPS issue. *Id.* ¶ 8.

[7] Novartis AG v. Union of India (Appeal No. 20539-549, 2009), *see also* Novartis AG v. Union of India, Order No. 100/2009 (Intellectual Prop. App. Board, June 29, 2009)

[8] Regardless of India's patent laws, a company is likely to seek patent protection in other countries in order to ensure its invention is considered "new." As patent applications generally become published, earlier patent applications on the same invention would bar a later-filed application in India from being considered new as the published application would have disclosed the same thing. As discussed in Chapter 1, an inventor can be denied a patent based on information that existed before the patent application was filed (prior art)—even if that information was created by the same inventor.

[9] Office of the U.S. Trade Representative, 2010 Special 301 Report, 26 (2010) (noting the United States has urged India to provide stronger protection, and that section 3d is of particular concern in limiting patentability of "potentially beneficial innovations, such as temperature-stable forms of a drug or new means of drug delivery"). Interestingly, although this law has been in place since 2005, the United States has not consistently mentioned it in its Special 301 reports. For example, in 2009, there was no mention of this. *See generally* Office of the U.S. Trade Representative, 2009 Special 301 Report (2009).

Also, no country has thus far formally challenged this provision from India within the WTO dispute settlement process since it became effective in 2005. Nonetheless, a review of TRIPS compatibility is warranted.

There are two related issues with respect to TRIPS compliance. The first is whether India's law complies with the TRIPS requirement to patent all "inventions." In particular, some have suggested India's law creates an additional hurdle to patentability beyond those mentioned in TRIPS. The second is whether the law—even if permissibly defining an invention—improperly discriminates against a field of technology.

a. Invention?

Is section 3d inconsistent with the requirement to patent "inventions?" Although some patent enthusiasts have suggested the only possible inventions excluded from patentability are those specifically mentioned in TRIPS,[10] the better view (and one widely accepted among scholars) is that nations may define what constitutes an invention so long as their definition does not discriminate against a field of technology. After all, the word *invention* is not defined, and it is generally accepted that undefined terms under TRIPS may be defined by member countries. Also, although TRIPS does list certain things that do not need to be patented, these categories are things that in fact *are* inventions, but nonetheless need not be patented.[11]

b. Discrimination against Field of Technology

The tougher issue is whether the provision discriminates against a field of technology: pharmaceuticals. Differential treatment may not necessarily rise to improper discrimination against a field of technology.[12] Importantly, a WTO panel has noted this requirement "does not prohibit bona fide exceptions to deal with *problems that may exist only in certain product areas*."[13] Thus, the issue is whether India's provision addresses a problem that only exists in the field of pharmaceuticals.

---

[10] *E.g.*, Roger Bate, *India and the Drug Patent Wars*, AEI HEALTH POLICY OUTLOOK, Feb. 2007, at 2, *available at* AEI Online, File No. 25566 (suggesting the only things that can be excluded from the scope of patentable subject matter are the specific ones listed in TRIPS article 27); Rajnish Rai, *Patentable Subject Matter Requirements: An Evaluation of Proposed Exclusions to India's Patent Law in Light of India's Obligations under the TRIPS Agreement and Options for India*, 8 CHI. KENT. J. INTELL. PROP. 41, 60–61 (2008).

[11] TRIPS states that "subject to the provisions of paragraphs 2 and 3, patents shall be available for any inventions" that meet the TRIPS requirements. TRIPS art 27(1) The noted paragraphs then talk about specific *inventions* that may be excluded. *E.g.*, TRIPS art 27(2) ("Members may exclude from patentability inventions, the prevention within their territory of the commercial exploitation of which is necessary to protect *ordre public* or morality. . . .").

[12] Panel Report, *Canada—Patent Protection of Pharmaceutical Products*, WT/DS114/R, ¶ 7.94 (Mar. 17, 2000) [hereinafter WTO Panel Report, *Canada Generics*].

[13] *Id.* ¶ 7.92 (emphasis added).

The critical issue, then, is whether section 3d is a bona fide exception that deals with a problem that only exists in the area of pharmaceuticals—so-called "evergreening." Arguably, this provision deals with a situation unique to the pharmaceutical industry as evergreening is only a problem discussed in the context of drugs.[14] However, companies tend to characterize the same practice as "life cycle management," suggesting they are legitimately managing their product and business market. Accordingly, they would be unlikely to consider India's proivision to be a bona fide exception.

India clearly was concerned about evergreening when it amended its patent laws to comply with TRIPS, and an argument can be made that section 3d does not unduly discriminate because it is a "bona fide exception." In particular, the Doha Public Health Declaration (which is a relevant interpretive document) seems to focus on pharmaceutical products for special treatment.[15] In addition, TRIPS article 8 also mentions public health in particular as an important social value to be considered with respect to patent rights.[16] Indeed, this was India's intent in creating section 3d—especially after an Indian commission concluded that a proposal to bar all drugs except new molecular entities unduly discriminate against technology.[17]

B. INVENTIVE STEP

The requirement that an invention have an "inventive step" is often considered one of the most important for ensuring that only deserving inventions are patented. In recent years, a number of studies have suggested this criterion is not applied stringently enough.[18] Although this can be a problem in any field, it may be particularly problematic with

---

[14] This issue is discussed in more detail in Chapter 1.

[15] *See* DANIEL GERVAIS, THE TRIPS AGREEMENT: DRAFTING HISTORY AND ANALYSIS 358 (3d ed. 2008) (noting that because Doha focuses on pharmaceutical products, that may suggest an area where "compelling public interests constitute bona fide reason for differentiation"); Amy Kapczynski, *Harmonization and its Discontents: A Case Study of TRIPS Implementation in India's Pharmaceutical Sector*, 97 Cal. L. Rev. 1571, 1598 (2009) ("exceptions to patentability adopted in order to protect public health might not be deemed to discriminate by field of technology, even if they have applications only to particular fields").

[16] TRIPS art. 8; *see also* Kapczynski, *supra* note 15, at 1598 (suggesting interpretation of a bona fide exception "should be informed by the areas given special solicitude under article 8").

[17] RA MASHELKAR ET AL., COMM. FOR REV. OF DRUGS REGULATORY SYSTEM IN INDIA, REPORT OF THE TECHNICAL EXPERT GROUP ON PATENT LAW ISSUES, ¶ 5.6 (2006) ("Granting patents only to NCEs or NMEs and thereby excluding other categories of pharmaceutical inventions is likely to contravene the mandate under Article 27 to grant patents to all 'inventions.' Neither Articles 7 and 8 of the TRIPS Agreement nor the Doha Declaration on TRIPS and Public Health can be used to derogate from this specific mandate under Article 27.").

[18] *E.g.*, NATIONAL RESEARCH COUNCIL OF THE NAT'L ACADS, PATENT SYSTEM FOR THE 21ST CENTURY 47 (Stephen A Merrill et al. eds., 2004). FED. TRADE COMM., TO PROMOTE INNOVATION: THE PROPER BALANCE OF COMPETITION AND PATENT LAW AND POLICY 10 (2003).

respect to patents that directly impact whether patients can afford needed medical treatment. To promote greater access to medicine, a more stringent definition of inventive step would be helpful.

India's current patent laws provide a definition of inventive step that is tougher than most other countries. In particular, India requires that an invention involve a "technical advance," "economic significance," or both, such that the invention is not obvious to someone with similar skill as the inventor. India appears to be the first country to suggest a primary factor in evaluating whether an invention possesses an invention is whether an invention might be of economic significance. How this will be assessed remains to be seen as the law does not define the term. Nonetheless, what is clear is that the definitions seem to be an attempt to make patentability more difficult.

In terms of TRIPS compliance, the analysis of India's new inventive step definition is fairly straightforward. Commentators uniformly agree that countries can define what constitutes an inventive step. Both before and after TRIPS, differing definitions of what constitutes an inventive step have been adopted by member countries. Thus, while India's law provides a new definition of inventive step, that does not make it inconsistent with TRIPS.

C. IMPLICATIONS OF PATENTABILITY RULES FOR INNOVATION AND ACCESS

Although more stringent patentability requirements may promote greater access to medicine, some suggest this approach has negative repercussions: in particular, that if the requirement is too stringent, there will be inadequate incentive to create new—and possibly useful—drugs. Even if unpatented inventions are theoretically available, that may be a moot point if there is inadequate incentive to create such drugs in the first instance.

There is a very important counterargument to the fear of decreased innovation. A significant point to remember is that although the patentability requirements of a single country (such as India) will impact what is patented in India, most companies that create new drugs are multinational ones. In other words, multinational companies' incentive to innovate will not solely be a function of Indian patent law. Rather, so long as some markets—particularly lucrative ones—provide an incentive to innovate, the more limited patentability options in India should not have much impact on incentives. Lucrative markets are known for having particularly lax standards on permitting patents on new variations of known compounds. In addition, even if other developing countries adopt laws that mirror India's, that is unlikely to have an impact on overall innovation as the major markets where companies make their profits will be largely untouched.

Some have also suggested that stringent patentability requirements may hinder the ability of Indian drug manufacturers to benefit from patent protection in India and thus fail to promote an incentive to develop products for the local economy. In particular, some suggest that Indian companies tend to be smaller than most multinational companies and less likely to be in a position to create new inventions that would pass section 3d.

However, Indian companies are not limited to obtaining patents in India: they can apply for and obtain patents in other countries with lesser requirements for patentability. Indeed, financial incentives encourage Indian companies to invent drugs for more profitable markets where drugs can easily be sold at a higher profit. Even before India introduced its new patent laws, some Indian manufacturers were beginning to gear their development efforts toward developing drugs for these wealthier markets—efforts that would be tied to patent standards in those countries.

One caveat: although Indian companies may seek patents in other countries, those countries will likely be wealthy markets with different interests than the Indian market. In other words, if an Indian patent is not readily available, an Indian company may be less likely to develop a drug that would predominantly affect the Indian population. However, even if an Indian patent is available, the slim profits from a relatively poor country such as India may not provide much incentive to develop a drug for the local population. To the extent there is an incentive to develop drugs for the local market, an Indian company would not necessarily be barred from obtaining a patent. After all, a new drug to treat a condition previously not known to be treatable because multinational companies ignored research in this area would likely not be barred as a mere variation, as there probably would be no prior drugs.

Overall, more stringent patentability requirements in India, as well as possibly more stringent requirements in other developing countries, is unlikely to substantially impact incentives for global companies to develop drugs. That has always been the case with global pharmaceutical companies and now appears to be equally true with Indian companies; after all, more profits always exist in markets that have more ability to pay. Although there is a clear need to better promote development of drugs for developing countries, this issue is a red herring in the context of whether limiting patent protection will impede innovation for those unprofitable drugs. For this chapter, it is merely important to acknowledge that heightened patentability standards of a few developing countries are unlikely to change global innovation for drugs—at least those that are predominantly profitable in wealthy countries. Moreover, any possible threat to innovation should be outweighed by the strong access considerations of heightened patentability standards. This is particularly true in India where a robust generic drug industry already exists and could potentially continue to be a source of low-cost drugs to many other countries.

III. Reinforcing Patentability Requirements

India also reinforces patentability requirements in several ways. First, it permits third parties to help show that a patent application or recently issued patent fails to meet one or more requirements of patentability; these are called pre-grant and post-grant oppositions. Second, India does not provide a presumption of validity to issued patents. In other words, a patent owner that sues another for infringing its patent is not entitled to any evidentiary presumption of validity, such that the proof needed to defeat a patent is no greater in litigation than in opposition proceedings. In addition, India permits invalid patents to be revoked. To help put these procedures in context, this section will first revisit the procedures for obtaining patents.

A. PATENT APPLICATION PROCEDURE

To obtain a patent, a patent application must be submitted to a national agency (such as the US Patent Office, the Japanese Patent Office, or the Indian Patent Office) that examines the application to assess whether it satisfies national patentability criteria. There is no global patent; accordingly, an inventor must apply for a patent in every country where rights are desired. Although nations have variations in their patent application procedures, the following summary reflects the core process for most nations, including India.

The goal of the patent office in examining an application is to assess whether the invention meets all the requirements of patentability. To achieve this objective, the patent office typically must first conduct an independent search of public information to find similar prior inventions—what is known as *prior art*. The patent office then compares the prior art to the invention in the application to assess whether the invention meets the patentability requirements. In other words, the patent office must assess whether the invention is actually *new* and has an *inventive step* compared to the prior art. These assessments require an understanding of the invention. Accordingly, a patent application is usually assigned to an examiner with expertise in the invention's field of technology. For example, a patent application for a new drug would likely be examined by someone with a background in chemistry.

Although issued patents are public documents, not all aspects of the patent application are public. In particular, at the time the patent application is filed, the application is not available to the public, and information is not even provided about what application has been filed. However, most patent offices publish patent applications eighteen months after filing to help disclose new technology and let potential competitors know what might be protected. Notably, the published application is typically not identical to a patent that issues from the application. In particular, although most of the text of a patent will be identical to the originally filed patent application, the most important part of a patent—the legal "claims" at the end that describe the scope of the patent rights—are often amended during the patent application process, especially in response to an initial assessment by the patent office that the claims fail to describe a truly new invention. Accordingly, although patent applications provide a rough idea of the technology being developed and what the applicant hopes to have rights to, the patent eventually issued may provide different rights.

In some countries, the patent examiner's assessment may be supplemented by the assistance of third parties. Some countries permit third parties to *oppose* a patent application or issued patent for failing to meet one or more criteria of patentability, although post-grant oppositions are more common in countries with strong patent protection, such as the United States and Japan.[19] Often the third party will bring previously unconsidered prior art to the attention of the patent office. Typically, these third parties are competitors who not only have a vested interest in whether a patent issues, but also good knowledge of the relevant prior art because they are peers in the same technical field. Most patent offices have only limited resources and capacity to evaluate whether an invention deserves a patent. However, peers in the field may have substantial expertise, as well as information not otherwise available to the patent office.

B. INDIAN LAW

As shown in Figure 9, India provides third parties with two opportunities to file an opposition.

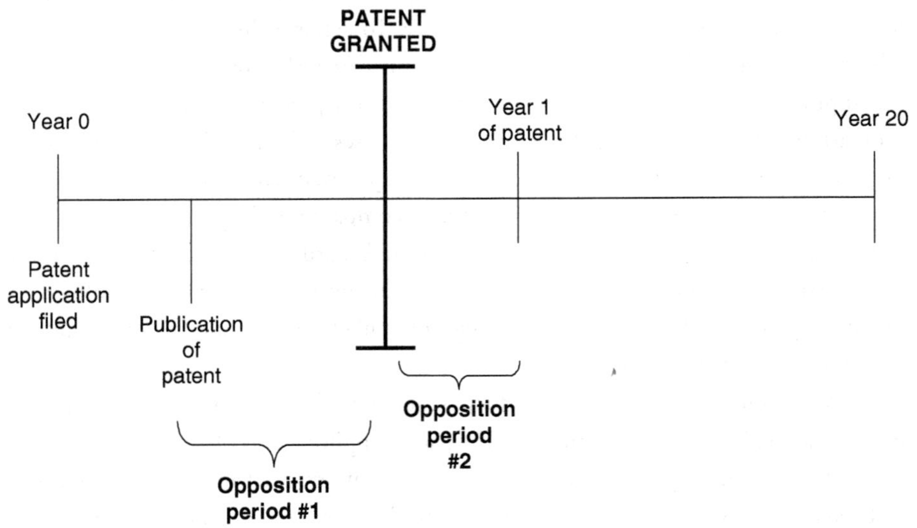

FIGURE 9 Indian Opposition Periods

1. Pre-Grant Opposition

India permits "any person" to oppose a pending patent application after it is published, but before a patent is granted, on the basis that the invention in the application is not patentable.[20] A potential competitor with a commercial interest is an obvious person

[19] *See, e.g.*, Jay Kesan, *Carrots and Sticks to Create a Better Patent System*, 17 BERKELEY TECH. L. J. 763, 781–82 (2002) (noting that Japan and Germany both abandoned pre-grant oppositions and only have post-grant oppositions today); Janice M. Mueller, *The Tiger Awakens: The Tumultuous Transformation of India's Patent System and the Rise of Indian Pharmaceutical Innovation*, 68 U. PITT. L. REV. 491, 568 (2007) (noting Brazil and Jordan permit pre-grant opposition).

[20] India Patents Act § 25.1.

who would oppose. In addition, an individual or an organization committed to the goal of promoting access to medicine could oppose an application. Pre-grant oppositions have in fact been filed by organizations dedicated to the interests of both HIV and cancer patients.[21]

Any ground of patentability may be raised in opposition proceedings; an application can be challenged for not being "new," lacking an "inventive step," or failing to comply with section 3d of the Indian patent act. The most common grounds for successful pre-grant oppositions thus far are lack of inventive step and failure to comply with section 3d.[22]

The majority of pre-grant oppositions have been successful thus far, with the rate of successful opposition being nearly 80 percent, comparable to the percentage of patents found invalid when challenged by generic companies in the United States.[23] Notably, filed oppositions may result in abandonment of patent applications.[24] Although these statistics are not predictive of future cases in India, let alone other countries that may adopt such procedures, they indicate that pre-grant oppositions may be a useful tool to promote access to medicine.

a. TRIPS Compatibility

As with all other Indian patent laws, the pre-grant opposition proceeding must be evaluated for TRIPS compatibility. There is no specific TRIPS provision regarding oppositions, only a general requirement regarding procedures for issuing patents. In particular, TRIPS requires patent examinations be done "within a reasonable time" to avoid "unwarranted curtailment" of protection.[25] In other words, TRIPS imposes the requirement of "reasonableness" for issuing patents to ensure the patent term is not unduly limited. The connection to the patent term exists because TRIPS requires the patent term to be calculated from the date of filing, such that if the patent examination time were unduly long, the actual term of a patent might be abbreviated. Importantly, the focus of this TRIPS provision is on the overall process for issuing *all* patents and not individual ones.

---

[21] *E.g.*, Novartis Challenges the Indian Patent Law: Cancer Patients and Health Groups Demand Withdrawal of Cases, BERNE DECLARATION (Sept 10, 2006) http://www.evb.ch/en/p25011414.html; *ABIA and SAHARA Joint Press Release on Patent Opposition for Tenofovir*, CAMPAIGN FOR ACCESS TO ESSENTIAL MED. (June 26, 2008), http://www.msfaccess.org/main/access-patents/abia-and-sahara-joint-press-release-on-patent-opposition-for-tenofovir/; Peter Ollier, *Roche Loses Patent Opposition in India*, MANAGING INTELL. PROP. (June 1, 2010).

[22] Kapczynski, *supra* note 15, at 1600, n.165.

[23] *Id.* at 1600 (based on information collected from www.i-mak.org/pharma-patent-decisions).

[24] GlaxoSmithKline abandoned applications on two of its HIV drugs, Combivir and Abacavir. *E.g.*, Feroz Ali Khader, *Transcending Differences: The Challenge for Pharmaceuticals in the Post-TRIPS Indian Patent Regime*, 13 J. INTELL. PROP. RIGHTS 424, 428 (Sept. 2008).

[25] TRIPS states "members may require, as a condition of the acquisition or maintenance of the intellectual property rights provided ... Compliance with reasonable procedures and formalities ... Where the acquisition of an intellectual property right is subject to the right being granted or registered, Members shall ensure that the procedures for grant or registration ... permit the granting or registration of the right within a reasonable period of time so as to avoid unwarranted curtailment of the period of protection." TRIPS art. 62.1–2.

Accordingly, even if individual patents took longer to issue, as long as the process overall was reasonable there should be no problem.

Whether pre-grant oppositions result in substantial examination delays is unclear at this point, as only limited information is available about the examination time for opposed applications versus those that are unopposed. However, even if opposed patent applications take longer to issue, it could still arguably be reasonable, as the word itself reflects a notably ambiguous and flexible standard. In addition, even in wealthy countries with substantial resources devoted to patent examination, the length of time applications take to issue seems to vary.[26] Accordingly, for a country with more limited resources (such as India), a reasonable time could potentially be longer than in wealthier countries.[27]

2. Post-Grant Opposition

Even after a patent is granted, it may still be opposed, although only within the first year of issuance and only by an "interested" party.[28] The grounds for challenging an issued patent are the same as with pre-grant opposition. In other words, an issued patent can still be opposed for failing to meet one or more of the patentability requirements. If the opposition is successful, the patent will be nullified.

However, the limitation of post-grant oppositions to "interested" parties may narrow the scope of possible opposers during this period. The interested person category includes those who are engaged in the same field, including researchers.[29] It is generally understood that a member of the public is not entitled to oppose a patent; rather, the common types of interested parties are those who have a commercial interest.[30] In addition, in one case, an Indian patent controller explicitly expressed doubt whether a nonprofit group dedicated to promoting access to medicine would be an interested party.[31]

There should be no TRIPS issue with respect to the post-grant opposition. Unlike pre-grant oppositions, a post-grant opposition would not impact the patent term as it would not delay the actual grant of a patent.

---

[26] For example, the average period of examination in the United States and Japan is roughly three years, versus almost four years in Europe, with a slight increase in the time in the latest two years for which statistics are available. The pendency in Europe increased from 42.8 months to 46.9 months whereas the pendency in Japan increased from 32.4 to 33.9 months, and the US pendency increased from 32 months to 33.5 months. *See* European Patent Office, Japan Patent Office & US Patent & Trademark Office, *Trilateral Statistical Report: 2007 Edition* 49 tbl.4 (2007); European Patent Office, Japan Patent Office & US Patent & Trademark Office, *Trilateral Statistical Report: 2008* Edition 49 tbl.4 (2008).

[27] Kapczynski, *supra* note 15, at 1604–05.

[28] India Patents Act § 25.2.

[29] *Id.* § 2.*l*(t).

[30] P. Narayanan, Patent Law, at 8-12–8-17 (4th ed. 2006).

[31] F. Hoffmann-La Roche AG v. Wockhardt Ltd, 2009 Dec. Chennai Pat. Off. Patent No. 198952, at 20-21, *available at* http://www.i-mak.org/pharma-patent-decisions (search within page text for "198952").

3. No Presumption of Validity for Issued Patents

Indian law also ensures that patentability requirements are satisfied by providing no presumption of validity to issued patents even after the opposition period has passed.[32] This rule has been explained as logical in that the patent office search is not considered exhaustive.[33] In addition, it is recognized that questions of novelty and inventiveness need to be assessed from the view of experts, including those who would normally oppose a patent.[34] However, given that many patents are not opposed, patent validity is considered to be appropriately assessed in litigation proceedings without the hindrance of any evidentiary presumption.

This is a marked difference from other jurisdictions that provide an evidentiary presumption of validity, making it more difficult to invalidate a patent after it is granted. For example, some countries consider issued patents to have a presumption of validity; although this can be challenged, the challenger must establish the patent is not valid based on a heightened evidentiary standard.[35] The presumption of validity has been explained as appropriate because as the patent office is an expert agency, it should be entitled to deference.[36]

a. TRIPS Compatibility

No specific TRIPS provision speaks to evidentiary presumptions regarding the validity of a patent. Rather, TRIPS provisions on enforcement indicate it is generally required to be "fair and equitable." The only additional requirement regarding proof is that parties to litigation be entitled to substantiate their claims and present evidence, but the standard of proof required is not mentioned.[37] Given the lack of any requirements regarding evidentiary presumptions to challenge patents, India's law is easily consistent with TRIPS.

IV. Patent Rights and Remedies

This section addresses how patent rights and remedies under Indian law—including limitations as to both rights and remedies—may promote access to medicine under TRIPS.

---

[32] The Patents Act, 1970, No. 39, § 13.4, Acts of Parliament, 1970 (India) (*amended by* The Patents (Amendment) Act, 2005, No. 15, Acts of Parliament 2005 (India)) [hereinafter India Patents Act] (patent examination does not "warrant the validity of any patent").

[33] NARAYANAN, *supra* note 30, at 1–19.

[34] *Id.*

[35] U.S. Patent Act, 35 U.S.C. § 282 (2010); *see also* Canada Patent Act, R.S.C. 1985, c. P-4, s. 43.2. Japan considered patents to have a presumption of validity until 2000, but now consider suits based on invalid patents to be an abuse of the patent right. Saiko Saibansho [Sup. Ct.] Apr. 11, 2000 Heisei 10 (O) 364 (Japan).

[36] *E.g.*, Brooktree Corp. v. Advanced Micro Devices, Inc., 977 F.2d 1555, 1574 (Fed. Cir. 1992); *see also* Mark Janis, *Reforming Patent Validity Litigation: The Dubious Preponderance*, 19 BERKLEY J.L. & TECH. 923 (2004).

[37] TRIPS art. 42. There is another provision specifying that evidence must be made available to support claims, but it similarly does not speak to burdens of proof.

The first part of this section explains how India's law is consistent with the minimum patent rights required under TRIPS. Then, the section addresses a myriad of limitations to patent rights under Indian law.

A. PATENT RIGHTS

TRIPS imposes two basic requirements for patent rights. First, the patent owner normally has the right to exclude all others from making, using, selling, or importing the patented invention during the term of the patent.[38] Second, these rights are to exist during the term of the patent, which must end no earlier than twenty years from the date the patent application was filed.

1. Infringement

India's law does in fact provide patent owners with the right to generally exclude others from the range of activities as set forth under TRIPS.[39] However, what is notable is that India does not provide *more* than TRIPS requires in contrast to some other jurisdictions. For example, in some countries, patent liability may exist for someone who contributes to the infringement of another (what is commonly known as *contributory infringement*).[40] In other words, if a patent exists on a method of using a compound to cure hair loss, the person who uses and thus violates the method would actually be the consumer. However, in the case of contributory infringement, the person who makes the compound used by the consumer could be considered liable. This would thus enable the patent owner to sue the maker of the compound, who is likely a company with more resources than an individual consumer; moreover, it would be more effective to stop a company from supplying consumers who infringe, rather than suing each individual consumer. Of course, from a public relations perspective, it is also preferable to sue a corporate entity rather than a consumer.

2. Patent Term

India also provides a patent term consistent with TRIPS—one that ends twenty years from the filing of the application.[41]

However, in contrast to some other jurisdictions, India does not provide *more* than required under TRIPS. For example, India does not extend the patent term in cases where either the patent examination process or the regulatory approval process (for a

---

[38] TRIPS art. 28.

[39] India Patents Act § 48.

[40] *E.g.*, 35 U.S.C. § 271(c) (2010).

[41] India Patents Act § 53.

patented drug) is lengthy. Accordingly, a drug patented in both India and the United States (which does provide for patent term extensions) is likely to have differing patent terms. Moreover, the patent term for patents issuing from mailbox applications may be particularly short; as the term is twenty years from filing (and mailbox applications may have been pending for up to five years before even being examined), patents issued from these applications are likely to have shorter terms than other patents. However, this is permissible under TRIPS, which only requires the patent term to end twenty years from the filing of the patent application.

B. LIMITATIONS ON PATENT RIGHTS AND REMEDIES

India provides an extensive assortment of limitations on patent rights and remedies. This section begins with specific exceptions to the usual patent right to exclude: regulatory review, compulsory licensing, and international exhaustion. After discussing these specific exceptions, the section addresses tht the ultimate "exception" to patent rights—their complete revocation. This section then discusses how courts may limit patent remedies even when patent rights exist; it concludes with a discussion of a complicated provisions under Indian law that attempts to limit both the rights and remedies for patent owners of so-called "mailbox" application patents.

1. Regulatory Review

Indian law has a provision that fosters regulatory review of generic applications during the patent term to ensure generics can quickly enter the market upon patent expiry. Some type of exception is common in many national laws, but there are differences among them.

India exempts from infringement any making, using, selling, or importing of a patented invention that is related to development of information required under any national law regulating the manufacture, sale, or import of any product.[42] In other words, an activity is exempt from infringement if it is relevant to obtain regulatory approval to sell a drug not only in India, but in any other country. The inclusion of other countries importantly permits Indian companies to continue to be exempt from infringement when preparing an application for approval in the US market where their products comprise a substantial proportion of generic drugs.[43]

---

[42] India Patents Act § 107A(a).

[43] *E.g.*, William Greene, *The Emergence of India's Pharmaceutical Industry and Implications for the US Generic Drug Market* (US Int'l Trade Comm'n Off. of Econ., Working Paper No. 2007-05-A, 2007); Hitesh Gajaria, *India to be amongst Top Three Generic Makers in the World*, EXPRESS PHARMA, Jan. 16–31, 2008; *see also* C.H. Unnikrishnan, *Indian Firms Overtake US in FDA Drug Filings*, LIVE MINT, Mar. 17, 2010.

a. *TRIPS Compliance*

As discussed in Chapter 3, a WTO panel previously found Canada's regulatory review provision to be a permissible "limited exception."[44] However, a brief consideration of whether India's provision differs from Canada's is necessary to evaluate TRIPS compliance. Canada's law was similar except that it did not create an exemption for imports.[45] Although India's provision is slightly broader, the overall similarity and strong support for a regulatory approval exception suggest this provision is consistent with TRIPS. After all, the panel noted that whether an exception was limited was not a function of how many of the patent owner's usual rights were impacted.[46] Rather, the exception was seen as limited in that the activity was confined to conduct necessary to comply with regulatory approval process.[47] This also applies to the Indian law.

2. Compulsory Licensing

India provides a substantial exception to the usual rights of a patent owner through its detailed and expansive infrastructure for issuing compulsory licenses. India provides extensive grounds for issuing a compulsory license that include traditional ones, such as failure of the patent owner to "work" the patent locally (meaning the patent owner is not making the patented product in India and only importing the patented invention). India also provides some unusual bases for issuing compulsory license that are particularly pertinent to the issue of access to medicine. In particular, India specifically permits a compulsory license to issue if the patent owner is not providing the patented medicine at a reasonably affordable price.

India's provisions on compulsory licenses are addressed in more detail in Chapter 5 (which focuses exclusively on compulsory licenses) as an illustration of how a country can enact compulsory license provisions consistent with TRIPS that promote access to medicine. The separate discussion is necessary because not only are TRIPS requirements on compulsory licenses lengthy and often confusing, but they often raise additional issues. As discussed in that chapter, a nation issuing a compulsory license can encounter retaliation from patent-owning companies as well as other nations.

---

[44] Panel Report, *Canada—Generic Medicines*, ¶ 7.32 *See also* TRIPS art. 30.

[45] Canada Patent Act, R.S.C. 1985, c. P-4, s. 55.1.1 ("It is not an infringement of a patent for any person to make, construct, use or sell the patented invention solely for uses reasonably related to the development and submission of information required under any law of Canada, a province or a country other than Canada that regulates the manufacture, construction, use or sale of any product.").

[46] Panel Report, *Canada—Generic Medicines* ¶ 7.32 (rejecting this argument, albeit with respect to a different regulatory provision that was ultimately found not adequately limited on other grounds).

[47] *Id.* ¶ 7.45.

3. International Exhaustion

Another important exception to patent rights permitted under TRIPS is that countries can recognize international exhaustion. Essentially, this permits countries to create an exception to the patent owner's usual right to exclude imports of the patented invention. Most countries that adopt the principle of international exhaustion consider a patent owner's rights exhausted after the first unrestricted global sale by the patent owner. For example, a patent owner who sold a patented drug in Brazil would not be able to restrict its importation into Country X if Country X recognized international exhaustion as an exception to what would otherwise be patent infringement. The principle behind international exhaustion and its significance to patent-owning pharmaceutical companies is explained in greater detail in Chapter 2. This current discussion focuses on India's law and its TRIPS compatibility.

India's patent laws push the envelope of international exhaustion.[48] In particular, India will consider a patent owner's rights exhausted by the first global sale of a product—without regard to whether that first sale was authorized by the patent owner, or even whether the product was patented.[49] The linchpin of India's exception is that the first global sale was by someone "duly authorized under the law to produce and sell or distribute the product."[50] A person may be duly authorized in a number of scenarios. This would encompass the traditional scope of international exhaustion where a product was first sold by the patent owner under patent protection in another country (such as Brazil) before being imported into India. In addition, if a company made a patented product under a compulsory license in Thailand (i.e., expressly without the patent owner's consent), it could still be imported. Moreover, a product could be made in an LDC where it is not patentable and then imported into India.

a. TRIPS Compatibility

The TRIPS issue is whether India's law improperly conflicts with the TRIPS requirement that patent owners have the right to exclude imports of a patented product.[51] Although there are exceptions to TRIPS patent rights, none of them address international exhaustion.[52] However, a footnote to the basic TRIPS requirement concerning right to exclude imports does so[53] by stating the right to exclude imports is subject to the provisions

---

[48] Another country with a broad international exhaustion regime is Kenya. Industrial Property Act (2001) Cap. 3 § 58 (Kenya).

[49] India Patents Act §107A(b) ("importation of patented products by any person from a person who is duly authorized under the law to produce and sell or distribute the product, shall not be considered as a infringement of patent rights").

[50] *Id.*

[51] TRIPS art. 28.

[52] *Id.* art. 30 ("limited exceptions" to patent rights); TRIPS art. 30 (compulsory license requirements).

[53] *Id.* art. 28, n.6.

of Article 6.[54] Article 6, in turn, excludes "exhaustion of intellectual property" from the scope of official WTO disputes unless a nation is applying different exhaustion rules to different WTO members—something that is not at issue here.[55] This patchwork of TRIPS provisions states that whether a patent owner's right to import is exhausted by a prior global sale is outside the boundaries of official WTO disputes. Essentially, member states are permitted to adopt—or not—an international exhaustion regime for patents, as that decision cannot be challenged under the WTO. Although TRIPS does not explicitly say so in a single provision, that is a proper interpretation of these assorted provisions (as confirmed by the Doha Public Health Declaration, which clearly states that member states are free to establish their own exhaustion regimes free of challenge).[56]

Although the above TRIPS discussion suggests nations are to interpret exhaustion principles as they desire, the issue here is whether India can interpret international exhaustion in a way that not only differs from most interpretations, but seems inconsistent with traditional policy concerning exhaustion. In particular, international exhaustion presumes the patent owner initially profited from an earlier global sale, but India's law deprives patent owners of rights even when they never obtained any such profits. It is unclear whether international exhaustion under TRIPS should apply to goods made under a compulsory license where the patent owner clearly does not obtain the full monetary benefit desired.[57] Given the diversity in views on the scope of international exhaustion, India should have at least a reasonable argument to defend its interpretation as an expansion on previously accepted principles (such as permitting international exhaustion of goods made under compulsory license). However, this provision is clearly controversial, and one Indian professor has suggested it goes beyond TRIPS.[58]

---

[54] *Id.* ("This right, like all other rights conferred under this Agreement in respect of the use, sale, importation or other distribution of goods, is subject to the provisions of Article 6").

[55] *Id.* art. 6. Technically, a country is permitted to discuss exhaustion issues at the WTO, but only if they relate to a violation of national treatment or MFN status. TRIPS art. 28, n.6.

[56] World Trade Organization, Ministerial Declaration on the TRIPS Agreement and Public Health of November 14, 2001, ¶ 5(d), WT/MIN(01)/DEC/2, 41 I.L.M. 755 (2002) [hereinafter Doha Public Health Declaration] ("The effect of the provisions in the TRIPS Agreement that are relevant to the exhaustion of intellectual property rights is to leave each member free to establish its own regime for such exhaustion without challenge").

[57] *E.g.*, UNCTAD, Resource Book on TRIPS and Development 106–07 (2005) (explaining that there is a split in view among TRIPS experts as to whether a compulsory license would be the basis for exhaustion, but suggesting it is nonetheless permissible). *But see* Correa, Trade Related Aspects, *supra* note 3, at 82 (noting that under some laws, international exhaustion of patent rights is not applicable if the product was not put on the market with the consent of the patent owner, such that a product made under a compulsory license would not be subject to exhaustion).

[58] Shamnad Basheer, *India's Tryst with TRIPS*, 1 Indian J. L. & Tech. 15, 31 (2005); Shamnad Basheer & Mrinalini Kochupillai, *TRIPS, Patents and Parallel Imports in India: A Proposal for Amendment*, 2 Indian J. Intell. Prop. L. 63 (2009) (suggesting the current provision violates TRIPS along with amendments to eliminate the problem); *see also* Nuno Pires de Carvalho, The TRIPS Regime of Patent Rights 6.19

3. Revocation

A patent may be revoked[59] under Indian patent law for a number of reasons.[60] Two grounds are of particular interest to the issue of access to medicine. First, a patent may be revoked if it fails to meet one of the elements of patentability. Second, a patent may also be revoked in a limited category of cases where a compulsory license was previously issued. Each of these will be addressed.

a. Patentability Requirements Unmet

Notably, a patent may be revoked for failing to meet one of the patentability requirements even after both opposition periods have expired. However, revocation can only be initiated by the government or an *interested party*, with the term here being understood as being limited to those who have a financial or commercial interest in the patented invention.[61] Accordingly, a group opposed to a patent only because it impedes access to medicine would not seem entitled to ask for revocation, even though they could have initially opposed the patent application (as *anyone* may oppose a patent application).

b. Post-Compulsory License

A patent that was previously subject to a compulsory license for at least two years may also be revoked. Grounds for revocation mirror some grounds for issuance of compulsory license: non-working of a patent, failure to satisfy the reasonable requirements of the public, or lack of availability of the patented invention to the public at a reasonably affordable price.[62]

India's law seems to provide a powerful ability to terminate patent rights when a patented drug is not made at a reasonably affordable price. Although this ground mirrors a provision for compulsory licensing, the impact on a patent revocation is much more severe: once the patent is revoked, the patent owner has no rights and is not entitled to even the minimal compensation provided under a compulsory license scheme. It remains to be seen whether India will have the political will to use—or, at least, threaten to use—this provision so as to promote lower-cost drugs.

(2d. ed. 2005) (asserting conclusively that there is no exhaustion when the first sale is pursuant to a compulsory license).

[59] There are basically two ways to revoke a patent: through petition or as a counterclaim to a suit for infringement. The revocation may be initiated by either the government or an interested party.

[60] A patent may also be revoked for false representation or failure to disclose relevant foreign applications. India Patents Act §§ 64.1(j), 64.1(m). In addition, a patent may be revoked when the invention fails to disclose the source of the biological material—an issue not relevant to conventional drug patents. India Patents Act § 64.1(p).

[61] NARAYANAN, *supra* note 30, at 15-09.

[62] India Patents Act § 85.

c. TRIPS Compatibility

An important question is whether patent revocation is permissible under TRIPS, as such a provision seems to entirely terminate the usual patent rights provided under TRIPS article 28. However, revocation of patents seemed to have been envisioned by the TRIPS negotiators as permissible at least in some cases as there is a specific provision dealing with revocation or forfeiture of patents. TRIPS requires only that an opportunity for judicial review exist for any decision to revoke or forfeit a patent.[63] TRIPS does not suggest what might be a permissible basis to revoke or forfeit a patent, thus allowing the countries to decide on the grounds.[64] Accordingly, revocation based on the failure to provide the patented product at a reasonably affordable price could be a powerful ground.

There is one ground that should be clearly permissible under TRIPS, albeit with some time constraints. In particular, TRIPS (through incorporation of the Paris Convention rules) mandates that if revocation is based on failure to work after a compulsory license was already granted, the revocation cannot take place until at least two years after the compulsory license issued. India's law clearly complies with this timing requirement.

C. LIMITED REMEDIES—INJUNCTIONS

TRIPS imposes requirements on enforcement of intellectual property rights such as patents. TRIPS requires that courts be capable of ordering both injunctions[65] and monetary damages to compensate for infringement.[66] In addition, TRIPS requires that courts have the authority to prevent infringement from occurring—what is commonly known as a preliminary, interlocutory, or interim injunction.[67]

However, although TRIPS requires that courts have the power to provide these remedies, TRIPS is silent on when courts should grant injunctions, as well as on how courts should decide what monetary damages are appropriate. Accordingly, these issues should be within the discretion of member states.

India has thus far provided one notable example of how a national court may elect to deny a preliminary injunction in light of concern about access to medicine. In *Roche v. Cipla*, the generic company Cipla was perceived to be audaciously making the patented drug during the patent term and simply waiting to challenge the patent as invalid during

---

[63] TRIPS art. 32 ("an opportunity for judicial review of any decision to revoke or forfeit a patent shall be available").

[64] *E.g.*, CORREA, TRADE RELATED ASPECTS, *supra* note 3, at 343.

[65] TRIPS art. 44. However, injunctions are not required if national laws limit remedies to monetary compensation and/or a declaratory judgment.

[66] *Id.* art. 45 (with the exception of compulsory licenses (in which case no injunction is necessary)).

[67] *Id.* art. 50.

litigation, yet the court declined to impose a preliminary injunction on Cipla.[68] In denying the injunction, the court noted the generic company had established a strong likelihood of success on the merits (challenging the drug as being inconsistent with the section 3d requirement for increased efficacy).[69] The court noted the patented drug Erlotinib, sold as Tarceva, provided life-saving treatment for lung cancer and that the defending generic manufacturer offered the drug at a substantially lower cost than the patent owner;[70] specifically, the generic version made by Cipla cost Rs 1600/month (US $36), which is roughly a third of patent owner's Roche price of Rs 4880/month (US $110).[71] A subsequent court that entertained Roche's appeal of this case declined to focus on the cost differential and instead dismissed the case.[72] However, that does not prevent another court (whether in India or elsewhere) from revisiting the issue.

The court's denial of a preliminary injunction is consistent with TRIPS. The most pertinent requirement under TRIPS is that judicial authorities have the *power* to order a party to cease and desist. As recently confirmed in a WTO panel proceeding, courts need not always exercise that power.[73] The only other TRIPS requirement of relevance is that enforcement procedures be "fair and equitable." Assuming the patent owner was given a full opportunity to present its case, the court's ultimate decision not to grant an injunction should be within its discretion.

The decision to deny a preliminary injunction where the defendant had shown a strong likelihood of success and there are public interest issues such as public health at stake is not necessarily surprising. However, injunctions are more typically denied where the defendant is the *only* supplier of the drug—not simply because the defendant provides a cheaper alternative to the patent owner.

What remains to be seen is whether Indian courts may similarly consider access-to-medicine issues with regard to permanent injunctions where a defendant is found to actually infringe a patent. As noted above, TRIPS requires national courts only to have the *power* to issue injunctions, but does not state when they must use it. In addition, there

---

[68] F. Hoffmann-La Roche Ltd. v. Cipla Ltd., (2008) I.A. 642 (India) [hereinafter Hoffman-La Roche v. Cipla (2008)] *available at* http://www.indiankanoon.org/doc/64813/; C.H. Unnikrishnan, *Cipla to Sell Copycat Version of Roche Drug*, LIVEMINT.COM (Jan. 11, 2008), http://www.livemint.com/2008/01/11001205/Cipla-to-sell-copycat-version.html Notably, Cipla was already granted marketing approval for its generic as India does not provide market exclusivity. *See* Hoffman-La Roche v. Cipla (2008), ¶ 7.

[69] Hoffmann-La Roche v. Cipla (2008), ¶ 82.

[70] In particular, the court noted that it "cannot be unmindful of the right of the general public to access life saving drugs which are available and for which such access would be denied if the injunction were granted." *Id.* ¶ 85.

[71] *Id.* ¶ 84.

[72] F. Hoffmann-La Roche Ltd. v. Cipla Ltd., (2008) FAO (OS) 188/2008 ¶¶ 72–73, ¶¶ 79–80 (Delhi H.C., Apr. 24, 2009) *available at* http://www.indiankanoon.org/doc/401740/; *see also* http://spicyipindia.blogspot.com/2009/08/breaking-news-supreme-court-dismisses.html.

[73] WTO Panel Report, *China—Measures Affecting the Protection and Enforcement of Intellectual Property Rights*, ¶ 7.236, ¶¶ 7.238–7.239, WT/DS362/R (Jan. 26, 2009).

is precedence beyond India for not automatically granting injunctions in every case of patent infringement; for example, the US Supreme Court rejected such an approach.[74] However, it would be notable if Indian courts permitted adjudicated infringers to continue to make infringing products subject to payment of damages on the grounds the infringer was making the product at a more reasonable cost to consumers. This should technically be permissible under TRIPS (unless it was considered to be an impermissible compulsory license). Whether this will occur remains to be seen.

D. LIMITED RIGHTS AND REMEDIES—MAILBOX APPLICATION PATENTS

India has also crafted exceptions to patent rights as well as remedies with respect to patents granted on inventions filed as mailbox applications during the transitional period when India was not yet required to provide patents on drug compounds.[75] These applications were filed from 1996 to 2005 and have only recently begun to be examined, with just a few issuing as patents.

Before discussing the Indian law, it is helpful to recall the situation in India during the transitional period. At that time, some Indian companies made low-cost generic versions of inventions subject to mailbox applications; as there was no patent protection, the generics could not infringe.[76] Some inventions filed as mailbox applications were patented in other countries; however, Indian generic companies could make generic versions of these drugs because patents did not exist in India.

India's current law essentially permits companies that were making low-cost generic versions of mailbox application inventions prior to January 1, 2005 to *continue* making now-infringing products, subject only to payment of a reasonable royalty.[77] Whereas other patent owners have the right to exclude others from making and selling their patented inventions, owners of mailbox inventions are deprived of this right. The mailbox invention patent owners are not permitted to sue these generic companies for infringement and thus cannot request a court to stop the continuing infringement—even though courts have such powers in other cases to do so.[78]

---

[74] Ebay v. MercExchange, 547 U.S. 388 (2006).

[75] During this period, India was required to accept applications for products (such as drugs) that would be evaluated once patents on products were permissible. In addition, the applications would be reviewed against prior art using the date of the mailbox application rather than the date the application was actually examined.

[76] Not all inventions subject to mailbox applications could be made by Indian companies. India was also required to provide an exclusive marketing right to mailbox inventions if the invention was previously granted a patent as well as regulatory approval in another country.

[77] India Patents Act § 11A(7).

[78] *Id.* ("no infringement proceedings shall be instituted against such enterprises [that have made significant investment and were producing and marketing the patented product prior to January 1, 2005]").

The owner of a patent issued from a mailbox application may thus face competition from a generic supplier for the entire patent term. This owner is entitled to receive a "reasonable royalty" to compensate for the infringement. However, this amount is unlikely to provide the patent owner with the same economic benefit of being the sole supplier of a patented drug. After all, the generic company will likely price its version at a lower cost and undermine the patent owner's sales and profits.

India's law also provides different treatment to owners of patents on mailbox inventions with respect to the time period when monetary damages may be accrued and how much money may be obtained. For most inventions, the patent owner can claim damages not only from the day of the patent issuance, but retroactively to the date of publication of the patent application. For owners of mailbox invention patents, however, monetary damages only accrue from the date of the patent grant. In addition, whereas most patent owners have the right to seek damages due to the infringing activity, or the defendant's profits,[79] the only possible monetary compensation for owners of mailbox invention patents are a "reasonable royalty," which is likely less.[80]

However, not every mailbox application that becomes an issued patent will face competition from a generic company. For example, the first pharmaceutical product patent granted in India was on a hepatitis C treatment for which no generic versions previously existed.[81] Accordingly, the patent owner Hoffman-LaRoche had complete exclusivity when it entered the Indian market in 2006.

Before addressing the issue of TRIPS compliance, a few important points will be reivisited.

Q: Can any patented drug be made in India as a generic, subject only to payment of a reasonable royalty?

A: No. Although some generic versions of patented drugs may be made (subject to payment of a reasonable royalty), this is only possible with respect to generic drugs that were made prior to 2005 and for which a patent application was filed before 2005.

Q: Can a generic company make a drug that was invented before 2005, but not made in India until after 2005?

A: No. For a generic company to make a patented drug, two separate things must be true: the drug must have been subject to a mailbox application filed before 2005, and the generic company must have marketed the drug before 2005.

---

[79] India Patents Act § 108 (noting that court may grant an injunction and either damages or an account of profits).

[80] The procedure by which the mailbox invention patent owner would obtain a reasonable royalty is unclear, as this owner is by statute precluded from filing a regular patent infringement suit.

[81] Jyothi Datta, *Roche Gets Product Patent on Hepatitis C*, HINDU BUS. LINE (Mar. 2, 2006), http://www.thehindubusinessline.com/2006/03/03/stories/2006030302040200.htm.

a. TRIPS Compatibility

The different treatment of mailbox invention patents raises a host of issues under TRIPS. First, the ability of the mailbox invention owner to exclude all others from the invention is hindered. This inability to exclude violates the basic TRIPS requirement that patent owners be entitled to do so—unless this treatment is permissible under a specific TRIPS exception. In addition, as all mailbox inventions are necessarily pharmaceuticals, the differing treatment of such patent owners raises the issue of discrimination on the basis of technology. To help visualize the differing treatment, the Table 7, summarizes treatment of mailbox invention patents versus other patents.

TABLE 7

Indian Patent Rights Mailbox vs. Non-Mailbox Patent Remedies

	Patent Owners (except mailbox inventions)	Mailbox Invention Patent Owners
Right to exclude all others from patented invention (except for compulsory licenses)	**Yes**	**No**
Right to sue all infringing entities	**Yes**	**No**
Available remedy	Money (damages or profits) Injunction	Reasonable royalty only
Relevant time period for assessing compensation	From date of patent application	From date of patent *grant*

b. Exception to Patent Rights

The next issue is whether India's mailbox provisions may be a permissible exception to the usual patent rights required by TRIPS. The two possible exceptions are whether the law is a permissible compulsory license (under article 31) or a "limited exception" to patent rights under article 30. However, as further explained below, neither of these exceptions seem to support the Indian law.

c. Article 31 compulsory license?

One issue is whether the inability to enjoin infringement coupled with capped damages of a reasonable royalty constitutes a compulsory license that complies with TRIPS. Although this provision is not listed under the compulsory license section of Indian patent law, it seems to function as one in that a patented invention is used without authority of the patent owner. Although the mailbox provision may be considered a compulsory license, it would not be one that complies with TRIPS. There are many TRIPS requirements for a compulsory license, but the mailbox provision readily fails several. For example, the provision provides automatic compulsory licenses for an entire class of inventions inconsistent with the TRIPS requirement to consider inventions on their own merits before imposing a compulsory license.[82] In addition, the provision runs afoul of the usual requirement that the patent owner be first contacted to secure a voluntary license—and none of the situations that permit waiver of prior negotiation are applicable.[83] Although there are other requirements regarding compulsory licenses under TRIPS, further discussion is not necessary, as this provision clearly fails several requirements, and *all* TRIPS requirements must be met to be consistent with TRIPS.

d. A "Limited Exception" under TRIPS?

The next issue is whether the limited rights for mailbox invention owners is consistent with the exception to patent rights under TRIPS article 30, which only permits "limited exceptions." Although this exception is but a single sentence, a prior WTO panel has interpreted it to require three separate and cumulative requirements:

(1) the exception is a narrow deviation from the usual patent rights
(2) the exception does not "unreasonably conflict" with "normal exploitation" of the patent; and
(3) the exception does not "unreasonably prejudice the 'legitimate interests of the patent owner'—taking into account the legitimate interests of third parties."

The mailbox provision is unlikely to pass all three elements of this test.

First, the exception is hard to characterize as "limited" when it violates the patent owner's right to exclude others from making as well as selling the patented invention. In fact, the mailbox provision seems far less limited than a provision the WTO panel previously found to be inadequately limited: an exception permitting a generic manufacturer to make unlimited quantities of a patented drug during the last six months of a patent term (to enable competition with the patent owner on the first day of patent expiry).[84] The mailbox provision similarly places no limits on the amount that may be sold by generic companies. In addition, as the six-month time period was not considered an adequate limitation, the entirety of the patent term would clearly provide no limit at all.

The mailbox provision would likely also fail to meet the second requirement: that it not be an "unreasonable" conflict with the "normal exploitation" of a patent. The prior WTO

---

[82] TRIPS art. 31(a).

[83] *Id.* art. 31(b). Although TRIPS permits waiver of initial consultation in some cases, none of those are applicable here—there is no national emergency, situation of extreme urgency, or public noncommercial use that applies to the continuing commercial production of drugs. *Id.*

[84] WTO Panel Report, *Canada Generics*, ¶ 7.34.

panel found such an unreasonable conflict to exist in any case where the patent owner's expected economic return from normal exploitation (i.e., total exclusivity) would be significantly impacted. Clearly the mailbox provision has a significant impact on a patent owner's expected economic benefit from being the sole supplier of a patented drug.

The last element of article 30 is the only one that arguably is satisfied. The issue here is whether the limited rights given to mailbox patent owners constitute unreasonable prejudice to those owners in light of the legitimate interests of third parties. The third parties at issue must be first identified; here they include not only the generic manufacturers but patients. Both manufacturers and patients have "legitimate" interests, although their interests are not identical. For example, manufacturers who have invested money in manufacturing drugs that were not previously patented have a legitimate interest in receiving a financial return on their investment. Patients who previously purchased drugs from such companies may have a legitimate interest in continuing to receive the same drug at the same low price to maintain their treatment needs.

The next step is to consider whether these interests outweigh the *legitimate* interest of the patent owner. What legitimate interest does the patent owner have? Notably, the panel clarified that a legitimate interest differs from a legal right already analyzed under the prior requirement.[85] So, the question is whether the patent owner in fact has a legitimate interest in being the sole supplier of the patented invention when that invention was previously in the public domain. This would be a question of first impression with respect to TRIPS jurisprudence.[86] It seems an argument can at least be made that the patent owner does not have a legitimate interest in entirely taking away an invention from those who were previously using it legally. This would be consistent with the doctrine of prior user rights that are commonly recognized in national domestic laws; this doctrine provides a complete defense to infringement of a later-patented method that was previously used in secret. The prior use doctrine is not identical in that it usually only operates as a defense—but it does reflect a policy decision to recognize reliance interests existing before a patent was granted.

Given that the mailbox provision seems inconsistent with not just one but two of the three requirements set forth by a prior panel, it seems impossible—at least at first glance—to argue the mailbox provision is a limited exception. However, the prior panel decision has been criticized by some commentators, and subsequent panels are not bound to follow the previous approach. Although panels often give deference to prior decisions, there is at least the possibility the mailbox provision could be consistent with TRIPS based on a new interpretation of article 30.

---

[85] *Id.* ¶ 7.68.

[86] The prior panel found a patent owner had no legitimate interest in a period of de facto exclusivity of several years following the patent term that would exist if regulatory authorities could not review applications to sell generic versions during the patent term. *Id.* ¶ 7.82.

An alternative interpretation of article 30 might possibly support the mailbox provision as an appropriate limited exception. An alternative approach would need to discard the prior approach of considering every clause to be a separate and cumulative element as the mailbox exception is not adequately limited based on that approach. In addition, an alternative interpretation would suggest that article 30 be read in light of the broader context of TRIPS, including not only the preamble, but the objects and principles. Notably, these provisions emphasize a need to balance the interests of users and producers as well as to recognize socioeconomic needs of countries. If balancing the interests of users and producers is an important goal in interpreting article 30, it could require an emphasis on the legitimate interests of third parties with respect to *all* clauses of article 30, rather than merely the last one. Alternatively, perhaps the same three requirements can be used, but more emphasis be given to the balance between the interests of third parties versus the patent owner. Yet another possibility is to consider all the requirements, but not strictly demand that they all must be fulfilled for a limited exception to exist. In any of these cases, the mailbox provision could be permissible.

e. Does the Provision Discriminate against a Field of Technology?

Even if the mailbox provision is an arguably permissible "limited exception" under TRIPS, it may still be ultimately impermissible if it discriminates against a field of technology. One WTO panel has stated that all exceptions to TRIPS rights must also comply with the nondiscrimination requirement. Although some commentators have suggested this interpretation is inappropriate, a consideration of the nondiscrimination requirement seems warranted.

Multiple factors raise an issue of discrimination. First, Indian law provides differing treatment with respect to when damages accrue (either during the patent application, or only upon the patent grant).[87] Second, the amount of damages (as well as whether injunctive relief is available) are also different.

This exception seems focused on a specific field of technology, as it only applies to inventions that were subject to mailbox applications—that is, pharmaceuticals. However, the law refers to inventions subject to mailbox applications, so an argument could be made that it does not discriminate on its face. Even if it did, the crux of the question is whether the differing treatment can be justified as a bona fide exception to deal with a problem that only exists in certain product areas. If articles 7 and 8, together with the Doha Public Health Declaration are interpreted broadly, an argument could be made that it is permissible. After all, Doha states that nations can and should be able to

[87] TRIPS does not actually require any rights or remedies be provided with respect to pending patent applications. However, TRIPS is notably a minimum standards agreement that permits members to provide more rights—so long as they do not violate other provisions, such as nondiscrimination against a field of technology. TRIPS art 1.1.

interpret TRIPS to ensure access to medicine for all. Also, article 8 notes that members may adopt measures to protect public health, as well as protect sectors of vital importance to socioeconomic development. Permitting drugs once available at low cost to continue to be available would be relevant to protecting public health. In addition, permitting Indian generic drug companies to continue to make drugs not only for India but other countries would seem consistent with promoting a sector of vital importance. Accordingly, an argument can be made that the differential treatment of pharmaceuticals patented pursuant to mailbox applications has some support in TRIPS. However, the question is whether these contextual provisions are stretched so far as to completely nullify the requirements of TRIPS.

V. Regulatory Protection of Drugs

A. OVERVIEW

Some countries provide market protection to drugs beyond the patent system through the regulatory system. India presently does not, but a brief explanation of the types of protection that countries may provide, together with what India does and its consistency with TRIPS, is helpful to providing a complete picture of access to drugs.

There are two different ways in which regulatory agencies can protect a drug from generic competition. First, a regulatory agency can limit when a generic applicant can seek approval to market a generic drug by restricting when generic applicants are entitled to rely on clinical data submitted by the originator of the drug in establishing bioequivalence to the previously approved drug. The original clinical data is considered exclusive to the originator for a number of years during which no generic applicant can rely on it; this is referred to as *data exclusivity*. The second way regulatory laws may limit generic entry into the market is when the proposed generic is linked to a patented drug. This is referred to as *patent linkage* in that the regulatory authorities will not approve a generic that is "linked" to an existing patent. Although both of the above methods of protecting drugs from generic competition are favored by the pharmaceutical industry, neither are expressly required by TRIPS, as will be further explained below.

B. DATA PROTECTION (NOT DATA EXCLUSIVITY)

Under TRIPS, all countries that require submission of clinical data as a prerequisite for obtaining approval to sell a new drug must protect this data from undefined "unfair commercial use." Although some believe this requires countries to bar generic companies from relying on this data for a certain number of years (data exclusivity), the term *unfair commercial use* does not necessarily preclude reliance on data (as discussed in more detail in Chapter 3). Rather, a country could bar use of data obtained through dishonest commercial practices. The "use" in this case would go beyond simple reliance and instead include a misrepresentation that the data submitted was generated by the applicant or resulted from the improper use or public disclosure of the data by a government employee after leaving government service.

India does not have a specific law prohibiting unfair commercial use of clinical data; accordingly, it will approve generic drugs based on the clinical data of a previously approved drug at any time. Those who evaluate applications for new drugs are not to disclose any information received in the course of their official duties.[88] In addition, although there is no official statute that addresses trade secrets or unauthorized disclosure of information, Indian courts have begun to develop case law on this issue.[89] Those who suggest India's existing law complies with TRIPS indicate that current law regarding torts, contract, equity, and laws of confidence are sufficient to meet the minimum requirements of TRIPS.[90] Nonetheless, even some advocates of public health have expressed concern that if courts are the primary means to comply with TRIPS, that might not only lead to ambiguity, but perhaps even judicial adoption of data exclusivity.[91]

a. TRIPS Compliance

An obvious question is whether India's existing regime complies with TRIPS. PhRMA clearly believes India's domestic regime fails to do so,[92] and the United States has previously suggested this as well,[93] although in recent years it has encouraged more protection without specifically suggesting India's law is inconsistent with TRIPS.[94] As noted above, TRIPS can be interpreted to permit data protection rather than data exclusivity.

---

[88] The Drugs and Cosmetics Rules, 1945, No.F.-28-10/45/(1), R. 53 (India), *available at* http://cdsco.nic.in/Drugs&CosmeticAct.pdf.

[89] *See* Shamnad Basheer, *Protection of Regulatory Data under Article 39.3 of TRIPS: The Indian Context*, INTELL. PROP. INST, at 36–37 (2006), http://papers.ssrn.com/sol3/papers.cfm?abstract_id=934269.

[90] *E.g.*, Tahir Amin et. al, *The Impact of Article 39.3 in India: A Practical Perspective*, INITIATIVE FOR MED., ACCESS & KNOWLEDGE, 19 (2006), http://www.i-mak-org.blogspot.com/; SATWANT REDDY & GURDIAL SINGH SANDHU, GOV'T OF INDIA, REPORT ON STEPS TO BE TAKEN BY GOVERNMENT OF INDIA IN THE CONTEXT OF DATA PROTECTION PROVISIONS OF ARTICLE 39.3 OF TRIPS AGREEMENT, 13 (2007), http://chemicals.nic.in/DPBooklet.pdf.

[91] *E.g.*, Tahir Amin et. al, *supra* note 90, at 20–21.

[92] *E.g.*, PHARMACEUTICAL RESEARCH AND MANUFACTURERS OF AMERICA, SPECIAL 301 SUBMISSION 2010, 48 (2010).

[93] OFFICE OF THE U.S. TRADE REPRESENTATIVE, 2003 SPECIAL 301 REPORT, 4-5 (2003) (TRIPS recognizes that the original applicant should be entitled to a period of exclusivity during which second-comers may not rely on the data the innovative company has created to obtain approval for their copies of the product.).

[94] OFFICE OF THE U.S. TRADE REPRESENTATIVE, 2010 SPECIAL 301 REPORT, 26 (2010) ("The United States also encourages India to provide protection against unfair commercial use, as well as unauthorized disclosure, of undisclosed test or other data generated to obtain marketing approval for pharmaceutical . . . products."); OFFICE OF THE U.S. TRADE REPRESENTATIVE, 2009 SPECIAL 301 REPORT, 18 (2009) ("The United States continues to urge India to improve its IPR regime by providing . . . protection against unfair commercial use of undisclosed test and other data generated to obtain marketing approval for pharmaceutical . . . products.").

In addition, TRIPS expressly permits member states to determine how to best implement its provisions.[95]

C. PATENT LINKAGE

TRIPS does not require nations to bar denial of generic drugs on the basis that the generic drug might infringe an existing patent. In other words, TRIPS does not bar patent linkage.

India does not presently bar approval of generic drugs that might infringe a patent, although this has apparently been considered.[96] Thus, a generic drug may be authorized for sale regardless of whether it might infringe a patent. Indeed, a generic company can seek marketing approval with the expectation that it will be sued for patent infringement. Although this may seem risky, a generic company may be willing to undertake this risk. After all, many litigated patents (including drug patents), are ultimately found invalid in litigation.

Although the Indian regulatory agency is not legally precluded from approving generic versions of patented drugs during the patent term, some pharmaceutical owners have tried to create patent linkage through litigation. Bristol Myers, maker of a cancer drug sold as Dasatinib, took the bold step of suing generic company Hetero for patent infringement even before the company had received approval to sell a generic version. The Delhi High Court granted Bristol Myers an ex parte injunction and suggested the Indian regulatory agency should not approve a generic version, but as the Indian regulatory authority was not actually a legal party to that decision, the court technically had no authority to control its actions.[97] Bayer also sought to use Indian courts to force the Indian regulatory agency to deny approval to generic versions of other patented drugs. For example, Bayer sued Indian generic company Cipla as well as the Indian regulatory agency; the suit involved an anti-cancer drug sold by Bayer as Nexavar and as Sorafenib by Cipla. However, this time the Delhi High Court as well as the Delhi High Court Division Bench declined to grant Bayer's request to create patent linkage via judicial fiat, and also imposed costs on Bayer for bringing frivolous litigation.[98] The Indian Supreme Court as initially accepted Bayer's request to address patent linkage, but ultimately dismissed the

---

[95] TRIPS art. 1.1.

[96] *See, e.g.*, Shamnad Basheer, *The Potency of a Middle Path*, LIVE MINT (July 9, 2008), http://www.livemint.com/articles/2008/07/09215015/The-potency-of-a-middle-path.html; Bhuma Shrivastava, *Drug Companies Oppose Bid to Link Marketing, Patents*, LIVE MINT (May 2, 2008), http://www.livemint.com/2008/05/02224246/Drug-companies-oppose-bid-to-l.html.

[97] Bristol Myers Squibb v. Hetero, Court Order, (2008) CS(OS) No. 2680 (Delhi H.C., Dec. 19, 2008) (*quoting* Judge Rajiv Sahai Endlaw) ("It is expected that the Drug Controller General of India while performing statutory functions will not allow any party to infringe any laws and if the drug for which approval has been sought by the defendants is in breach of the patent of the plaintiffs, the approval ought not be granted.").

[98] Bayer v. Cipla, (2008) WP(C) No. 7833/2008 (Delhi H. C.,Nov. 20, 2008), *reversed by* Delhi High Court (Aug. 18, 2009, Judge S. Ravindra Bhat), *aff'd* by Delhi High Court Division Bench (Feb. 9, 2010, Judge S Muralidhar).

request. Accordingly, Indian courts have yet to hold that regulatory authorities must recognize patent linkage.

a. TRIPS Compatibility

TRIPS does not require patent linkage, although it is definitely desirable to the pharmaceutical industry. The only TRIPS provision that addresses what a regulatory agency does is the one on protection of clinical data. That provision applies without regard to whether the drug is patented. Accordingly, there is no problem with TRIPS compliance no matter what the Indian courts elect to do on this issue.

VI. Concluding Thoughts

An important question is the significance of India's law for the future of access to medicine. There are actually two views on this. To those dedicated to promoting access to medicine, retaining India's existing patent provisions is considered essential—both to preserve India's status as a global supplier of low-cost generic drugs and to potentially serve as an example for other countries with similar goals. However, to patent owners, India's patent laws could be seen as an audacious attempt to undermine TRIPS. In particular, as in the view of the pharmaceutical companies, a major goal of TRIPS was to provide patents on drugs and drug compounds, the aspects of Indian law that limit patentability or patent rights for drugs would seem wholly unreasonable.

It remains to be seen what impact India's laws will have on the global supply of generic drugs. First, it is unclear the extent to which India can continue to make generic versions of drugs that are patented elsewhere as India's patent laws are relatively new and in danger of being modified. Second, even if some drugs are available as generics in India, those low-cost drugs may lack a substantial export market if other countries consider those same drugs to be patented. In other words, even if a drug can be made generically at low cost in India, that is of little utility to other countries that cannot import the Indian drug without violating patent rights. Each of these issues will be discussed in turn.

A. WILL GENERICS DRY UP IN INDIA?

Although India's patent laws are designed to promote access to medicine through restrictive patent laws and broad exceptions to patent rights, it remains to be seen whether it will succeed in this mission. India's most controversial law (the section 3d provision limiting patentability of compounds that are similar to prior art unless they have improved "efficacy") has the potential to limit patent scope *if* the patent office and courts uniformly require therapeutic efficacy. Even if they do, India's ability to limit patent protection of drugs may be curtailed if a WTO panel finds its law inconsistent with TRIPS. Similarly, India may face political pressure to modify its patent laws, including but not limited to section 3d.[99] In addition, it is not only patent laws that are at issue. Even if India's patent laws remain unchanged, if India adopts data exclusivity pursuant to a free trade agreement, that would substantially limit generic drugs as such exclusivity would block competition from even unpatentable drugs.

B. MARKET FOR INDIA'S GENERIC DRUGS?

Even if India's patent laws enable it to make generic versions of drugs patented elsewhere, it is unclear whether the country will continue to be a global supplier of low-cost generics. India can definitely make and export drugs that were not patentable before 2005. However, over time, those drugs will be less relevant to patients. This is most acutely seen with HIV patients who commonly must switch to newer treatments as they acquire resistance to old ones. However, the issue is not limited to HIV drugs: newer drugs always offer the potential for improvements that patients consider desirable. Even if India continues to produce generic versions of newer drugs, India may be limited in its ability to export them.[100] In particular, if other countries patent these drugs, they may not be able to legally import these low-cost versions. Although these countries could theoretically import such drugs through either an expansive interpretation of international exhaustion or compulsory licenses, both of these approaches would likely be highly scrutinized by pharmaceutical companies as well as other countries.

Ironically, India's public health-oriented patent laws may prompt patent owners—and the countries that promote their interests—to be more aggressive in forcing LDCs to adopt patent rights prematurely. This is common even among countries that are considered more powerful than LDCs. For example, Brazil readily modified its patent laws soon after TRIPS was concluded, rather than taking advantage of the full transition period for developing countries as did India. Considering Brazil has occasionally resisted the

[99] Pharmaceutical companies have aggressively opposed India's section 3d law. As noted earlier, when Novartis's patent application for a new version of its cancer drug sold as Glivec was opposed, the company directly challenged India's patent law as both unconstitutional and inconsistent with TRIPS. Although Novartis has not prevailed thus far, it has consistently remained unconcerned by public pressure to drop the case and is continuing to pursue it before the Indian Supreme Court. *See, e.g.*, Eur. Parl. Declaration (PE 386.0760) 2 (2007), *availableat*http://www.europarl.europa.eu/sides/getDoc.do?pubRef=-//EP//NONSGML+WDECL+P6-DCL-2007-0022+0+DOC+PDF+V0//EN&language=EN (requesting Novartis drop its suit); Letter from Anne Ferreira et al., European Parliament, to Daniel Vasella, Novartis CEO (Feb. 9, 2007), http://www.cptech.org/ip/health/c/india/meps02092007.html; Press Release, India Department of Commerce, Norwegian Minister Urges Novartis to Withdraw Case Against India (Apr. 30, 2007), http://commerce.nic.in/PressRelease/pressrelease_detail.asp?id=2018; *see also* Tove Iren S. Gerhardsen, *Opposition Gains Support against Novartis Patent Lawsuit in India*, INTELL. PROP. WATCH (Feb. 15, 2007), http://www.ip-watch.org/weblog/2007/02/15/opposition-gains-support-against-novartis-patent-lawsuit-in-india/.

[100] Although there is an existing procedure to enable export of low-cost generics to poor countries, the procedure is for *patented* drugs made under compulsory license, such that it would not apply to drugs made in India that are not patented at all. This is described in more detail in Chapter 7.

demands of pharmaceutical industry as well as wealthy countries, this modification suggests that countries with less political leverage are more likely to comply with requests to modify their patent laws.

Theoretically, an LDC that adopts patent protection for drugs might still be able to import some unpatented drugs for a short time. As patent laws only apply prospectively, drugs that were already patented in other countries before an LDC recognized patent protection on drugs would be prior art and thus unpatentable in the LDC. Essentially, this is why drugs that were unpatentable in India before 2005—when it first recognized product patents on drugs—can continue to be made as generics. However, once the LDC begins to grant patents on drugs, imports of patented drugs would be a patent violation. There are possible exceptions, such as a compulsory license or a broad view of international exhaustion; however, such actions would likely prompt pressure and possibly retaliation from drug companies and wealthy countries.

C. WILL INDIA HAVE ANY FOLLOWERS?

Even if some developing countries were theoretically not approached to enter agreements with TRIPS-Plus requirements, it is still unclear whether they would be inclined to follow India's lead. India has provided a clear template, and there are many who believe that limited patent rights taking full account of TRIPS flexibilities are the best policy option for developing countries. However, the issue is not one of optimal policy, but practical, political, and tactical realities. Beginning with TRIPS, developing nations were required to adopt laws that were arguably not in their interest. In addition, as further discussed in Chapter 8, even nations that provide more protection than India may face pressure to accede to higher standards of patent protection in exchange for coveted access to wealthy markets.

India provides not only an example of how to fashion patent laws to promote access to medicine, but how to provoke the ire, controversy, and expense of litigation with wealthy companies. It remains to be seen whether India has the political will to maintain the patent laws it designed. Moreover, considering its political strength and resources, if India is unable to maintain its patent laws in the face of international pressure, it is unlikely any other developing countries would be able to adopt or maintain similar laws. This is especially true as many other developing countries are more directly reliant on financial support from other countries.

D. OTHER UNINTENDED CONSEQUENCES

India's aggressively creative interpretation of TRIPS to promote access to medicine may have some unintended consequences. Patent owners may see India's patent laws as a template of what they wish to *prohibit* in other developing countries through rules promulgated under free trade agreements. These agreements that require more protection than TRIPS (such that they are called "TRIPS-Plus") are discussed in more detail in Chapter 8. Patent owners were interested in raising global standards of patent protection even before India enacted its laws intended to promote access to medicine. However, India's 2005 amendments to its patent laws may have increased the desire and intensity for more TRIPS-Plus agreements to prevent other nations from following India's lead.

India could also potentially face other forms of retaliation or negative consequences. Pharmaceutical companies could charge India higher prices for drugs than other countries, to the extent that India can no longer make the drugs locally. In addition, even if pharmaceutical companies are willing to sometimes voluntarily license their drugs to permit generic manufacture for poor countries, such companies may decide not to license Indian manufacturers.

In conclusion, although India's patent laws provide an important example of the existing flexibilities under TRIPS, it remains to be seen what impact those laws will have in the global sphere. A number of potential issues may limit India's ability to provide generic drugs globally. In addition, it remains to be seen whether other developing countries will consider India's laws a useful model. Regardless of these uncertainties, India's laws should provide an interesting and important contrast to the TRIPS-Plus measures discussed in subsequent chapters.

5 Compulsory Licenses under TRIPS: An Introduction

This chapter aims to bring clarity to the contentious and confusing question of what types of compulsory licenses are permissible under TRIPS. There is an abundance of misinformation not only in the popular press, but also offered by some academics. A detailed discussion of the reasons for confusion and misinformation as well as the underlying issues they likely mask are discussed in Chapter 6. This chapter first clarifies the TRIPS rules on compulsory licenses and, in particular, distills the dozen complex provisions regarding compulsory licensing into more logical categories for explanation. The chapter also provides an illustration of how a country can craft a compulsory licensing law to promote access to medicine while simultaneously complying with TRIPS. The chapter concludes with a discussion of issues beyond the scope of TRIPS that countries must consider in connection with the benefits and costs of issuing compulsory licenses. This chapter provides an essential background to understanding a procedure involving a "waiver" of one of the usual TRIPS requirements to help provide low-cost drugs to poor countries, as described in more detail in Chapter 7.

By the conclusion of this chapter, a reader should be able to answer the following questions:

- What is a compulsory license?
- Are only the poorest countries permitted to issue compulsory licenses under TRIPS?
- Does TRIPS permit a country to issue a compulsory license on a drug that is not available at a reasonable cost?
- If TRIPS permits compulsory licenses, why should a country be circumspect in issuing them?

I. Introduction 127
II. TRIPS Requirements 127
  A. What Can be Licensed 128
    1. Basic Rule 128
    2. Possible Grounds for Compulsory License 130
  B. Procedural Requirements 133
    1. Timing 133
    2. Possible Prerequisite 133
    3. License Parameters 135
  C. Patent Owner Rights and Recourse 137
    1. Monetary Compensation 138
    2. Challenges 138
  D. Special Situations 139
    1. Blocking Patents and Dependent Patents 139
    2. Anticompetitive Use 140
  E. Concluding Thoughts 140
III. Implementing TRIPS: India as an Example 141
  A. What Can be Licensed—Notable Grounds 142
    1. Invention Not Available at a "Reasonably Affordable Price" 143
    2. "Reasonable Requirements of Public" Unsatisfied 143
  B. Procedural Requirements 144
    1. Timing 144
    2. Prior Negotiation 145
    3. Scope and Duration 145
  C. Patent Owner Rights and Recourse 146
    1. Amount of Compensation 146
    2. Additional Protection 147

D. Unaddressed Issues under TRIPS 147
1. General Framework 147
2. Who May Seek 148
3. Who Authorizes 148
IV. Constraints beyond TRIPS 149
A. Drug Company Retaliation 151
B. National Retaliation 151
C. Broader Innovation and Access Concerns 152
V. Conclusion 154

I. Introduction

Before addressing what TRIPS permits with respect to compulsory licenses, it is important to first define the term. A compulsory license permits a nation (or a third party authorized by the nation) to use a patented invention without permission of the patent owner in exchange for payment of a government-determined royalty. The license is *compulsory* in that it is forced upon the patent owner. In addition, the patent owner is required to accept the government's compensation—which is likely far less than the patent owner could obtain in a free market. A compulsory license, however, never "breaks" a patent, as the patent is still valid and in force. Thus, the patent owner can continue to exclude all others besides the government-authorized licensee from the patented invention. However, with a compulsory license in place, the patent owner no longer has exclusive control over the market for its patented invention. Although the patent owner can sell its patented invention at its preferred price, cheaper versions sold by the government-authorized licensee will likely undercut the patent owner's sales and profits.

Although patent-owning pharmaceutical companies tend to suggest that compulsory licenses should not be utilized, they are in fact permissible under TRIPS. In addition, it is noteworthy that during the negotiation of TRIPS, the parties discussed—but rejected—proposals to eliminate compulsory licenses. The inclusion of rules that explicitly contemplate compulsory licenses can be seen as part of the package negotiated by developing countries to minimize the harm of granting patents in the first instance.

II. TRIPS Requirements

Any nation may issue a compulsory license consistent with TRIPS because TRIPS imposes no limitations on the type of country that can utilize this procedure. A country of any level of economic status (as well as any type of government) is equally entitled to use a compulsory license under TRIPS.[1]

Although TRIPS permits any nation to issue a compulsory license, TRIPS does not require any nation to impose them.[2] Rather, TRIPS sets conditions with which all nations must comply *if* they elect to provide compulsory licenses. Before discussing the specific requirements, it is important to clarify that although any country can issue a compulsory license for *domestic use*, there are different requirements concerning what countries can issue a license to export to other countries. This chapter deals with the usual—and less complex—situation where a nation issues a compulsory license to predominantly address its own needs, whereas Chapter 7 addresses when a nation may issue a license primarily to export the product to another country.

TRIPS requires that every country that permits compulsory licenses comply with certain procedural requirements. As will be further explained in this section, TRIPS does not dictate what subject matter can be licensed, but instead imposes procedures intended to ensure that compulsory licenses are fairly issued with reasonable consideration of patent owner interests. For example, the patent owner must not only be provided compensation, but also an opportunity to challenge a decision to issue a compulsory license.

The TRIPS provision on compulsory licensing is the most lengthy and complex patent provision. To promote clarity and address misinformation, this section first begins with the most contentious issues. In particular, this section first clarifies what subject matter may be subject to a compulsory license, followed by what countries may issue them. After clarifying these key issues, the section explains the basic procedures that must be followed to issue a TRIPS-compliant compulsory license, followed by the appropriate scope of a license and rights of the patent owner. This section concludes with a discussion of special situations, such as compulsory licenses for "blocking" patents and anticompetitive use.

A. WHAT CAN BE LICENSED?

1. Basic Rule

A fundamental and recurring point of confusion concerns what type of patents can be subject to a compulsory license. The short answer is that under TRIPS, *any* patent may

---

[1] This is in stark contrast to suggestions by patent owners that only LDCs can utilize compulsory licenses or that a military-based government is less justified in using them (as discussed in Chapter 6).

[2] *See* TRIPS: Agreement on Trade-Related Aspects of Intellectual Property Rights art. 31, Apr. 15, 1994. Marrakesh Agreement Establishing the World Trade Organization, Annex 1C, The Legal Texts: The Results of the Uruguay Round of Multilateral Trade Negotiations 320 (1999), 1869 U.N.T.S. 299 [hereinafter TRIPS] ("Where the law of a Member allows for other use of the subject matter of a patent without the authorization of the right holder . . . .").

be subject to a compulsory license. Although there are a dozen subparts to the relevant TRIPS provision concerning compulsory licensing (article 31), *none* of these contain any provision that limits the type of inventions that may be licensed.[3] This is in marked contrast to earlier drafts of TRIPS that had only limited categories of permissible subject matter for compulsory licensing. Although some patent proponents may be tempted to claim that the prior drafts limit the scope of subject matter, that is in complete opposition to a proper interpretation of TRIPS according to customary principles of international law. In particular, prior negotiating history can confirm but not contradict the actual TRIPS text. The 2001 Doha Public Health Declaration lends additional support to this conclusion.[4] Scholars generally consider this document an appropriate interpretive tool.[5] Importantly, the Declaration clearly affirms "each member has the right to grant compulsory licenses and the freedom to determine the grounds upon which such licenses were granted."[6]

Of particular importance to the issue of access to medicine is that any and all drug patents may be subject to compulsory license under TRIPS because no subject matter is excluded. Of course, nations need not permit any compulsory licenses at all, but, if they do, drug patents are definitely within the range of inventions that may be subject to them. This is an important point to emphasize as drug companies as well as patent proponents repeatedly state that drug patents should not be subject to compulsory licenses and tend to minimize or ignore that TRIPS considers such licenses to be perfectly legal.

---

[3] A few provisions of article 31 mention specific subject matter: semiconductor technology and dependent patents. TRIPS art. 31(c) (stating that in the case of semiconductor technology, the scope and duration shall only be for public noncommercial use or to remedy a practice determined to be anticompetitive); TRIPS art. 31(l) (stating that additional conditions apply). However, these areas are singled out for *additional* requirements; the provisions do not purport to suggest these are the only subjects appropriate for licensing. To the contrary, as these are the only instances where subject matter is mentioned, they support the general rule that inventions in all areas of technology may be licensed.

[4] World Trade Organization, Ministerial Declaration on the TRIPS Agreement and Public Health of November 14, 2001, WT/MIN(01)/DEC/2, 41 I.L.M. 755 (2002) [hereinafter Doha Public Health Declaration].

[5] Many scholars consider the Declaration to be a subsequent agreement between the parties. *E.g.*, CARLOS CORREA, IMPLICATIONS OF THE DOHA DECLARATION ON TRIPS AGREEMENT AND PUBLIC HEALTH 45 (WHO 2002), *available at* http://www.who.int/medicines/areas/policy/WHO_EDM_PAR_2002.3.pdf; Denis Borges Barbosa et al., *Slouching Towards Development in International Intellectual Property*, 2007 MICH. ST. L. REV. 71, 131–32 (2007); Steve Charnovitz, *The Legal Status of the Doha Declarations*, 5 J. INT'L ECON. L. 207, 211 (2002); Susy Frankel, *WTO Application of the Customary Rules of Interpretation of Public International Law to Intellectual Property*, 46 VA. J. INT'L L. 365, 400–01 (2005–2006). Others view the Declaration to be binding, but based on a different rationale. *E.g.*, HOLGER HESTERMEYER, HUMAN RIGHTS AND THE WTO: THE CASE OF PATENTS AND ACCESS TO MEDICINES 281–82 (2007) (suggesting that the Declaration is an authoritative interpretation of TRIPS pursuant to WTO, art. IX(2)); Carmen Otero Garcia-Castrillon, *An Approach to the WTO Ministerial Declaration on the TRIPS Agreement and Public Health*, 5 J. INT'L ECON. L., 212, 212 (2002) (noting the Declaration constitutes a supplementary means of interpretation).

[6] Doha Public Health Declaration ¶ 5(b).

However, although drug patents are not immune from compulsory licenses, there can be no automatic licenses of drugs. Prior to TRIPS, some countries (such as India and Canada) had provisions that automatically granted compulsory licenses on inventions relating to pharmaceuticals or food, with a set royalty rate.[7] Although it is theoretically possible that every drug patent could be subject to compulsory license in a post-TRIPS world, that is highly unlikely given the TRIPS requirement that nations must evaluate *individual* patents to determine if a license is appropriate, as well as determine the proper remuneration.[8] Accordingly, a country may not automatically issue compulsory licenses for an entire class of technology, such as all drugs. This is consistent with the remainder of article 31 requirements, which focus on what steps must be taken to issue (as well as challenge) a specific license. Moreover, this meaning is also confirmed by the negotiating history that shows countries declined to adopt India's proposal to automatically issue licenses for patents relating to food and drugs.[9]

2. Possible Grounds for Compulsory License

Although TRIPS permits nations to issue compulsory licenses on any patent and to decide on the appropriate grounds for them, a brief discussion of possible grounds (including ones that have historically been recognized) may be useful. The pharmaceutical industry tends to suggest that compulsory licenses are extreme and unprecedented. However, they in fact have a long history; at the time TRIPS was concluded, most nations had provisions authorizing them.[10]

a. Failure to Work

The oldest basis for issuing a compulsory license is for "failure to work" the invention in the local market. This term refers to a patent owner that does not make the patented invention locally. Historically, a patent was considered a privilege granted with the expectation

---

[7] *E.g.*, Canada Patent Act, R.S.C. 1985, c. P-4, § 39 (repealed 1993); The Patents Act, 1970, No. 39, § 87, Acts of Parliament, 1970 (India). *See also* JEROME H. REICHMAN WITH CATHERINE HASENZAHL, NON-VOLUNTARY LICENSING OF PATENTED INVENTIONS: HISTORICAL PERSPECTIVE, LEGAL FRAMEWORK UNDER TRIPS, AND AN OVERVIEW OF THE PRACTICE IN CANADA AND THE UNITED STATES OF AMERICA 28 (UNCTAD-ICTSD 2003) (noting that in the past, Canadian laws broadly allowed compulsory licenses to manufacture drugs as a way to increase public access to low-cost drugs).

[8] The official TRIPS requirement is that "authorization for such use" (i.e., the compulsory licenses) "shall be considered on its individual merits." TRIPS art. 31(a).

[9] Negotiating Group on Trade-Related Aspects of Intellectual Property Rights, including Trade in Counterfeit Goods, *Standards and Principles Concerning the Availability Scope and Use of Trade-Related Intellectual Property Rights*, Communication from India, ¶ 15, MTN.GNG/NG11/W/37, (July 10, 1989) (proposing licenses of right for food, pharmaceuticals, and chemicals, with no opportunity for administrative or judicial review).

[10] *See, e.g.*, CARLOS M. CORREA, TRADE RELATED ASPECTS OF INTELLECTUAL PROPERTY RIGHTS: A COMMENTARY ON THE TRIPS AGREEMENT 313 (2007).

that the patent owner would make the patented invention locally.[11] In most countries, a patent owner who did not make the product locally and only imported the product would be considered to fail to work the patent and thus abuse the patent right. One of the goals of patents that is seen as violated by failure to work is that of promoting transfer of technology. Many nations permitted compulsory licenses where a patent owner had not locally made the patented invention several years after the patent was granted (unless there were legitimate reasons); however, a legitimate reason did not include importing the product. Although this may seem extreme, it is actually a more moderate position than some earlier laws that required the patent owner to entirely forfeit the right to a patent if it was not used locally.[12]

There is an open dispute concerning whether compulsory licenses are permissible for lack of local working of a patented invention. Some countries clearly believe that TRIPS bars local working. For example, the United States initiated a suit against Brazil on this ground, although the case settled before a WTO panel could provide definitive guidance.[13] Commentators are divided on this issue,[14] with the disagreement focusing not on the TRIPS provisions concerning compulsory license—which are silent as to what constitutes permissible grounds—but on a more general provision concerning nondiscrimination. In particular, TRIPS requires that patent rights shall be available without discrimination, whether products are imported or locally produced.[15] Some, such as the

---

[11] *E.g.*, Paul Champ & Amir Attaran, *Patent Rights and Local Working under the WTO TRIPS Agreement: An Analysis of the U.S.-Brazil Patent Dispute*, 27 YALE J. INT'L L. 365, 370–71 (2002).

[12] Forfeiture or cancellation of a patent for lack of working existed in patent laws dating back to the Venetian Patent Act of 1474. *See, e.g.*, Champ & Attaran, *supra* note 11, at 371. However, the Paris Convention of 1883 prohibited automatic forfeiture in cases where the patentee imported the invention and the 1925 revision of the Paris Convention clarified that patent forfeiture could result only after a compulsory license had been ineffective to remedy failure to work. Paris Convention for the Protection of Industrial Property art. 5, Mar. 20, 1883, 21 U.S.T. 1583; Paris Convention for the Protection of Industrial Property art. 5(2), Mar. 20, 1883, *as revised* at The Hague, Nov. 6, 1925, 21 U.S.T. 1583, 828 U.N.T.S. 305 [hereinafter Hague Revision to Paris Convention].

[13] Request for Establishment of a Panel by the United States, *Brazil – Measures Affecting Patent Protection*, WT/DS199/3 (January 9 2001); Notification of Mutually Agreed Solution, *Brazil – Measures Affecting Patent Protection*, WT/DS199/4, G/L454, IP/D/23/Add.1 (July 19, 2001).

[14] Professor Daniel Gervais believes that local working is no longer permissible, but his interpretation is based in part on some countries having defined working of a patent to include imports; the national laws he cites seem more appropriate to support a conclusion that countries may, but need not, consider local working as a ground for compulsory license, and have the ability to self-define what constitutes local working. DANIEL GERVAIS, THE TRIPS AGREEMENT: DRAFTING HISTORY AND ANALYSIS 340, 395 (3d ed. 2008). Conversely, other commentators believe that local working is clearly permissible. Carlos M. Correa, *Patent Rights*, *in* INTELLECTUAL PROPERTY AND INTERNATIONAL TRADE: THE TRIPS AGREEMENT 314–15 (Carlos Correa & Abdulqawi Yusuf eds., 2nd ed. 2008); Bryan Mercurio & Mitali Tyagi, *Treaty Interpretation in WTO Dispute Settlement: The Outstanding Question of the Legality of Local Working Requirements*, 19 MINN J. INT'L L. 275 (2010); Amy Kapczinski, *Harmonization and Its Discontents: A Case Study of TRIPS Implementation in India's Pharmaceutical Sector*, 97 CALIF. L. REV. 1571, 1611–12 (2009); Champ & Attaran, *supra* note 11, at 383–90. Others believe the issue is ambiguous. REICHMAN WITH HASENZAHL, *supra* note 7, at 13–14.

[15] TRIPS art. 27(1).

United States and PhRMA, interpret this provision to bar compulsory licenses on grounds of local working. However, that interpretation is not squarely supported by the actual text. The nondiscrimination clause seems to affirm the patent owner's right to exclude others from making, selling, or importing the patented invention without speaking to whether the patent owner would lose its right to exclude others if the patent owner itself failed to make the product locally.

TRIPS should be read as only providing *additional procedural* requirements for compulsory licenses but not removing any prior bases for issuing them. Additional support lies in the Paris Convention, an international agreement that is expressly incorporated by reference into TRIPS; this agreement expressly considers lack of local working as an abuse of patent rights that is a ground for issuing compulsory licenses.[16] Permitting compulsory licenses to issue on the ground of lack of local working would thus be consistent with the fundamental cannon of treaty interpretation that favors interpretations that give full effect to clauses. As member states are bound to the relevant Paris Convention clause on non-working, the nondiscrimination clause should be interpreted so as to not nullify this requirement.[17] This is especially true because TRIPS does not contain any language expressly repealing local working; accordingly, the existing default that lack of local working may give rise to a compulsory license should continue to be the norm. This conclusion is also supported by the fact that a proposal to expressly exclude local working from the grounds of compulsory licenses was not adopted.[18]

b. Public Interest

Compulsory licenses were later expanded to promote a variety of public interest goals. In other words, nations elected to grant compulsory licenses in certain areas where the public interest in broader access to the patented invention was considered to outweigh the patent owner's right to exploit its exclusive rights. Traditionally stated public interest grounds include government use, dependent or "blocking" patents (on improvements to

---

[16] Paris Convention for the Protection of Industrial Property art. 5(A), Mar. 20, 1883, as revised July 14, 1967, 21 U.S.T. 1583, 828 U.N.T.S. 305 [hereinafter Paris Convention]; TRIPS art. 2(1)(requiring all WTO members to comply with Paris Convention articles 1 through 12).

[17] For a detailed discussion of a proper reading of TRIPS article 27 with the Paris Convention, see Mercurio & Tyagi, *supra* note 14, at 296-325. This interpretation is also supported by other TRIPS provisions that suggest that abuse of patent rights may need to be policed. *E.g.*, TRIPS art. 8(2) ("appropriate measures . . . may be needed to prevent the abuse of intellectual property rights by right holders").

[18] *Draft Agreement on the Trade-Related Aspects of Intellectual Property Rights*, Communication from the United States, art. 27, GATT Doc. No. MTN.GNG/NG11/W/70 (May 11, 1990) (proposing that compulsory licenses only be permissible to "remedy an adjudicated violation of competition laws or to address, during its existence, a declared national emergency."). However, this issue is complicated by the fact that the proposal was for the TRIPS provision explicitly governing compulsory licenses—article 31—whereas current debate focuses on the nondiscrimination clause of article 27. Nonetheless, because the nondiscrimination language in article 27 was added in part when parties could not agree to language under article 31 regarding local working, the history of article 31 could be relevant to interpreting article 27.

a prior invention), and anticompetitive practices. However, a compulsory license on the basis of public interest is certainly not limited to these areas; a license necessary to lower the price of a patented medicine would also be for public interest. TRIPS does not limit the grounds that countries may consider to issue compulsory licenses, such that public interest remains a possible consideration. In addition, nations are not limited to these common grounds—they are free to create new grounds for issuing licenses. TRIPS does impose additional requirements when a license is issued for blocking patents and anticompetitive practices (as later discussed in this chapter).

B. PROCEDURAL REQUIREMENTS

Now that it is clear that any country is permitted to issue a compulsory license with respect to any technology, this section discusses some procedural requirements that do place some limitations on such licenses. This section discusses three primary types of procedural parameters: when a license is permissible, whether prior negotiation with the patent owner is required, and the parameters of the actual license.

1. Timing

TRIPS is silent on when a compulsory license may be issued during the patent term. This means that a nation can generally issue one at *any* time during that term—subject to one exception. A compulsory license issued for failure to work the invention cannot be issued until three years after the patent has issued (and only if the patent owner cannot establish a legitimate reason for non-working), according to the pertinent Paris Convention provision that applies to all WTO member states. What this means is that for compulsory licenses issued on any basis *except* failure to work, the license may be issued at any time during the patent term.

2. Possible Prerequisite

a. General Rule of Prior Negotiation

In most cases, TRIPS requires that before a compulsory license issues, the proposed user must first attempt to obtain a voluntary license from the patent owner.[19] In other words, TRIPS envisions that compulsory licenses should generally not be imposed unless there is a breakdown in private negotiations. This is an important requirement, as patent owners frequently reduce prices on their own accord when a compulsory license is threatened.

However, there is ambuiguity as to exactly how hard the proposed user must try to secure a license. TRIPS requires that efforts be made to obtain voluntary authorization

---

[19] TRIPS art. 31(b) ("Such use may only be permitted if, prior to such use, the proposed user has made efforts to obtain authorization from the right holder on reasonable commercial terms and conditions and that such efforts have not been successful within a reasonable time.").

from the patent owner on "reasonable commercial terms and conditions" and for a "reasonable period of time."[20] However, patent owners and consumers are likely to have opposite views of what is considered "reasonable." For example, when Thailand recently imposed its licenses, it asserted it had negotiated for months, but patent owners expressed great surprise because they did not know that a compulsory license was being contemplated. TRIPS does not explicitly require a patent owner be told that a compulsory license is under consideration for the negotiation requirement to be satisfied. However, the Thailand example suggests it may be prudent to expressly articulate this possibility, rather than requesting lower rates without any mentioned implications. This is especially important because (as later discussed in this chapter) nations may suffer political and economic consequences from issuing compulsory licenses.

b. Waiver of Prior Negotiation

The usual prerequisite of prior negotiation with the patent owner may be waived under three separate circumstances. In particular, a country can issue a compulsory license without prior negotiation in the case of a "national emergency," circumstances of "extreme urgency," or in cases of "public noncommercial use," so long as the patent owner is notified "as soon as practicable."[21] In each of these situations, the patent owner is provided no opportunity to dissuade the nation from imposing a compulsory license. However, even if prior negotiation is waived, all other requirements remain, including compensation to the patent owner.

Notably, these categories do not limit the situations in which compulsory licenses may be issued. In other words, compulsory licenses can be issued even if there is no national emergency, because as discussed in the previous section, TRIPS does not limit the grounds for compulsory licenses. However, the existence of a national emergency (or extreme urgency or public noncommercial use) will remove the procedural hurdle of having to engage in prior negotiation with the patent owner.

Of great importance is defining the situations under which prior negotiation is not necessary. For example, what constitutes a "national emergency"? TRIPS does not dictate how a nation assesses whether an emergency exists. There is not even a requirement of a formal declaration. Accordingly, a nation should be permitted to simply note in the license itself that it is issued based on a national emergency. In addition, the lack of a definition provides nations with discretion to decide what constitutes an appropriate emergency. This approach is not only consistent with standard TRIPS interpretation, but also explicitly endorsed in the unanimously concluded 2001 Doha Public Health Declaration. The Declaration reflects the agreement of the WTO community that a national emergency or situation of extreme urgency exists in the case of HIV/AIDS,

---

[20] *Id.*

[21] *Id.*

tuberculosis, malaria, or other epidemics.[22] However, these are not the only possible situations; the Doha Public Health Declaration plainly states that "each member has the right to determine" what constitutes a national emergency or situation of extreme urgency, and that the listed conditions are only illustrative.[23] On a practical level, a nation may be more vulnerable to challenge for issuing a compulsory license without prior negotiation on the basis of a national emergency that is *not* specified in the Declaration. However, this is no different than with any other undefined term under TRIPS.

A trickier question is what constitutes "public noncommercial use." This has never been defined, and it was subject to some controversy when Thailand issued a series of licenses on this basis. Although many key TRIPS terms are undefined, this particular one is critical with respect to potentially opening up a wide range of situations where compulsory licenses can be imposed without prior negotiation. The word *public* could refer either to a license that is for the public *benefit*, or a license that is given to a public (as opposed to private) entity. Similarly, *noncommercial* could refer to use by a public institution, or to the use being not-for-profit.

3. License Parameters

TRIPS sets forth a number of parameters concerning the scope of a compulsory license. In general, a compulsory license must be issued predominantly for domestic use. In order to limit the impact of such licenses on the patent owner, they must be nonexclusive and nonassignable. Licenses are further restricted in that the scope and duration of individual licenses must be "limited to the purpose for which it was authorized." Once the purpose for which the license was authorized ceases to exist (and is unlikely to reoccur), the license must be subject to termination. Each of these parameters are further discussed below.

a. Domestic Use

A compulsory license must be "predominantly" for the domestic market where issued—unless it was issued to address anticompetitive conduct[24] or pursuant to a very limited exception further discussed in Chapter 7. Although TRIPS does not define *predominantly*, it would seem to require that at least a majority of goods made under the license be for domestic use. In other words, if Brazil issued a compulsory license to make a patented cancer drug, the drug must be primarily distributed to Brazilian citizens.

---

[22] Doha Public Health Declaration ¶ 1.

[23] *Id.* ¶ 5(b).

[24] TRIPS art. 31(f) ("predominantly for the supply of the domestic market"); *id.* art. 31(k) (waiving dominant use requirement when compulsory license issued for anticompetitive conduct).

b. Nonexclusive and Nonassignable

The license must be nonexclusive and nonassignable. The requirement of nonexclusivity means that the patent owner is free to authorize other entities to make and sell the patented drug. The requirement that the license itself be nonassignable prevents the creation of a market in compulsory licenses with independent value that might lead to greater interest in seeking compulsory licenses.

c. Scope and Duration

TRIPS also imposes two requirements on the scope of the license. First, the scope and duration of the license be limited to the "purpose for which it was authorized." Second, it must be subject to termination. Termination will be addressed first as it is the simpler issue.

TERMINATION

Under TRIPS, a compulsory license must be subject to termination if the circumstances that led to the authorization cease to exist (and are unlikely to reoccur). For example, if a license were issued for a patented drug to treat a viral condition that has abated, that license must be subject to termination. However, some licenses are not likely to be subject to termination. This will notably be the case when the condition is chronic or a nation continues to be unable to pay the retail cost of the drug.

SCOPE AND DURATION LIMITED TO AUTHORIZED PURPOSE

The other requirement on compulsory license scope is that the scope and duration of the license must be limited "to the purpose for which it was authorized." Notably, this requirement does not say that compulsory licenses should be issued only in limited circumstances (contrary to some misinformation propagated by patent proponents).[25] Rather, the limitation is with respect to the purpose. There are two possible interpretations of what it means to limit the license to its authorized purpose. First, it could mean the scope of the license cannot be expanded after the license is issued; for example, if the initial license was for 50,000 doses, this amount cannot be later doubled. An alternative interpretation would be that the scope and duration of the license must be tailored to the purpose authorized. Importantly, however, it is up to a nation to declare its authorized purpose. For example, a license on an HIV drug that lasts for the duration of the patent could be considered limited in duration in a nation with an HIV epidemic; as HIV is a condition that is not curable, the need would presumably be indefinite. Similarly, if a

[25] A different exception to patent rights explicitly states that it provides a "limited exception" to patent rights. TRIPS art. 30. Moreover, the negotiating history indicates parties rejected proposed language that would have mandated all member states to "minimize" the grant of compulsory licenses overall. *See* Status of Work in the Negotiating Group: Chairman's Report to the GNG, Part III, § 5A.2, GATT Doc. MTN.GNG/NG11/W76 p.19 (July 23, 1990).

compulsory license of an HIV drug was considered necessary to enable a cash-strapped government to satisfy a legal mandate to provide free HIV treatment to all citizens, a license might be reasonably limited if it covered the poorest citizens.

This interpretation may seem overly broad and even outlandish to patent owners. However, it is consistent with the actual text of TRIPS. Notably, TRIPS does not impose any criterion for assessing the scope and duration; for example, it does not require the license be limited to a "legitimate" purpose.

Whether a compulsory license of the heart treatment Plavix for lower-income Thai citizens could satisfy this requirement presents a difficult question. The stated purpose of the license is to expand the accessibility of an effective treatment to prevent undue morbidity and mortality. The license scope was not limited to a specific number of patients, but instead referred to patients covered by government health insurance, who are typically the lowest-income citizens. Drug companies have vigorously protested this license as being improper when there are alternative generic treatments to Plavix (as well as raising some irrelevant grounds[26]). However, nothing under TRIPS article 31 requires that nations use unpatented alternatives when possible. In addition, although other WTO agreements use language such as "necessary,"[27] this language is notably absent from article 31, which suggests it would be improper to impose such an additional requirement.

The above illustrations help to show that almost any license can meet the requirement that it be "limited in scope and duration" to the stated purpose. Although this conclusion may seem very broad, it is subject to important caveats. First, this is only *one* of the many requirements of TRIPS. Second, this is also what parties agreed to during the TRIPS negotiations.

C. PATENT OWNER RIGHTS AND RECOURSE

A compulsory license is by definition imposed on a patent owner—not something a patent owner would seek out. However, TRIPS does contain provisions aimed at protecting the patent owner's interests. First, TRIPS mandates the patent owner be compensated for use of the invention. Second, TRIPS requires that patent owners have the opportunity to challenge both the decision to issue a compulsory license and the compensation provided.

---

[26] Patent owners improperly complained that heart disease was not a national emergency. *E.g.,* Roger Bate, *Thailand's Patent Attack*, N.Y. SUN, Feb. 13, 2007. However, this is irrelevant because the license specifically stated that it was issued based on public noncommercial use.

[27] *See, e.g.*, General Agreement on Tarrifs and Trade, art. XX. Oct. 30, 1947, 61 Stat. A-11, 55 U.N.T.S. 194.

1. Monetary Compensation

Although patent owners are deprived of the opportunity to choose their own compensation, they are provided some compensation under TRIPS. In particular, TRIPS requires the patent owner be paid "adequate remuneration" that considers the "economic value of the authorization."[28]

There is currently no established meaning of these phrases. As with other key terms, what constitutes "adequate" remuneration is not defined under TRIPS. Patent owners would likely prefer the amount be equivalent to the prevailing market rate. However, if that were the case, a compulsory license would seem of little utility as such licenses are generally issued to lower the price of drugs. In addition, the plain meaning of *adequate* would suggest it means something sufficient on a minimal level rather than *optimal*, such that it should likely be less than the prevailing market rate. In addition, it should arguably be less than the prevailing *local* market rate in the country where the compulsory license is authorized for a license to have any effect in reducing drug prices. Although some have tried to promote guidelines for what royalty amounts should be considered adequate, as with all other undefined terms, the issue ultimately lies within the scope of national discretion.[29]

2. Challenges

The patent owner must be able to challenge two aspects of a compulsory license. First, the patent owner must be afforded an opportunity to challenge the initial decision to issue a compulsory license. In addition, the patent owner is entitled to challenge any national decision concerning remuneration for a compulsory license.

Of important note is that although a patent owner must be granted an opportunity to challenge a license, the required opportunity is within the judicial system of the nation granting the license.[30] Although some may suggest this is improper as a nation is unlikely to find its own license invalid, this is what countries negotiated for under TRIPS. Moreover, if the judicial review process is in any way inappropriate, member states can always bring a challenge within the WTO dispute settlement system. As noted earlier, individual stakeholders (such as patent owners) have no standing in that system. However, patent owners can petition their home country to act on their behalf and have effectively done so in the past when their interests were at stake.

---

[28] TRIPS art. 31(h).

[29] *E.g.*, World Health Organization, Remuneration Guidelines for Non-Voluntary Use of a Patent on Medical Technologies, Health Economics and Drugs 5 (2005).

[30] Technically, review is subject to judicial "or other independent review" to take into account the fact that not all members have judicial systems. TRIPS art. 31(i)–(j).

D. SPECIAL SITUATIONS

1. Blocking Patents and Dependent Patents

As noted earlier, a traditional ground for issuing a compulsory license arises in the case of a blocking patent; in other words, a license is sometimes necessary to use a subsequent patent without infringing an earlier patent.[31] Many nations have used compulsory licenses to ensure that new innovations can be used. A compulsory license to the later patent owner is considered to maximize social welfare by ensuring the subsequent invention can be used.

To understand this ground, it is important to first understand why use of a second patent would infringe. Recall that a patent only confers a right to *exclude* and not an affirmative right to use; this means a patent owner could be precluded from using the patented invention if it infringes another's patent. This could happen because there is no patentability requirement that the invention not violate an existing patent. Rather, the patentability rules are that the invention be capable of industrial application, new, and have an inventive step. Accordingly, it is possible a patent will issue for an invention that cannot be used without violating another patent unless permission to use the prior patent is obtained. Because the prior patent "blocks" use of the later patent, it is often referred to as a "blocking patent." Similarly, because legal use of the later patent depends on the ability to use the first patent (whether through voluntary permission or a compulsory license), the later patent is referred to as a "dependent patent."

It may be helpful to illustrate how one patent could block the use of another. Imagine a world where a three-legged stool is a new invention, such that it is granted a patent. Now imagine that someone invents a stool with *four* legs and a back. In other words, the new invention is a chair. If the chair is patented as new (and does not lack an inventive step based on the prior stool),[32] the owner of the chair patent will be able to preclude others from making chairs. However, because a chair has all of the elements of a stool, making the chair would violate the stool patent owner's right to exclude others from making a stool.

TRIPS imposes several additional requirements when a compulsory license is issued in the case of a related or blocking patent. First, TRIPS states that this is only permissible if the second patent involves an "important technical advance of considerable economic significance" relative to the first patent. So, for example, to be entitled to a license to use the (blocking) stool patent, the owner of the (dependant) chair patent would need to show that the chair was an important advance. Second, the owner of the stool patent

---

[31] This is sometimes referred to as a situation where an improvement or "dependent" patent cannot be used without infringing another's dominant or blocking patent.

[32] A chair (stool with an additional leg for stability and a back for support) might be considered lacking an inventive step. However, for the purpose of explaining blocking patents with a readily understandable invention, this section assumes this is not a problem.

must be entitled to a cross-license of the chair patent. Third, the compulsory license must be nonassignable.[33]

2. Anticompetitive Use

Another special situation involves a compulsory license issued not for any of the public interest grounds previously discussed, but as a remedy to address officially determined anticompetitive action.[34] In such a case, certain requirements are actually waived. In particular, where a compulsory license is issued as a remedy for anticompetitive conduct, there is no need for prior negotiation with the patent owner, and the license need not be predominantly for domestic use. In addition, the amount of remuneration may be limited by the fact that a nation can take into consideration the need to correct anticompetitive practices. For example, the patent owner might be required to provide a compulsory license with no royalty as part of the penalty for the anticompetitive action.

E. CONCLUDING THOUGHTS

Although TRIPS article 31 lists many requirements, it actually does not spell out every possible option. For example, TRIPS does not dictate who within a given member country has authority to issue a compulsory license, or who can seek a compulsory license. Instead, national law governs these issues. For example, nations must decide whether a license could be issued as a matter of executive authority, or whether there is a special agency, committee, or other organization that would be dedicated to this task. Similarly, nations must determine whether anyone beyond the government can seek a compulsory license. Each of these issues will be addressed in the following section, which uses India's domestic laws as an example of a domestic compulsory license regime. Before discussing those laws, some important TRIPS requirements will first be reviewed.

Q: Is compulsory licensing of patents permissible under TRIPS?

A: Yes! There are restrictions on *how* licenses are issued, but they are definitely permissible.

Q: What countries may impose compulsory licenses?

A: Any country! TRIPS does not impose any limitations on which countries may utilize compulsory licenses.

---

[33] TRIPS art. 31(l).

[34] TRIPS art. 31(k). This provision permits states to define what constitutes anticompetitive conduct. Nonetheless, TRIPS contemplates that anticompetitive conduct can be an abuse of patent rights. *See* TRIPS art. 40 (suggesting that licenses may constitute anti-competitive conduct); *see also* TRIPS art. 8(2) (suggesting that "abuse" of rights must be addressed).

Q: Are only HIV drugs subject to compulsory licenses?

A: No! TRIPS does not impose any limitations on the type of subject matter that can be licensed; however, nations could self-impose such limits.

Q: Are compulsory licenses only appropriate in the case of a national emergency, such as an epidemic?

A: No. Whether there is a national emergency is only pertinent to whether the patent owner must first be consulted in an effort to voluntarily negotiate a license. If there is a national emergency, there is no need to first negotiate with the patent owner. However, a national emergency is not a prerequisite to issuance of a compulsory license. In addition, prior negotiation with the patent owner can also be waived if the license is for public noncommercial use.

Q: Does the WTO review whether each compulsory license issued by a nation is valid?

A: No. The WTO dispute resolution procedure is available if a country believes that another country is issuing licenses that fail to comply with TRIPS. However, TRIPS does not require each compulsory license to be subject to WTO review.

Q: Does TRIPS require all compulsory-licensed drugs to be specially labelled to protect the patent owner from undesired imports into a different market?

A: No. Only compulsory licenses of drugs made for export pursuant to a complex procedure (described in Chapter 7) need be specially labelled. There is no general requirement for licensed drugs that serve the domestic market.

Q: Can a patent owner contest a compulsory license?

A: Yes! A patent owner can challenge the issuance of a compulsory license (as well as the royalty rate associated with the license) through the judicial system of the nation imposing it.

III. Implementing TRIPS: India as an Example

This section highlights aspects of India's compulsory licensing provision to provide an example of how nations can use compulsory licensing to best promote short-term access to medicine in a manner consistent with TRIPS.[35] Examining the actual laws of one country is helpful in seeing how TRIPS can be implemented because TRIPS alone is a poor template for domestic implementation: it lacks definitions of key terms and does not address important issues, such as who can issue a compulsory license. Interestingly, although India has one of the more detailed provisions on compulsory licenses and

[35] A number of the compulsory licensing provisions were actually enacted before TRIPS was concluded, yet have greater relevance now that India must provide patents on drug compositions.

government use,[36] there has been recent discussion about whether additional guidelines or clarification are necessary.[37] In addition, how India's laws will be applied remain to be seen, since no compulsory licenses have issued since India modified its laws to comply with TRIPS.

For ease of readability, this section discusses Indian patent laws on compulsory licenses in roughly the same order as the prior section on TRIPS requirements. However, not every TRIPS requirement will be discussed. Rather, this section will focus on notable provisions as well as examples of how India has provided more clarification than TRIPS. This section begins with some novel Indian grounds for issuing compulsory licenses, then addresses procedural requirements and protection for the patent owner. The section concludes with issues unaddressed by TRIPS; namely, who can seek a license, who authorizes licenses, and additional criteria to consider in granting licenses.

A. WHAT CAN BE LICENSED—NOTABLE GROUNDS

India provides a number of grounds for compulsory licenses. Some grounds are expressly mentioned in TRIPS, such as anticompetitive acts, dependent patents (that require use of a "blocking patent"), and cases of national emergency, extreme urgency, or public non-commercial use. This section does not provide a comprehensive discussion of each basis. Rather, it aims to focus on the most notable grounds for compulsory licenses under Indian law and evaluates whether they comply with TRIPS. As indicated earlier, nations generally have free reign to decide on what is appropriate. However, that discretion is somewhat constrained by other TRIPS requirements that must always be satisfied; for example, the basis for a compulsory license may be impermissible if it discriminates against different WTO members or unduly discriminates against subject matter.

The focus here is on notable grounds for issuing compulsory licenses that have not been previously addressed in the earlier TRIPS discussion. For example, this section discusses whether compulsory licenses should issue if a patented drug is either not available at a "reasonably affordable price" or if the "reasonable requirements of the public" are unsatisfied. Each ground will first be explained, then followed with an analysis of its TRIPS compatibility.

---

[36] The government use provision will not be discussed in detail here. However, it is notably broader than the compulsory license provisions. In particular, any invention in a patent or patent application may be used by the government, or a party authorized by the government. The Patents Act, 1970, No. 39, § 87, Acts of Parliament, 1970 (India) (*as amended by* The Patents (Amendment) Act, 2005, No. 15, Acts of Parliament 2005 (India)) [hereinafter India Patents Act].

[37] INDIAN DEPARTMENT OF INDUSTRIAL POLICY AND PROMOTION, DISCUSSION PAPER: COMPULSORY LICENSING ¶2 (2010), *available at* http://dipp.nic.in/; *see also* Ramnath Subbu, *Pharma Units Await Clarity On Compulsory Licensing*, HINDU, Aug. 29, 2010, http://www.thehindu.com/business/Industry/article602336.ece; Madhur Singh, *India May Issue Compulsory Licenses to Control Drug Prices*, PAT. TRADEMARK AND COPYRIGHT LAW DAILY (BNA), Aug. 27, 2010.

1. Invention Not Available at a "Reasonably Affordable Price"

A notable ground for access to medicine in India is the ability to issue a compulsory license where the patented invention is not available at a "reasonably affordable price." The Indian law does not define this phrase. Nonetheless, this wording, together with general principles applicable to all the Indian compulsory licensing provisions, suggests the term should be defined to ensure that patented medicines are affordable to Indian citizens.[38]

Although countries are free to decide what grounds are appropriate for compulsory licenses, one possible issue that might be raised with respect to this provision is whether it unduly discriminates on the basis of technology. Notably, this ground applies to all inventions that are not available at reasonably affordable prices; it could apply equally to patents on drugs and patents on automotive parts. Accordingly, there is no explicit discrimination. There could potentially be an issue of discrimination if this provision were used to issue licenses only on patented drugs. However, even in that case, the action could be defended on the ground that differential treatment of drugs is justified because of a unique problem in that area of technology; namely, that inadequate access to drugs—unlike other inventions—may have deadly consequences.

2. "Reasonable Requirements of Public" Unsatisfied

India also permits a compulsory license to be granted if the "reasonable requirements of the public" are "unsatisfied" with respect to the patented invention.[39] Although this phrase is ambiguous, there are half a dozen specific situations in the statute, some with further subsections that define the term. These provisions fall into two general categories of abuse of the patent right: inadequate working of the patent and unfair license conditions.

a. Inadequate Working

Although lack of working of a patented invention is already its own basis for a compulsory license, this provision focuses on situations where the patent is *partially* worked. For example, a compulsory license may issue if the patented invention is not worked on a commercial scale to an adequate extent.[40] A compulsory license is also appropriate if importations of the patented product by either the patent owner or a third party (who would infringe the patent) hinder local working. In addition, a compulsory license may issue if local working is minimized.[41]

---

[38] For example, one principle is that patents are granted to make the benefit of the patented invention available at reasonably affordable prices to the public. Similarly, other provisions emphasize that patents do not impede protection of public health and do not prohibit government measures to protect public health.

[39] India Patents Act §§ 84(1)(a), 84(7) (providing further explanation of the term).

[40] *Id.* § 84(d).

[41] *Id.* § 84(e).

b. Abuse of Patent Power

Several provisions also permit compulsory licensing of a patented invention for some type of abuse of the patent through unfair license terms. For example, a compulsory license is permissible if the patent owner refuses to grant a license on reasonable terms, such that a local industry is negatively impacted.[42] In addition, a compulsory license is also appropriate if the patent owner imposes conditions on a license that improperly protects unpatented materials;[43] although not all countries consider this grounds for a compulsory license, there is broader recognition that requiring the purchase of patented and unpatented components may constitute misuse of patent power.[44] A compulsory license is also appropriate if a patent owner previously granted a voluntary license but imposed inappropriate restrictions, such as requiring the licensee to agree not to challenge the validity of the patent; this would be an abuse of the patent power because patents are generally subject to challenge to ensure the public is not burdened by invalid ones.[45]

c. TRIPS

All of the situations where the "reasonable requirements of the public" are unsatisfied easily satisfy TRIPS requirements. As noted earlier, TRIPS does not limit the grounds for issuing compulsory licenses. In addition, none of these provisions seem to unduly discriminate either between different WTO members or different types of technology.

B. PROCEDURAL REQUIREMENTS

Indian patent law serves as an example of how countries can provide additional clarification or even more limitations on the use of compulsory licenses beyond TRIPS. In particular (as will be discussed here), Indian patent law provides more detail on (1) when compulsory licenses may be issued, (2) the usual period of prior negotiation required with the patent owner, and (3) the scope and duration of the license.

1. Timing

India's law offers an example of how countries may decide to differentiate when different types of compulsory licenses issue. As noted earlier, TRIPS permits compulsory licenses to issue at any time unless the license is imposed for non-working of the invention (in which case it must be at least three years after patent issuance as required by the Paris Convention).

---

[42] *Id.* § 84(7)(a)(i).

[43] *Id.* § 84(7)(b).

[44] *See, e.g.,* Morton Salt Co. v. G.S. Suppiger Co., 314 U.S. 488, 491 (1942).

[45] India Patents Act § 84(7)(c).

Although some may disagree with the time period chosen by India, the types of situations under which a license is immediately available as opposed to being subject to a waiting period are distinct. For example, India permits a compulsory license to issue at any time in the case of a national emergency—a situation that would seem to be properly prioritized as being more important than the rights of a patent owner in a nonemergency situation. India gives more deference to the patent owner in less urgent situations. For example, India will not grant a compulsory license on the basis of an unreasonably priced drug until at least three years after the patent grant. Although a patent owner is likely to think three years of sales at its desired price is too short, those interested in promoting access to medicine likely think that three years is an unnecessary delay. Either timeline is permissible under TRIPS. In addition, other timelines could be imagined, such as a longer duration of patent owner exclusivity before a compulsory license is imposed.

One aspect of timing that is left open under India's patent laws is the time for a decision to be made on a compulsory licensing application. Recently, some have suggested that it would be helpful to have specific timelines under which the Controller should act on applications.[46] A timeline is permissible, but not required under TRIPS. The benefit of a timeline is that it can give parties certainty. However, any such timeline would need to be chosen with care to ensure that it would not unduly rush a thoughtful consideration of factors.

2. Prior Negotiation

India provides more detail than TRIPS with respect to the amount of prior negotiation that a proposed licensee must first engage in with the patent owner. TRIPS requires the proposed licensee seek a license under reasonable commercial terms and conditions and that the effort be unsuccessful within a "reasonable period of time" (which is not defined in TRIPS). India specifies that the reasonable period should normally be no more than six months.[47] This number thus provides guidance to proposed licensees as well as patent owners. Some nations may prefer to choose a different time frame, but determining an actual number is likely helpful to all parties with respect to providing predictability.

3. Scope and Duration

In addition, India also provides a presumptive term for compulsory licenses to last the remainder of the patent unless a shorter term is consistent with public interest. This is probably compatible with TRIPS so long as the term is evaluated in every case. After all,

---

[46] *E.g.*, NATCO Pharma Limited Hyderbad, Feedback on Compulsory Licensing- Discussion Paper II (30 Sept. 2010) (NATCO's answer to question twelve, suggests 2-8 weeks), http://www.dipp.nic.in/ipr-feedback/Feedback_NATCO2_30September2010.pdf.

[47] India Patents Act § 84(6)(iv).

TRIPS states that the scope and duration of a compulsory license must be limited to the stated purpose. So, as long as there is some connection between the term of the license and the stated purpose, this requirement should be satisfied.

C. PATENT OWNER RIGHTS AND RECOURSE

1. Amount of Compensation

India also elaborates on the TRIPS requirement that patent owners be paid "adequate remuneration" that takes into account the "economic value of the authorization." India's law seems to expand further on this requirement. In particular, the law requires the remuneration be "reasonable, having regard to the nature of the invention, the expenditure incurred by the patentee in making the invention or in developing it and obtaining a patent... and other relevant factors."[48] At the same time, India's law suggests the Controller aim to ensure the compulsory license provides a "reasonable profit" to the licensee[49] and that the patented product be made available to the public at "reasonably affordable prices."[50]

However, there is a different standard of remuneration when a compulsory license is based on a national emergency, situation of extreme urgency, or public noncommercial use. In these cases, the Controller is directed to ensure that products made under the compulsory license are available to the public at the lowest prices consistent with the patent owners deriving a "reasonable advantage" from their patent rights.[51]

There is a possible question with respect to whether the "adequate remuneration" required by TRIPS permits two different standards of remuneration based on different grounds for issuing a license. TRIPS seems to contemplate only a single standard of remuneration. However, it is possible the two different standards may both take into consideration the economic value of the authorization. Notably, TRIPS does not require the remuneration be *entirely* based on the economic value, as that factor need only be taken into account. Arguably, each provision does take the value of the authorization into account—while also considering what price is necessary to provide the patented product to the public at a reasonable price. Although India's compensation is likely consistent with TRIPS, it is not the only method of calculating remuneration. A country could alternatively set a maximum ceiling to provide clarity and avoid litigation about whether an amount is reasonable.[52] Indeed, there has been recent discussion in India about fixed

---

[48] India Patents Act § 90(1)(i).

[49] *Id.* § 90((1)(ii).

[50] *Id.* § 90(1)(iii).

[51] *Id.* § 92(1)(ii).

[52] With respect to an adequate amount of compensation, some suggest that 4% is a reasonable rate.

royalty rates, which have elicited criticism of a possible TRIPS problem in that the royalty might not adequately consider the economic value of the authorization.[53]

2. Additional Protection

India goes beyond TRIPS in providing additional procedural protection to the patent owner. TRIPS only requires the patent owner's interests be protected by the owner's ability to challenge the decision to issue a compulsory license as well as the amount of the royalty.[54] India also provides the patent owner with notification of a compulsory license application and the opportunity to oppose that application.[55] If an opposition is filed, both the applicant and opposer have an opportunity to be heard before a decision is made on issuing a compulsory license. The notification and opposition procedures are not required by TRIPS. However, whether other countries decide to provide similar opportunities to patent owners may be a function of several factors. For some countries, providing such opportunities may be considered important for an equitable process. An additional consideration may be whether providing such a procedure to patent owners would minimize general objections to compulsory licenses, or only result in expected opposition and additional delay.

D. UNADDRESSED ISSUES UNDER TRIPS

1. General Framework

India provides a set of governing principles for all its different compulsory licenses at the beginning of its section on compulsory licenses. These principles are likely ones that would resonate with other countries desirous of maximizing the use of TRIPS flexibilities to promote access to medicine. For example, India's provisions emphasize that patents should not only promote innovation, but also domestic industry, rather than profits of patent owners.[56] In addition, the impact of patents on affordable medicine is heavily

[53] *E.g.,* Views and Suggestions of the Pharmaceutical Research and Manufacturers of America (PhARMA) and the European Federation of Pharmaceutical Industries and Associations (EFPIA) in Response to the Invitation of the Government of India on Compulsory Licensing Ministry, p. 7, *available at* http://www.dipp.nic.in/iprfeedback/Feeback_PHRMA_29September2010.pdf.

[54] TRIPS art. 31 (i)–(j).

[55] Notification only occurs if the Controller is satisfied that a prima facie case is set forth. India Patents Act § 87(1). However, this notification is waived in the case of national emergency, situation of extreme urgency, or public noncommercial use. *Id.* § 92(2)

[56] *Id.* § 83(a) ("patents are granted to encourage inventions and to secure that the inventions are worked in India on a commercial scale and to the fullest extent that is reasonably practicable without undue delay"); *Id.* § 83(b) ("they [patents] are not granted merely to enable patentees to enjoy a monopoly for the importation of the patented article").

stressed, with multiple provisions emphasizing that patented goods should be available at reasonably affordable prices to the public.[57]

None of these governing principles need be adopted to comply with the TRIPS requirements on compulsory licenses, which refer only to the procedure under which compulsory licenses are granted. However, these principles are not only permissible, but consistent with other TRIPS provisions. For example, TRIPS article 7 states that protection of rights should contribute not only to technological innovation, but to the transfer and dissemination of technology in a manner conducive to social and economic welfare. Although patent rights may promote innovation, the information disclosed in a patent has traditionally been considered important to disclose new technology.

2. Who May Seek

Although TRIPS clearly contemplates that someone will make a decision on compulsory licenses, it does not specify whom that will be. India's law makes it clear that any person may seek a compulsory license on the ground that the patent is not worked or the patented invention is not available to the public at a reasonable price.[58] Moreover, the law specifically states that even an existing licensee may seek a license on grounds of non-working, unreasonable price or the reasonable requirements of the public not being satisfied.[59] Although an existing licensee may seem unlikely to seek a license, if the terms of the license are viewed as too onerous, a compulsory license would be preferable—especially as the Controller has the authority to cancel an existing license or order it be amended.[60]

3. Who Authorizes

Indian law clarifies that the Controller of Patents makes the determination on compulsory licenses; it also provides a number of guidelines for the Controller to follow in addition to the general guidelines for all compulsory licenses.

In evaluating applications for compulsory licenses, the Controller is to consider two issues.[61] First, patented inventions are to be worked to the fullest extent practicable in

---

[57] *Id.* § 83(d) ("patents granted do not impede protection of public health and nutrition and should act as instrument to promote public interest specially in sectors of vital importance for socio-economic and technological development in India"); *id.* § 83(e) ("patents granted do not in any way prohibit Central Government in taking measures to protect public health");*id.* § 83(g)("patents are granted to make the benefit of the patented invention available at reasonably affordable prices to the public").

[58] India Patents Act §§ 84(1)–(2).

[59] *Id.* § 84(2).

[60] *Id.* § 88(2).

[61] *Id.* § 89. These factors are not applicable to compulsory licenses issued on grounds of national emergency, extreme urgency, or noncommercial use. *Id.* § 92(3).

India; in other words, local and commercial manufacturing of the patented invention is favored. Second, the interests of the patent owner are not to be unfairly prejudiced.

In addition to the above issues, the Controller is to always consider the general guidelines that focus on what the patentee (or licensee) has done to make full use of the invention, as well as the ability of the applicant to use the invention to public advantage.[62] For example, if the patent owner has imported the patented invention and sold it at a high price to the public, that might be considered as not making full use of the invention or not using it to the public advantage. Other factors, however, focus more on whether the compulsory licensee applicant will be well positioned to effectively use the license. For example, the Controller is to consider the ability of the applicant to use the invention to "public advantage" and whether the applicant has adequate funding to properly use the invention; these factors could address whether the potential licensee would be able to make the patented invention such that the public could benefit.

While the Indian law is helpful in clarifying who has authority to issue licenses and providing some parameters for consideration, it may not necessarily be an ideal model. Those interested in promoting the fastest access to low-cost medicine may consider India's laws to give the Controller undue discretion to deny a compulsory license. Moreover, whether the Controller would have adequate legal expertise to decide the many relevant issues is questionable. An independent panel that includes members with legal and economics expertise might be better suited for this role.[63]

IV. Constraints beyond TRIPS

Although TRIPS seems to provide substantial leeway to countries to issue compulsory licenses, their practical ability to do so is constrained by factors beyond TRIPS. In particular, nations that issue—or attempt to issue—compulsory licenses may face retaliation from patent owners, as well as countries that are home to patent owners such as the United States and EU Member States.

A. Drug Company Retaliation

A potentially troublesome problem with issuing compulsory licenses is that patent-owning companies may retaliate by not selling patented drugs in the license-issuing nation.

[62] *Id.* § 83; *see also id.* § 84(6).

[63] The panel proposal is suggested to not only provide a better result, but may also be necessary as a matter of constitutional law in India since the Controller would otherwise be engaged in essentially judicial functions even though the Controller is a civil servant that lacks the independence of a judiciary. Shamnad Basheer & Prashant Reddy, *Constitutional Infirmities in the Compulsory License Scheme* (Oct. 1, 2010), http://www.dipp.nic.in/ipr-feedback/Feedback_ShamnadBasheer(II)_01October2010.pdf>.

The retaliation occurs outside of TRIPS and actually outside of the patent system. For example, after Thailand issued a compulsory license on Abbott's HIV drug Kaletra, Abbott announced it was withdrawing its applications to sell seven new drugs in Thailand, including its new HIV drug marketed as Aluvia that should have been well-suited to Thailand's climate. Abbott's action is believed to be the first known retaliation by a drug company in response to compulsory licenses.[64] This action prompted substantial criticism, calls for boycotts, and protests at Abbott's annual shareholder meeting.[65] Abbott eventually decided to seek registration of Aluvia, but has thus far not changed its position on the other drugs.[66]

While patent owner Abbott eventually withdrew its threat regarding Aluvia, the more important issue to consider is the impact of a withdrawal of drugs from the market. In particular, while new (and patented) drugs generally will not be marketed before the patent owner has first filed clinical data concerning safety and efficacy in a developed country, that is not necessarily the situation in a developing country. Some developing countries will grant approval to manufacturers to sell a generic version of a drug based on a showing of bioequivalence to a drug first marketed elsewhere, such as in a developed country. Alternatively, developing countries may grant approval based on bioequivalence to published data concerning an existing drug. There may still be some hurdles to overcome since to establish bioequivalence, a company must be able to first reverse engineer commercial production of the drug and may also have limited access to necessary active ingredients. However, if these issues can be overcome, the threat of withdrawal substantially diminishes.

In addition, retaliation is not limited to withdrawal of drugs from the market, as it can also be done through pricing. For example, a drug company may provide its lowest prices only to countries that do not issue compulsory licenses. Indeed, after Thailand issued its series of compulsory licenses and Abbott faced public pressure about not registering its new drug Aluvia, Abbott announced it would offer discounted prices on Aluvia to developing countries—but only where its patents were "respected."[67]

The possibility of drug company retaliation may have a chilling effect on the use of a compulsory license. After all, what good is a compulsory license for a drug for a few

---

[64] *See, e.g., U.S. Drugmaker Abbott, Thailand Face Off in AIDS Drug Patent Stalemate*, INT'L HERALD TRIB., June 6, 2007, at 2; James Hookway & Nicholas Zamiska, *Harsh Medicine: Thai Showdown Spotlights Threat to Drug Patents*, WALL ST. J., Apr. 24, 2007, at 1 (noting that some referred to Abbott's action as the "nuclear option").

[65] *E.g.*, Nirmal Ghosh, *Battle Rages Over Thai Actions on AIDS Drugs*, STRAITS TIMES, May, 28, 2007, *available at* http://www.tillekeandgibbins.com/Publications/pdf/interview_thai_actions_on_aids_drugs.pdf; Bruce Jaspen, *AIDS Controversy Dominates Abbott's Labs' Annual Meeting*, CHIC. TRIB., Apr. 27, 2007.

[66] *E.g.*, Nicholas Zamiska, *Abbott's Thai Pact May Augur Pricing Shift*, WALL ST. J., Apr. 23, 2007, at A3.

[67] Press Release, Abbott, Abbott Agrees with WHO Direct-General to Expand Access to Kaletra/Aluvia (Apr. 10, 2007) http://www.cptech.org/ip/health/c/thailand/abbott04102007.pdf.

thousand citizens if it means that all citizens will be denied half a dozen other drugs? Clearly, the patent holder holds all the cards in terms of what drugs it will withhold.

B. NATIONAL RETALIATION

Another very real implication of compulsory licenses is that countries may be subject to unilateral economic sanctions (or at least political pressure) imposed by individual countries even if a license is TRIPS-compliant. Economic implications are a major problem as any cost savings from issuing a compulsory license may be dwarfed by more substantial economic sanctions.

Thailand provides a useful illustration: Thailand's compulsory licenses were noted by the United States as an issue in the 2007 and 2008 "Special 301" reports[68] (which are usually a first step towards trade sanctions).[69] Thailand was not alleged to have violated any specific provision of TRIPS, but a country may be listed in the Special 301 report for having "inadequate" protection of intellectual property rights despite compliance with existing international agreements.[70] In addition, the United States suggested that Thailand could be removed from the "Watch" list if it promised to increase protection of intellectual rights.[71] In addition, Thailand lost the preferential treatment for Thai imports pursuant to the Generalized System of Preferences after it issued compulsory licenses, although there is some dispute as to whether this is directly a result of the licenses.[72]

Technically, WTO member states can not exercise unilateral retaliation against another member state for failure to comply with TRIPS. As noted in Chapter 3, member states are to exclusively use the Dispute Settlement Understanding for such disputes.[73]

---

[68] *See* 2008 Special 301 Report, at 36–37; 2007 Special 301 Report, at 27.

[69] The Special 301 report lists countries with perceived inadequate or ineffective levels of intellectual property protection. 19 U.S.C. § 2242(a). In addition, the report is not an independent assessment by a government agency; the statute explicitly permits the USTR to consider information submitted by interested persons. 19 U.S.C. § 2242(b)(2)(B) (permitting input from interested parties in determining priority foreign countries); *see also* PETER DRAHOS WITH JOHN BRAITHWAITE, INFORMATION FEUDALISM 94–99 (2002) (discussing close cooperation between USTR and companies in determining countries to include on Special 301 list). Accordingly, a major drug company and patent owner can not only only directly retaliate against a country such as Thailand, but may also suggest that the United States impose additional economic sanctions.

[70] 19 U.S.C. § 2242(d)(4) (2000) (noting that a foreign country may be determined to be denying "adequate and effective protection of intellectual property rights, notwithstanding the fact that the foreign country may be in compliance" with TRIPS).

[71] Pennapa Hongthong, *US Action Plan Must be Opposed, Groups Tell Government*, NATION, May 9, 2007.

[72] *E.g.*, Jon Fernquest, *More Expensive Thai Exports Effects of Lost Thai GSP Privileges?*, BANGKOK POST ECONOMIC BUSINESS BLOG (June 21, 2007, 10:58 AM), http://www.readbangkokpost.com/business/2007_currency_crisis/more_expensive_thai_exports_ef.php.

[73] Understanding on Rules and Procedures Governing the Settlement of Disputes, art. 23(2)(a), Apr. 15, Marrakesh Agreement Establishing the World Trade Organization, Annex 2, Legal Instruments—Results of the Uruguay Round, 33 I.L.M, 1125 (1994) [hereinafter DSU].

However, by not alleging a clear violation of TRIPS, the United States may be attempting to avoid this requirement. A developing country that believes that the United States is improperly using the Special 301 process as retaliation could bring a dispute before the WTO and a prior WTO panel ruling gives an indication that such a dispute may be successful.[74] However, there have been no such challenges to the Special 301 process since the conclusion of WTO/TRIPS. From a practical perspective, developing countries may be reluctant to do so. Indeed, countries have been reluctant to even engage in permissible compulsory licenses under TRIPS.[75] Notably, although Thailand recently issued compulsory licenses and was well aware of possible negative political consequences, it declined to issue a compulsory license on the HIV drug ddI in 1997, after the United States threatened to impose economic sanctions.[76] The most recent compulsory licenses were seen as necessary by the World Bank to address an inevitable budget gap as more HIV patients need second line treatments, but even the World Bank recognized that issuing such licenses would require strong political resolve.[77]

C. BROADER INNOVATION AND ACCESS CONCERNS

Although TRIPS permits compulsory licenses, there remains another issue that is likely an undercurrent to drug company and national retaliation. In particular, some suggest that if compulsory licenses were widely used, innovation would suffer.[78] Although this seems

[74] Panel Report, *United States - Sections 301-310 of the Trade Act of 1974*, para 7.89, WT/DS152/R (Dec. 22, 1999). In addition, a violation of the DSU could theoretically entitle a party to engage in self-help by suspending its obligations under TIRPS as compensation for TRIPS pursuant to the Vienna Convention on the Law of Treaties. Vienna Convention on the Law of Treaties, art. 60, May 23, 1969, 1155 U.N.T.S. 331, 8 I.L.M. 679; *see also* Jerome H. Reichman, *Compulsory Licensing of Patented Pharmaceutical Inventions: Evaluating the Options*, 37 J. L., MED. & ETHICS 247 (2009) (suggesting that this is a possible, yet thus far "underappreciated" remedy).

[75] An exception is Brazil – in May, 2007, the President of Brazil issued a compulsory license on an AIDS drug patented by Merck in a televised ceremony after a series of unsuccessful attempts to obtain price reductions from Merck. *E.g.*, Jon Cohen, *Brazil, Thailand Override Big Pharma Patents*, 316 SCIENCE 816 (2007); *see also* Alastair Stewart, *Brazil to Break Patent on Merck AIDS Drug*, WALL ST. J., May 5, 2007 (noting that Brazilian President Lula Da Silva threatened additional licenses if drug prices were not affordable).

[76] *See, e.g.*, SUSAN SELL, PRIVATE POWER, PUBLIC LAW 151 (2003).

[77] ANA REVENGA ET AL., THE WORLD BANK, THE ECONOMICS OF EFFECTIVE AIDS TREATMENT: EVALUATING POLICY OPTIONS FOR THAILAND 15, 36, 169 (2006), *available at* http://web.worldbank.org/WBSITE/EXTERNAL/COUNTRIES/EASTASIAPACIFICEXT/EXTEAPREGTOPHEANUT/EXTEAPREGTOPHIVAIDS/0,contentMDK:21024879~pagePK:34004173~piPK:34003707~theSitePK:503157,00.html.

[78] *See, e.g.*, Aide Memoire: Compulsory Licenses in Thailand on Pharmaceuticals under Patent Protection, ¶ 4 (Feb. 25, 2008), *available at* http://www.keionline.org/misc-docs/1/swiss2thailand_cl.pdf; Richard P. Rozek & Renee L. Rainey, *Broad-Based Compulsory Licensing of Pharmaceutical Technologies—Unsound Public Policy*, 4 J. WORLD INTELL. PROP 463 (2001); Richard Epstein & Scott Kieff, *Questioning the Wisdom and Frequency of Compulsory Licenses* (U. of Chicago Sch. Law, Olin Law & Economics Working Paper, Paper No. 527, 2010, *available at* www.law.uchicago.edu/files/file/527-rae-pharma-patents.pdf; Letter from Peter Mandelson, E.U.

plausible, given that compulsory licenses are generally an issue for developing countries that make up only a fraction of the market, the monetary impact will likely be small.[79] In other words, so long as most wealthy markets do not impose compulsory licenses, there should be no impact on innovation; in fact, some have gone so far as to suggest that sales in developing countries are essentially a "windfall" to companies who readily recoup their costs in first-world countries.[80] The limited studies to date suggest that such licenses have no negative effect on research and development.[81] But one caveat to these studies is that they do not replicate the precise issue at dispute here: compulsory licenses issued to address public health, as most studies instead focus on licenses issued as penalties for antitrust violations. However, there may be other evidence that so long as there is no widespread use of compulsory licenses among the dominant global markets (i.e., the United States and Europe), innovation may not be negatively impacted. For example, Canada previously imposed compulsory licenses for drugs, but neither Canada's innovation nor global innovation has clearly increased since Canada ceased issuing such licenses. In general, there are no definitive studies that patents or increased patent rights in fact promote innovation; what exists are assertions of a correlation between strong patent rights and research[82]—and correlation is not causation. Even if compulsory licenses *might* negatively impact global innovation, there could be an argument that modest and especially only hypothetical harm to innovation is a reasonable trade-off in promoting access to life-saving medicines. The pharmaceutical industry may even recognize this when it suggests limiting compulsory licenses to emergencies. Although countries can legally issue compulsory licenses in other cases as well, the controversy and misunderstanding about nonemergency uses may reflect a lack of consensus on whether such uses should be proper, without regard to what was agreed to under TRIPS. In addition, even if there is a fear of reduced innovation, any reduction need not be permanent. After all, compulsory

---

Trade Comm'r to Krirk-krai Jirapaet, Thai Minister of Commerce, ¶ 4 (July 10, 2007), *available at* http://www.wcl.american.edu/pijip_static/documents/mandelson07102007.pdf?rd=1.

[79] *E.g.*, Reichman with Hasenzahl, *supra* note 7, at 6 ("the customary assertion of some economists that the use of compulsory licensing will depress investment in research and development requires careful and skeptical evaluation"); Colleen Chien, *Cheap Drugs at What Price to Innovation: Does Compulsory Licensing of Pharmaceuticals Hurt Innovation?*, 18 Berk. Tech. L.J. 853, 893–94 (2003); WHO & WTO, WTO Agreements and Public Health: A Joint Study by the WHO and the WTO Secretariat 91 (2002).

[80] Reichman with Hazenshal, *supra* note 7, at 6; *see also* Frederick Abbott & Jerome Reichman, *The Doha Round's Public Health Legacy: Strategies for the Production and Diffusion of Patented Medicines Under the Amended TRIPS Provisions*, 10 J. Int'l. Econ. L. 921, 971 (suggesting that as long as companies recoup costs in OECD markets, developing countries should pay only marginal costs of production plus a five percent royalty to provide "generous compensation").

[81] *See, e.g.*, Chien, *supra* note 79, at *passim*; Donald McFetridge, *Intellectual Property, Technology Diffusion and Growth in the Canadian Economy*, *in* Competition Policy and Intellectual Property Rights in the Knowledge Based Economy 65, 88-90 (Robert D. Anderson & Nancy T. Gallini eds., 1998).

[82] *See* Chien, *supra* note 79, at 883–92.

licenses under TRIPS must be for specific patents, and if broader use of compulsory licenses presents a problem, the issue could be revisited.

Pharmaceutical companies may nonetheless be opposed to compulsory licenses on grounds beyond purported innovation concerns. First, they may fear that lower costs in some markets will prompt pressure for lower costs in more profitable ones. Second, they may be concerned that drugs made under a compulsory license might be subject to parallel importation into the wealthy markets where they would undercut expected profits (as discussed in Chapter 2). Although most wealthy markets have laws in place that would restrict parallel imports, market segregation is so important to pharmaceutical companies that they may have a heightened sensitivity to the issue of compulsory licenses.

In addition, even if drug companies do not directly retaliate by withdrawing drugs from the market, there are other ways patent-owning drug companies can indicate displeasure with a country that issues compulsory licenses. For example, a drug company can minimize research on issues of concern to developing countries. Although the current amount of research is negligible in comparison to conditions impacting richer countries, this may make any such research especially important to preserve. After all, most developing countries are not in a position to develop drugs for their own needs. Accordingly, consideration is still pertinent of the impact of the developing countries' actions on those who may develop useful drugs—even if only for a small portion of the global pipeline.

V. Conclusion

The ability to actually use TRIPS-consistent compulsory licenses may be seriously hindered by issues beyond TRIPS. In particular, the threat of retaliation by drug companies and major trading partners may make countries hesitant to issue them. This situation may seem particularly unfair to developing countries because many of them entered into TRIPS assuming the agreement would end unilateral trade sanctions.[83] Although formal unilateral trade sanctions do not necessarily occur, the threat of such sanctions still looms large for many countries.

Nonetheless, some countries seem to be testing the waters of compulsory licenses. In April 2010, Ecuador issued its first ever compulsory license on the patented HIV combination therapy lopinavir/ritonavir, sold under the trademark Kaletra by Abbott Laboratories.[84] One week after the license issued, the cost of the licensed drug was almost

---

[83] *See, e.g.*, Frederick M. Abbott, Commentary, *The International Intellectual Property Order Enters the 21st Century*, 29 VAND. J. TRANSNAT'L L. 471, 472–73 (1996); Barbosa et al., *supra* note 5, at 124; James Thuo Gathii, *Rights, Patents, Markets and the Global AIDS Pandemic*, 14 FLA. J. INT'L L. 261, 316–17 (2001–2002).

[84] Catherine Saez, *Ecuador Grants First Compulsory License, for HIV/AIDS Drug*, IP WATCH MICROBLOG (Apr. 22, 2010, 6:24 PM), http://www.ip-watch.org/weblog/2010/04/22/ecuador-grants-first-compulsory-licence-for-hivaids-drug/.

30 percent less than what the patent owner had previously charged.[85] In addition, despite pressure from governments and pharmaceutical companies, Thailand has extended two of its compulsory licenses on HIV/AIDS drugs.[86]

However, the long-term prospects for widespread continued use of compulsory licenses are not rosy. In particular (as will be explained in Chapter 8), a number of developing countries have entered into agreements requiring them to limit the use of compulsory licenses. Some agreements also impose additional restrictions beyond the patent realm that may make compulsory licenses moot because the company authorized to make the licensed drug is barred from obtaining regulatory approval to sell the drug.

For countries interested in using compulsory licenses to reduce the cost of patented drugs, the following issues under TRIPS should be borne in mind:

- compulsory licenses *are* permissible under TRIPS
- any country may issue a compulsory license
- any invention may be subject to a compulsory license
- individual nations decide appropriate grounds for compulsory licenses
- nations can clearly issue licenses in the event of a national emergency, but no emergency is ever necessary for a license to issue
- individual nations decide on appropriate compensation to the patent owner
- individual nations decide whether a license and the compensation for a license are appropriate—subject only to a procedure to enable a patent owner to challenge these decisions

---

[85] *Ecuador to Expand Access to Medicines*, PUBLIC CITIZEN (April 22, 2010), http://www.citizen.org/pressroom/pressroomredirect.cfm?ID=3116.

[86] *Compulsory Licensing for Two AIDS Drugs Extended by Thailand Health Authority*, PHARMALETTER, (Apr. 22, 2010), http://www.thepharmaletter.com/file/95883/compulsory-licensing-for-two-aids-drugs-extended-by-thailand-health-authority.html.

6 Compulsory License Case Study: An Introduction to Competing Patent Perspectives

This chapter builds upon the basic rules for compulsory licenses under TRIPS (Chapter 5) to consider one reason TRIPS-compliant compulsory licenses remain controversial. It introduces two competing perspectives of patents as a possible social schema that can help anticipate (or at least better understand) not only controversy, but misstatements concerning the TRIPS rules for compulsory licenses. This chapter addresses the following issues:

- How are compulsory licenses viewed from the perspective of a country with inadequate resources to pay the retail price for drugs?
- How are compulsory licenses viewed from the patent owner perspective?
- How do competing perspectives of patents promote misunderstandings of TRIPS requirements for compulsory licenses?

I. Introduction 158
II. The Competing Patent Perspectives 160
A. Patent as Mere Privilege: A Moldable Tool 161
B. Patent as an Uber-Right 163
C. Social Science Support for Perspectives 166
III. A Case Study in Competing Perspectives 167
A. Background 167
B. Privilege (Thailand) Perspective 170
C. Uber-Right (PhRMA) Perspective 173
IV. Impact of Perspectives on Compulsory Licensing Controversy 179
A. TRIPS Article 31 179
B. Distorted TRIPS Interpretations Fostered by an Uber-Right Perspective 180
1. Mythical Emergency Limitation 180
2. Mythical Exclusion of Some Countries 183
C. Non-issues Raised versus Real Issues in Need of Resolution 184
1. No Emergency versus Public Noncommercial Use 184
2. No Research versus Adequate Remuneration 186
3. Factual Dispute on Prior Negotiation versus Defining Prior Negotiation 187
4. Factual Dispute Regarding Frequency of Compulsory Licenses 188
V. Examining Alternate Explanations 189
A. A Fake Phenomenon? 189
B. All About Property 190
C. Genuine Confusion 191
VI. Conclusion 193

I. Introduction

This chapter uses one of the most controversial topics—compulsory licenses—to expose not only differing views of patents, but the *impact* of such views on how compulsory licenses are viewed. Public discussions of compulsory licenses not only include quick dismissals of opposing views, but outright hostility and name-calling. Patent-owning pharmaceutical companies are called greedy corporations that place profits above life,[1] while public health advocates are decried as anti-property activists and patent hooligans.[2] The one commonality between these positions is that each side believes only one view can be

[1] For example, a website critical of patent owner Abbott's actions against Thailand was called "abbotttsgreed.com." Although the website is no longer live, an archived version is available at http://web.archive.org/web/20080424010611rn_2/www.abbottsgreed.com.

[2] *See, e.g.*, Editorial, *Bangkok's Drug War Goes Global*, WALL ST. J. ASIA, Mar. 7, 2007, at 13; Ronald A. Cass, Op-Ed., *Patent Remedy*, WALL ST. J. ASIA, Aug. 28, 2007, at 13 [hereinafter Cass, *Patent Remedy*].

true. However, considering there are intelligent people on both sides—including attorneys who should understand the law—the polarized positions raise a question of whether there is an alternative phenomena at issue. The seeming deadlock in views suggests there is a need to look for new insight to better understand and ultimately get beyond the deadlock.

This chapter suggests there are two competing perspectives of patents that each provide a different lens through which to view whether compulsory licenses are consistent with TRIPS. Compulsory licenses are an excellent vehicle to analyze these perspectives as the issue is recognized as controversial, yet a correct understanding of the law should be straightforward. The perspectives thus provide an explanation for not only heated controversy, but misstatements concerning TRIPS beyond simplistic suggestions of malice or ill intent. Recognizing these perspectives is critical because each serves as a powerful prism that impacts how laws such as TRIPS are created and understood—not only with respect to compulsory licenses, but for other issues involving access to medicine, as will be subsequently explored in other chapters.

The basic thesis is that there is a spectrum of views on patents benchmarked by two distinct and seemingly irreconcilable perspectives. One perspective of patents is that they are a mere privilege granted by a nation and thus inherently subject to limitations to accommodate other societal goals, such as access to medicine. This view is referred to as the *privilege view* of patents. The alternative perspective views patents as a type of super-property right that should seldom, if ever, be subject to exception. This is referred to as an *uber-right view* in contrast to the traditional conception of property rights that necessarily include limitations and exceptions.[3] The privilege view of patents can find some roots in the historical genesis of patents because patents were once a privilege conveyed by the Crown, often with restrictions on where the patent could be used.[4] However, the uber-property view may reflect a more modern conception by some.

These differing perspectives are important to understanding why TRIPS has been misinterpreted with respect to compulsory licenses. Some clear rules, such as what subject matter may be licensed, have been distorted to reflect the desired perspective of patents. In addition, both perspectives are responsible for raising a number of nonissues. The extent of the distortions and nonissues suggests the distortions may be explained by a more fundamental phenomenon than rhetoric and issue framing (although these can be

---

[3] The term "uber-right" is used to emphasize a view of patent rights as very strong. However, there is no intent to imply a connection with a particular political ideology.

[4] *See, e.g.*, SUSAN SELL & CHRISTOPHER MAY, INTELLECTUAL PROPERTY RIGHTS: A CRITICAL HISTORY 53–54 (2006); Oren Bracha, *The Commodification of Patents 1600–1836: How Patents Became Rights and Why We Should Care*, 38 LOY. L.A. L. REV. 177, 183–86 (2004); Paul Champ & Amir Attaran, *Patent Rights and Local Working under the WTO TRIPS Agreement*, 27 YALE J. INT'L L. 365, 370–73 (2002).

very powerful in shaping the creation and interpretation of law).[5] The competing perspectives thus provide a vehicle for an enriched understanding of the compulsory licensing controversy.

II. The Competing Patent Perspectives

This section outlines each patent perspective and then discusses the supporting social science literature for the existence of such perspectives. The goal of this section is not to identify a single one as necessarily superior. Rather, the goal is to highlight the existence of these distinct perspectives as a means to better understand global debates concerning access to medicine.

The perspectives approach outlined here bears some similarities to existing intellectual property theories, yet is nonetheless distinct. In particular, different types of patent rights are often referred to as promoting either a "liability" or "property" regime.[6] A compulsory license is seen as a liability rule in that the patent right is taken and the owner compensated afterwards; the person granted the license incurs a liability, while the patent owner has no control in protecting unauthorized use of his property. A property regime, on the other hand, would suggest that a patent right be more absolute, subject to cases of extreme necessity; in other words, the patent owner can prevent unauthorized use of her right under a property regime. The privilege perspective thus bears some similarity to a liability regime whereas the uber-right perspective bears similarity to the property regime. However, an important distinction is that debates about liability versus property rights are generally made in the context of discussing what is economically efficient and usually promotes one view – either the liability or property view. This dichotomy does not necessarily explain intransigent debates and positions concerning how patents impact access to medicine. The patent perspectives theory, on the other hand, can be conceived as a type of unconscious bias through which the world is viewed and resistant to change. In other words, the patent perspectives complement existing theories in intellectual property and add further nuance.

[5] The power of rhetoric and issue-framing is widely recognized with respect to the development of international laws that promote patent rights. *E.g.*, SUSAN K. SELL, PRIVATE POWER, PUBLIC LAW (2003); Amy Kapczynski, *The Access to Knowledge Mobilization and the New Politics of Intellectual Property*, 117 YALE L. J. 804 (2008). On the other hand, the patent perspectives suggested here may be the backdrop for the development of rhetoric and issue-framing; unlike those issues, the perspectives are suggested to operate largely on a sub-conscious level.

[6] *See, e.g.*, Robert Merges, *Of Property Rules, Coase, and Intellectual Property*, 94 COLUM. L. REV. 2655 (1994).

A. PATENT AS MERE PRIVILEGE: A MOLDABLE TOOL

On one side of the spectrum of views is the concept that patents are a tool inherently subject to limitations. Although many—including those with an uber-right view—may concur that patents are a utilitarian tool to promote innovation by providing an incentive or reward for new innovations, those who view patents as a privilege do not necessarily place innovation as a priority over all other social interests. Rather, promoting innovation is only one goal among other competing societal goals that inherently contemplate the need for balance. Regardless of whether patents are legally considered property rights, they may be viewed as a special type of property to those who see patents as a privilege. As stated by Professor Baker (who is also associated with Health Global Access Project, which is dedicated to eliminating barriers to access to HIV treatment), "[P]atents are not 'property' in the traditional sense—they are government granted rights that are intended to balance the interests of innovators and the public at large, and which are granted by governments with many express and implied conditions. . . ."[7]

The UN Commissioner of Human Rights supports the idea of patents as a privilege that must be "subject to limitations in the public interest."[8] In particular, the UN has suggested that certain human rights, such as the right to health, are "inalienable and universal" ones that must be recognized over state-granted rights, such as patents.[9] Moreover, the UN Commissioner has suggested that to the extent there is an "actual or potential conflict," patent rights should yield to the right to public health.[10]

If patents are conceived as a tool to promote innovation as one among many societal goals, exceptions to ensure the patent purpose is served would seem reasonable. In particular, although patents are assumed to provide an incentive to innovate, to the extent they fail to provide appropriate incentives (or actually interfere with additional innovation), modifications are necessary. Accordingly, under the view of patents as privilege, a nation would limit or craft exceptions to typical patent remedies if doing so would promote greater innovation, such as use of patented inventions by researchers.

---

[7] Posting of Brook K. Baker, *Pharma's Seven Deadly Lies about Thai Compulsory Licenses*, to IP DISPUTES IN MEDICINE (Feb. 1, 2007), http://www.cptech.org/blogs/ipdisputesinmedicine/2007/02/pharmas-seven-deadly-lies-about-thai.html.

[8] U.N. Office of High Comm'r for Human Rights, Sub-Comm'n on the Promotion & Prot. of Human Rights, *Intellectual Property Rights and Human Rights*, ¶ 1, U.N. Doc. E/CN.4/Sub.2/RES/2000/7 (Aug. 17, 2000) [hereinafter Resolution 2000/7].

[9] U.N. Econ. & Soc. Council, Sub Comm'n on Promotion and Prot. of Human Rights, *The Impact of the Agreement on TRIPS on Human Rights*, ¶ 14, U.N. Doc. E/CN.4/Sub.2/2001/13 (June 27, 2001) (suggesting that rights under TRIPS, which are state-granted rather than inalienable, should, where appropriate, bow to the more universal human rights, such as the right to health); *accord* Resolution 2000/7, *supra* note 8, ¶ 3 (reminding "all governments of the primacy of human rights obligations over economic policies and agreements").

[10] *See* Resolution 2000/7, *supra* note 8, pmbl. ¶ 3.

In the area of health care and access to medicine, the idea that patents are a privilege has particularly dramatic consequences. One possible view is that "the lives of patients have to come before the patents of drug companies,"[11] such that substantial modifications of patent rights are appropriate—particularly if a nation places a premium on the right to health through programs such as universal access to essential drugs, including HIV treatment for all citizens. Accordingly, possible limitations on a patent right so as to accommodate public health interests might include limits on the scope of patentability, shorter patent terms, and compulsory licenses.

Holders of the view of patents as a privileged tool do not necessarily advocate abolishing all patents. However, consideration of patents as a privilege injects more hesitation into whether they should be granted in the first instance. For example, countries such as India and China could be seen as embracing a view of patents as privilege in their national approaches to patents prior to TRIPS because they provided patents on *methods* of making drugs, but not on the drugs themselves. The rationale of providing a patent incentive to increase new methods of creating drugs (but not monopoly rights on needed drugs) seems to fit with a view of patents as a tool to promote progress: one to be balanced against other social interests, such as broad access to low-cost drugs. Another example of a nation viewing patents as privilege would be Canada; until a trade agreement with the United States required Canada to change its laws,[12] Canada broadly granted compulsory licenses on drugs to increase access to medicine. However, even after abandoning automatic compulsory licenses, some aspects of Canada's patent laws reflected a privilege view. For example, Canada permitted generic manufacturers to not only make, but stockpile patented drugs during the patent term to enable true competition with the patent owner (and lower costs) on the very day of patent expiry).[13]

The view of patents as a privilege is also consistent with the historical requirement of many nations that a patent owner "work" the patent locally. Although there is currently some question as to whether such a requirement would be consistent with TRIPS,[14] a number of countries historically limited or entirely eliminated patent rights if a patent owner failed to use the patent in the nation that granted the patent. The earliest patents required the patent owner to make the product locally to help transfer the technology to local citizens. Even when countries moved away from automatic forfeiture of patents for failure to work patents locally, they continued to permit penalties for lack of local

[11] Press Release, Campaign for Access to Essential Medicines, MSF Welcomes Move to Overcome Patent on AIDS Drugs in Thailand, (Nov. 30, 2006), www.accessmedmsf.org/prod/publications.asp?scntid=30112006945525&contenttype=PARA

[12] North American Free Trade Agreement, art. 104, U.S.-Can.-Mex., Dec. 17, 1992, 32 I.L.M. 289 (1993).

[13] Canada eliminated the stockpiling exception after a WTO panel found it inconsistent with TRIPS. *See* WTO Panel Report, *Canada—Patent Protection of Pharmaceutical Products*, WT/DS114/R (Mar. 17, 2000).

[14] This is more fully addressed in Chapter 5.

working; in particular, compulsory licensing was introduced as a less onerous punishment than complete patent forfeiture.[15]

Holders of this view would take a cautious approach toward requiring patent rights globally. Because patents are seen as a tool to promote innovation, and there is literature to support that patents alone are not adequate to do so, increasing patents globally would not be encouraged.[16] Rather, those who believe that patents are a privilege would advocate allowing each nation to decide whether to grant patents, as well as to determine the scope of such patents. Accordingly, those who view patents as a privilege would be opposed to an international agreement that mandates patents. However, to the extent an international agreement was required, those who view patents as privilege would likely be more amenable to one that permitted national discretion to recognize competing interests. Indeed, some of the language in TRIPS can be seen as the handiwork of those who view patents as privilege: articles 7 and 8 talk about balancing patent rights, and article 30 explicitly considers the rights of third parties in limiting the rights of the patent holder.

B. PATENT AS AN UBER-RIGHT

The alternative perspective views patents not as a privilege, but rather, as a *privileged* property right, or an *uber*-right that is stronger than other property rights. All rights have exceptions, including property rights. For example, the traditional property right may be limited by the nuisance doctrine, as well as by takings. Nonetheless, those who subscribe to the uber-right view likely see any possible limitation on patent rights as extremely suspect. Although they recognize that property rights may be limited, they nonetheless analogize legal exceptions to patent rights (such as compulsory licensing) to stealing. They suggest that limits on patents should be exercised with caution because the nature of the patent right is based entirely on the right to exclude.

Those who see patents as an uber-right believe that patents deserve a special status among other rights because they provide the necessary reward to fuel innovation that benefits all of society. The high cost of patented drugs are acknowledged, but most often in the context of emphasizing the high costs and long path toward drug discovery.[17]

---

[15] *Compare* Paris Convention for the Protection of Industrial Property, art. 5, Mar. 20, 1883, 25 Stat. 1372 (imposing local working requirement) *with* Paris Convention for the Protection of Intellectual Property art. 5, *as revised at* The Hague, Nov. 6, 1925, 47 Stat. 1789 (permitting revocation for non-working if compulsory license was inadequate); *see also* Michael Halewood, *Regulating Patent Holders: Local Working Requirements and Compulsory Licenses at International Law*, 35 OSGOOD HALL L. J. 243, 285, 266–67 (1997).

[16] *E.g.*, COMM'N ON INTELLECTUAL PROP. RIGHTS, INTEGRATING INTELLECTUAL PROPERTY RIGHTS AND DEVELOPMENT POLICY 79–89 (3d ed. 2003) (suggesting increased patent rights may not be optimal for developing countries).

[17] *See, e.g.*, Amir Attaran, & Lee Gillespie-White, *Do Patents for Antiretroviral Drugs Constrain Access to AIDS Treatment in Africa?*, 286 J. AM. MED. ASS'N 1886 (2001); Robert C. Bird, *Developing Nations and the Compulsory License*, J. L. MED & ETHICS 209, 216 (2009).

To the extent that high costs of patented drugs may impede access, proponents of uber-patent rights suggest the problem is one of poverty, not of patent protection.[18] In addition, uber-right sympathizers are inclined to suggest that problems related to the exclusivity of patent rights may pale in comparison to other factors that increase the cost of medicine, such as tariffs and taxes that some countries impose on imported drugs.[19]

Although they recognize that access to medicine is limited during the patent term by prices set by patent owners, they suggest that patents benefit all. For example, Fred Hassan, Chairman and CEO of major pharmaceutical company Schering-Plough, suggests patent protection is responsible for low cost generics:

> IP protection for pharmaceutical innovation creates a wonderful cycle. It rewards and incentivizes the huge investments needed to create new medicines. Then, on expiration of the patent, the innovation becomes freely available to all. Generic drugs are thus the direct result of IP-fueled innovation. They would not exist without IP. And without IP, we would not see new advances in medicines that in turn would become generic drugs.[20]

Similarly, Professor Martin Adelman has suggested that the question of access to medicine often overlooks the simple fact that "without patents there would be far fewer drugs around for people to access. One cannot have access to something that does not exist."[21]

The uber-right perspective has also invoked human rights norms to support their position. In particular, they argue based upon language in the Universal Declaration of Human Rights as well as the International Covenant on Economic, Social and Cultural Rights (ICESCR) that inventors have a right to benefit from the fruits of their invention.[22]

---

[18] *E.g.*, Richard P. Rozek, *The Effects of Compulsory Licensing on Innovation and Access to Health Care*, 3 J. World Intell. Prop. 889, 896–99 (2000) (pointing to other barriers to access).

[19] *E.g.*, Roger Bate et al., Still Taxed to Death: An Analysis of Taxes and Tariffs on Medicines, Vaccines and Medical Devices (AEI-Brookings Joint Center for Regulatory Studies, Publication 05-04 2005), *available at* http://regulation2point0.org/wp-content/uploads/downloads/2010/04/php88.pdf.

[20] Fred Hassan, Chairman & CEO, Schering-Plough Corp., Keynote Address at U.S. Chamber of Commerce 5th Annual Intellectual Property Summit: Fueling Innovation: To Be Our Best for a Better World (Oct. 8, 2008), http://www.phrma.org/about_phrma/ceo_voices/fueling_innovation_to_be_our_best_for_a_better_world. This quote also illustrates a false dichotomy—the choice presented is patents versus no patents, without consideration of the possibility of promoting both innovation and access to cheaper generic medications.

[21] Martin J. Adelman, *Compulsory Licensing of Drugs: TRIPS Context* 1, Paper Presented at ATRIP Annual Meeting in Tokyo, Japan (Aug. 4, 2003), *available at* http://www.atrip.org/upload/files/activities/tokyo2003/s02-Adelman_art.doc.

[22] *See* Universal Declaration of Human Rights, G.A. Res. 217A (III), at 71, art. 27(2), U.N. GAOR, 3d Sess., 1st plen. mtg., U.N. Doc. A/810 (Dec. 12, 1948) [hereinafter UDHR] ("Everyone has the right to the protection of the moral and material interests resulting from any scientific, literary or artistic production of which he is the author."); International Covenant on Economic, Social and Cultural Rights, G.A. Res. 2200A (XXI), art. 15(1), U.N. GAOR, 21st Sess., Supp. No. 16, U.N. Doc. A/6316 (Dec. 16, 1966) [hereinafter ICESCR].

Although neither document refers explicitly to either patents or intellectual property rights, both include a clause about how everyone should enjoy the benefits of scientific progress and benefit from protection of interests from any scientific production of which he or she is the author.[23] The basic argument is that if patent rights are minimized, the author is deprived of the protection of his or her interests.[24]

If patents are considered a special right, they should be given more protection than other rights. For example, although injunctions are equitable remedies that typically require a court to consider and balance a variety of factors (including harm to the plaintiff, defendant, and public interest), an uber-right perspective may take a more extreme approach. In particular, the uber-right view would hold that if a patent is found to be infringed, a permanent injunction should be nearly automatic.[25]

The uber-right perspective would want to limit any and all exceptions to patent rights. Compulsory licenses would be considered anathema to an uber-right perspective and only permissible in the narrowest of circumstances.[26] In particular, a compulsory license would likely be considered valid only if the patent owner was incapable of providing an adequate supply of the patented product.[27]

Extending the global reach of patents would also be consistent with patents as an uber-right. In particular, some of the arguments made in support of mandatory international minimum rights for patents reflect this perspective. Prior to the conclusion of TRIPS, some suggested the need for strong patent laws to promote innovation and prosperity. Companies contended that increased patent rights would spur innovation in every community. In addition, they argued that strong patent systems would improve the economic status of nations by promoting foreign direct investment.

Holders of the uber-right perspective would argue for strong international patent norms with minimal exceptions. They would not want exceptions to patent rights, and would want to minimize as much as possible exceptions such as compulsory licenses.

---

[23] UDHR, *supra* note 22, art. 27(2); ICESCR, *supra* note 22, art. 15(1).

[24] However, the same article has been read to support the perception of patents as privilege—that consumers are entitled to enjoy the results of scientific progress in drug discovery such that they have actual access to medicine, not merely theoretical access based upon economic conditions beyond their control.

[25] This view has been espoused for years by the Federal Circuit, presumed in part on the finite term of the patent court. *E.g.*, W.L. Gore & Associates, Inc. v. Garlock, Inc., 842 F.2d 1275, 1281 (Fed. Cir. 1988); H.H. Robertson Co. v. United Steel Deck, Inc., 820 F.2d 384, 390 (Fed.Cir.1987). The US Supreme Court, however, has held that this is an incorrect reading of the law. Ebay v. Mercexchange, 547 U.S. 388, 392–93 (2006). Even so, that may not necessarily change existing perspectives.

[26] *See, e.g.*, Rozek, *supra* note 18, at 889, 890 (stating "compulsory licensing is a policy that should be used rarely, if at all"); *id.* at 904 (stating compulsory licensing "is not intended as a general component for a system that protects IPRs").

[27] *E.g.*, Adelman, *supra note* 21, at 2 (suggesting TRIPS requires "adequate remuneration" that is equivalent to actual damages such that compulsory licensing "is only useful when the patent owner is unwilling or unable to provide a sufficient supply of a needed patented drug").

Moreover, to the extent prevailing global norms seem to permit too many exceptions, those who believe in uber-rights would want to change those norms, such as by negotiating free trade agreements that create stronger patent rights with narrower exceptions.

Those who believe in patents as an uber-right tend not to see a conflict between strong patent rights and public health. Their view is that patent rights necessarily promote innovation, which benefits all of mankind. Possible future breakthroughs in medicine are emphasized as spurred by patents and probably considered to outweigh short-term access problems.

C. SOCIAL SCIENCE SUPPORT FOR PERSPECTIVES

The perspectives presented here may be viewed as consistent with social science literature that documents imperfect information processing. In particular, such research suggests that people rely on prior schemas that are essentially unconscious biases in receiving and understanding new information.[28] A schema has been defined as a "mental structure which contains general expectations and knowledge of the world."[29] Although schemas are recognized as essential to enable individuals to cope with the large amounts of information present in everyday life, they also have been identified as a possible source of bias. Indeed, one example of a schema is a stereotype.[30] Schemas are used by all and can result in cognitive biases and imperfect decision making regardless of education or sociocultural background.[31] They are often discussed in connection with fundamental beliefs or stereotypes, and have provided a useful method for gaining new insight into the operation of existing law and policy.[32]

Although patent perspectives are not a traditional stereotype, the two polar views seem to nonetheless have similarities with stereotypes or fundamental beliefs. Each of the patent perspectives explained above can be seen as a schema through which individuals receive and understand information. Research on the existence and power of schemas suggests that attempts to interpret TRIPS will be done through the lens of the preexisting perspective. In fact, as will be further explored in this chapter, these perspectives may

---

[28] *See, e.g.*, Ronald Chen & Jon Hanson, *Categorically Biased: The Influence of Knowledge Structures on Law and Legal Theory*, 77 S. CAL. L. REV. 1103, 1125 (2004); *see also* SUSAN T. FISKE & SHELLEY E. TAYLOR, SOCIAL COGNITION 97–99 (2d ed. 1991); ZIVA KUNDA, SOCIAL COGNITION: MAKING SENSE OF PEOPLE 17–19 (1999).

[29] MARTHA AUGOUSTINOS & IAIN WALKER, SOCIAL COGNITION: AN INTEGRATED INTRODUCTION 32 (1995).

[30] *E.g.*, Chen & Hanson, *supra* note 28, at 1125–26.

[31] *E.g.*, Cass Sunstein, *Behavioral Analysis of the Law*, 64 U. CHIC. L. REV. 1175, 1183 (1997).

[32] *E.g.*, Jerry Kang, *Trojan Horses of Race*, 118 HARV. L REV. 1489, 1494-95 (2005)(showing how relaxation of ownership requirements by the FCC may exacerbate implicit racial bias); L. Song Richardson, *When Human Experimentation is Criminal*, 99 J. CRIM. L. & CRIMINOLOGY 89, 94 (2008)(suggesting that implict bias results in more lenient treatment of doctors who commit criminal acts in the course of scientific study).

be sufficiently tenacious that they result in interpretations inconsistent with the plain text of TRIPS. Research shows that people will maintain preexisting beliefs out of proportion with actual correctness, including subsequent evidence revealing that the initial information was incorrect.[33] The general theory is that people have an inherent need to ignore, discredit, or rationalize inconsistent information to avoid the discomfort of cognitive dissonance.[34] This theory could thus explain some statements concerning compulsory licensing that seem to lack firm support in the text of TRIPS, as later discussed.

III. A Case Study in Competing Perspectives

This section illustrates competing perspectives of patents through the lens of recent conflicts concerning compulsory licenses issued by Thailand. The section begins with a chronological background of Thailand's actions. Then, the section presents separate vignettes to present the story of Thailand versus PhRMA. The vignettes were not created by either party, but should represent their beliefs as the information is distilled from a ninety-page "white paper" issued by Thailand, as well as numerous news articles and corporate press releases.[35]

A. BACKGROUND

Since 2002, Thailand, a lower middle-income country, has offered universal access to medicine to its citizens, including access to anti-retroviral drugs (ARVs) for patients with HIV since 2003.[36] International agencies, including the World Health Organization and

---

[33] *See, e.g.*, RICHARD NISBETT & LEE ROSS, HUMAN INFERENCE 169 (1980); Craig A. Anderson et al., *Perseverance of Social Theories: The Role of Explanation in the Persistence of Discredited Information*, 39 J. PERSONALITY & SOC. PSYCHOL. 1037, 1041–42 (1980); Kari Edwards & Edward E. Smith, *A Disconfirmation Bias in the Evaluation of Arguments*, 71 J. PERSONALITY & SOC. PSYCHOL. 5, 18 (1996); Arthur G. Miller et al., *The Attitude Polarization Phenomenon: Role of Response Measure, Attitude Extremity, and Behavioral Consequences of Reported Attitude Change*, 64 J. PERSONALITY & SOC. PSYCHOL. 561, 564 (1993).

[34] LEON FESTINGER, A THEORY OF COGNITIVE DISSONANCE 2–3 (1957).

[35] *E.g.*, THAI MINISTRY OF PUBLIC HEALTH AND THAI NATIONAL HEALTH SECURITY OFFICE, FACTS AND EVIDENCES ON THE TEN BURNING ISSUES RELATED TO THE GOVERNMENT USE OF PATENTS ON THREE PATENTED ESSENTIAL DRUGS IN THAILAND 96 (Vichai Chokevivat ed., 2007), *available at* http://www.moph.go.th/hot/White%20Paper%20CL-EN.pdf [hereinafter WHITE PAPER]; THAI NATIONAL HEALTH SECURITY OFFICE, THE TEN BURNING QUESTIONS ON THE GOVERNMENT USE OF PATENTS ON THE FOUR ANTI-CANCER DRUGS IN THAILAND 4 (2008), *available at* http://www.moph.go.th/hot/Second_white_paper_on_the_Thai_CL_%5BEN%5D.pdf [hereinafter TEN BURNING QUESTIONS ON CANCER DRUGS].

[36] Thai National Health Security Act B.E. 2245; *see also* Thai Constitution B.E. 2250, § 51 (noting that "a person shall enjoy an equal right to receive standard public health").

the World Bank, have recognized Thailand's success in containing its AIDS crisis.[37] However, Thai HIV patients are similar to HIV patients globally; they eventually need newer treatments as they build resistance to older drugs. However, newer treatments tend to be patented and much more expensive.[38]

A looming budget problem for Thailand's continued treatment of its HIV patients had previously been forecast, with the idea of compulsory licensing explicitly mentioned by the World Bank as a possible mechanism to provide cost-effective treatment. At the same time, the World Bank noted that such an approach would require "high-level political commitment."[39]

After years of negotiation with patent owners failed to yield price cuts beyond the level of currency appreciation, Thailand issued compulsory licenses to achieve its mandate of providing access to essential drugs. The licenses were issued to cover only Thai citizens who were supported by government-funded insurance, not the small percentage of Thai citizens who are capable of paying the premium patent prices for the drugs.[40] Accordingly, wealthy Thai citizens were required to purchase their own drugs at the price set by patent owners.

Thailand issued three compulsory licenses over a period of several months. Two were for patented drugs used in the treatment of HIV (first-line treatment Efavirenz sold by Merck as Stocrin and Abbott's second-line treatment sold as Kaletra); the third was for a heart disease medication sold by Sanofi-Aventis as Plavix.[41]

To defend these three compulsory licenses, in February 2007, Thailand issued a ninety-page white paper, titled "Facts and Evidence on the Ten Burning Issues Relating to the Government Use of Patents in Thailand," including supporting documents.[42] Although the white paper was meant to support Thailand's position, some statements likely induced additional concern. In particular, the document telegraphed Thailand's intent to consider

---

[37] World Health Organization, External Review of the Health Sector Response to HIV/AIDS in Thailand 35–36 (2005), *available at* http://www.searo.who.int/LinkFiles/News_and_Events_ThailandProgrammeReviewNEW.pdf; Ana Revenga et al., World Bank, The Economics of Effective AIDS Treatment: Evaluating Policy Options for Thailand 169 (2006), *available at* http://web.worldbank.org/WBSITE/EXTERNAL/COUNTRIES/EASTASIAPACIFICEXT/EXTEAPREGTOPHEANUT/EXTEAPREGTOPHIVAIDS/0,contentMDK:21024879~pagePK:34004173~piPK:34003707~theSitePK:503157,00.html. Similarly, Doctors Without Borders described Thailand as having "one of the gold standard treatment programs for the developing world." Amy Kazmin & Andrew Jack, *Thailand Breaks Patent for AIDS Drug to Cut Costs*, Fin. Times, Nov. 30, 2006, at 9.

[38] *See, e.g.*, Médicins Sans Frontières, Will the Lifeline of Affordable Medicines for Poor Countries be Cut?: Consequences of Medicines Patenting in India 2–3 (2005), http://msf.fr/drive/2005-02-01-msf.pdf.

[39] Revenga et al., *supra* note 37, at 169–70.

[40] White Paper, *supra* note 35, at 1–2, 6.

[41] Efavirenz is an effective first-line treatment for HIV that has fewer adverse side effects than generic HIV drugs; most developed countries treat all new patients with Efavirenz.

[42] White Paper, *supra* note 35.

issuing additional licenses on up to 15 percent of patented drugs with a license being warranted when the market price was considered too high to achieve Thailand's universal access to essential drugs.[43]

Companies and countries began to react. Switzerland, home to patent owner Novartis, published an "Aide Memorie" in late February 2008 in which it acknowledged that TRIPS provides for compulsory licenses, but emphasized they should be "used only as a last resort" given that "patents are of paramount importance for the research and development of new pharmaceuticals."[44] Abbott responded to the compulsory license of its Kaletra drug by withdrawing several drugs from the Thai marketplace in March 2007, including a heat-stable form of Kaletra particularly well suited for the Thai climate.[45] The USTR elevated Thailand to Priority Watch status on its Special 301 list, which may lead to unilateral trade sanctions as well as pressure to enter into a regional trade agreement that increases patent protection.[46]

Thailand continued to explore additional compulsory licenses amidst criticism. In June 2007, Thailand established two exploratory committees to consider possible compulsory licenses on cancer medications. At the same time, Thailand was pressured to refrain from issuing additional compulsory licenses by EU Trade Commissioner Peter Mendelssohn and by the U.S. Ambassador to Thailand, Ralph Boyce.[47]

After initially promising negotiations with patent owners stalled, Thailand issued licenses on four cancer drugs in January 2008 on the eve of a change in government

---

[43] *Id.* at 10. Thailand attempted to stem concern by noting that "life-style" drugs, such as those used to treat baldness, acne, or erectile dysfunction would not be considered for compulsory licenses.

[44] Aide Memoire, Compulsory Licenses in Thailand on Pharmaceuticals under Patent Protection, Feb. 25, 2008, *available at* http://www.keionline.org/misc-docs/1/swiss2thailand_cl.pdf [hereinafter Swiss Aide Memorie].

[45] *See, e.g.*, Ambika Ahuja, *Thailand: Abbott Laboratories Won't Introduce New Drugs in Thailand Due to Breaking of Patent*, Associated Press, Mar. 14, 2007, http://www.aegis.com/news/ap/2007/AP070327.html; *Putting Meaning Back into TRIPS*, Bangkok Post, Mar. 22, 2007, *available at* http://www.aegis.org/news/bp/2007/BP070308.html; Nicholas Zamiska, *Abbott Escalates Thai Patent Rift*, Wall St. J., Mar. 14, 2007, at A9. In addition, patent owner Sanofi-Aventis threatened legal action against the Indian drug company Cadila that was to supply low-cost versions of Plavix to Thailand. Rupali Mukherjee, *Pharma Firms under US Pressure to Stop Generic Sale,* Times India, Feb. 13, 2008, *available at* http://timesofindia.indiatimes.com/articleshow/2777692.cms; C.H. Unnikrishnan, *Sanofi in Talks with Thai Govt to Protect its Patent Right,* LiveMint.Com, Mar. 13, 2008, http://www.livemint.com/2008/03/13164842/Sanofi-in-talks-with-Thai-govt.html.

[46] Office of the U.S. Trade Representatives, Special 301 Report 27 (2007) [hereinafter 2007 Special 301 Report]; Office of the U.S. Trade Representatives, Special 301 Report 36–37 (2008) [hereinafter 2008 Special 301 Report].

[47] Letter from Peter Mandelson, E.U. Trade Comm'r to Krirk-krai Jirapaet, Thai; Minister of Commerce (July 10, 2007), *available at* http://www.wcl.american.edu/pijip_static/documents/mandelson07102007.pdf?rd=1 [hereinafter Peter Mandelson July 10 Letter]; Letter from Ralph Boyce, U.S. Ambassador, to Surayud Chulont, Thai Prime Minister (July 20, 2007), *available at* http://lists.essential.org/pipermail/ip-health/2007-August/011610.html.

administration.[48] Thailand asserted these licenses were necessary because cancer is currently the number one cause of death in Thailand, and most effective cancer treatments are patentedand thus effectively inaccessible to most citizens.

On February 7, 2008, on his first day of taking office, the new Thai Public Health Minister announced he would reevaluate the decision to issue licenses on the cancer drugs.[49] Ultimately, Thailand decided not to revoke any of the compulsory licenses despite being told that their continued imposition threatened to impact Thailand's international trade. Thailand has also resisted the suggestion that it promise to forego the option of compulsory licenses in the future, stating that to do so would be considered a "neglect of duty or failure to exercise the rights established by the law to safeguard public interest and public health."[50]

B. PRIVILEGE (THAILAND) PERSPECTIVE

Here is the argument that would likely be made by those who view patents from a privilege perspective:

Thailand is committed to the health of its citizens, even if meeting this goal requires taking bold steps in the international arena and potentially prompting the ire of more powerful countries.[51] Thailand is required to provide all Thais with access to essential medicines, which includes drugs to treat HIV.[52] Thailand's commitment to healthcare has been recognized by international agencies such as the World Bank and the World Health

---

[48] On January 4, 2008, licenses were issued regarding Letrozole (a breast cancer medicine made by Novartis AG), Docetaxel (the breast and lung cancer drug by Sanofi-Aventis), Erlotinib (a drug for treating lung, pancreatic, and ovarian cancer by Roche), and Imitinab (a cancer drug sold by Novartis as Glivec). TEN BURNING QUESTIONS ON CANCER DRUGS, *supra* note 35, at 22–29.

[49] *See, e.g.*, Ambika Ahuja, *Thailand to Review Decision to Break Patents on Cancer Drugs*, INT'L HERALD TRIB., Feb. 9, 2008, *available at* http://www.iht.com/articles/ap/2008/02/09/asia/AS-GEN-Thailand-Cancer-Drugs.php; Editorial, *Bangkok's Drug War, Round Two*, WALL ST. J. ASIA, Feb. 27, 2008 at 13.

[50] TEN BURNING QUESTIONS ON CANCER DRUGS, *supra* note 35, at 13.

[51] The Thai Ministory of Public Health noted:

> [A]t least two thirds of our economy depends on exporting of our goods and services.... If the US government applies retaliation measures on our exports which results in 10 percent reduction of exports to the US market, it will mean a one to 1.2 percent loss of economy and several hundred thousands job losses. So this is a very sensitive issue. Unless there is very important need for the people supported by solid evidence, we will not make these decisions.

WHITE PAPER, *supra* note 35, at 16.

[52] Since 2002, Thailand has had a mandate to achieve universal access to essential medicines for all of its citizens, with access to ARVs included since 2003. *See, e.g.*, *id.* at 1–2.

Organization.[53] Thailand believes in a public interest orientation to patent rights with the view that patients must come before profits.[54]

Increasing drug costs have outstripped Thailand's ability to provide universal access to all necessary drugs. As recognized by the World Bank and the World Health Organization, Thailand needs to use the flexibility granted by TRIPS to deal with the situation of HIV patients developing tolerance to cheaper first-line ARVs. For patients who cannot obtain second-line treatment, AIDS will become a death sentence rather than just a long-term disease. Until Thailand's bold action, many patients were denied these essential drugs because they are generally patented and priced beyond reach. For example, Abbott set the annual price of its Kaletra drug—a good second-line drug—at $2200 per year per Thai patient when the average annual wage is only $1600 per year.[55] Heart disease and cancer are also major causes of death in Thailand and no less serious than AIDS. Without compulsory licenses on cancer drugs, patients and their families will suffer either severe economic hardship (including bankruptcy) or certain death.

Thailand carefully considered the implications of granting compulsory licenses before doing so. Thailand did in fact attempt to negotiate with each patent owner for *years* prior to the actual licenses, but with respect to reducing prices, it was to no avail.[56] Although Thailand could have continued to attempt such negotiations, past experience indicates that companies are much more willing to negotiate after a compulsory license has issued rather than before.[57] And Thailand did not even need to undertake such negotiations: TRIPS clearly states that prior negotiations may be skipped not only when there is a public health crisis, but when the license is for public noncommercial use, as is clearly true in Thailand.[58]

Contrary to the bluster of pharmaceutical giants, compulsory licenses are an essential part of national and international laws. The WTO rules allow member states to issue compulsory licenses to protect public health according to their own criteria, with the Doha

---

[53] WORLD HEALTH ORGANIZATION, *supra* note 37, at 35–36; REVENGA ET AL., *supra* note 37, at 169.

[54] WHITE PAPER, *supra* note 37, preface ("We are convinced and committed to the view that 'public health interest and the life of the people must come before commercial interest.'").

[55] *E.g.*, Paul Cawthorne, *Access to Drugs: The Case of Abbott in Thailand*, 7 LANCET: INFECTIOUS DISEASES 373, 373–74 (2007); *see also* Open Letter, Christopher Fournier, Int'l President, Médecins Sans Frontières, MSF Response to *Wall Street Journal* Editorial on Compulsory Licenses in Thailand, Mar. 14, 2007, *available at* http://doctorswithoutborders.org/publications/article.cfm?id=1957&cat=openletters (noting that second-line ARV cost "at best 5 times the price of current first-line treatments and in countries like Thailand, as much as 22 times!").

[56] WHITE PAPER, *supra* note 35, at 5; *see also* Robert Weissman, Op-Ed, *Compulsory Licenses are the Right Medicine*, NATION (Bangkok), Feb. 23, 2008 ("There is no question that the government undertook extensive negotiations before issuing the recent licenses on cancer products.").

[57] WHITE PAPER, *supra* note 35, at 6.

[58] Each license clearly stated that it was issued for noncommercial use to protect public health and to provide wider access to medicines.

Public Health Declaration clearly affirming that each country has the "freedom to determine the grounds upon which such licenses are granted."[59] No emergency is required, nor is compulsory licensing limited solely to ARVs. In fact the Declaration—which was unanimously adopted by all WTO member states in 2001—states that members "agree that the TRIPS Agreement does not and should not prevent Members from taking measures to protect public health." Accordingly, it is puzzling that legal licenses have been improperly characterized as theft and breaking of patents.

Thailand is opening up a new market for pharmaceutical companies rather than compromising existing markets. In particular, the compulsory licenses simply enable more Thai citizens to have access to essential drugs; those who can afford the retail price will continue to pay it to the patent owner.[60] Even though the compulsory licenses provide patent owners with less profit than their preferred price, the patent owners will still receive greater net profit than without the compulsory license.

Thailand should not be punished for being more committed to the health of its citizens than to the profits of multinational pharmaceutical companies. In addition, Thailand's budget is not reduced by means of the compulsory licenses; such licenses simply increase the number of citizens who receive access, rather than decreasing financial costs to Thailand. In fact, in some cases, more money will be spent. In addition, the Thai drug market is less than 0.5 percent of the global drug market, so the effect of compulsory licenses on the profits of pharmaceutical companies should be insignificant.[61]

Thailand's ability to use compulsory licenses to provide its citizens with access to essential drugs is internationally recognized. Organizations including the Clinton Foundation and Doctors without Borders have supported Thailand's actions; as President Clinton stated, "no company will live or die because of high price premiums for AIDS drugs in middle-income countries, but patients may."[62] Moreover, the WHO has stated that it supports use of TRIPS flexibilities by developing countries to ensure access to affordable drugs.[63] In addition, over twenty US congressmen supported the Thai position in a letter

---

[59] Doha Public Health Declaration, ¶ 5(b).

[60] White Paper, *supra* note 35, at 17 ("[I]t should be reiterated that the government use of patent does not touch on the out of pocket payment market, the current market of the patented drugs. The Government Use only opens new market for those who never have access to these drugs before").

[61] *Id.* ("[T]he size of the Thai drug market is less than 0.5 percent of the global drug market. It is even less for the market of patented drugs. So, there should not be significant effect on the market and return of the research based drug companies.").

[62] Marwaan Macan-Markar, *Thailand: Holding Big Pharma's Feet to the Fire*, IPS, May 17, 2007, http://www.aegis.org/news/ips/2007/IP070508.html; *see also* Letter from Peter Piot, Executive Dir., UNAIDS, to Mongkol Na Songkhla, Thai Minister of Pub. Health (Dec. 26, 2006), *reprinted in* White Paper, *supra* note 35, at 84 (commending Thailand for its strong and steadfast efforts to provide access to antiretroviral treatment).

[63] Letter from Margaret Chan, WHO Dir.-Gen., to Mongkol Na Songkhla, Thai Minister of Pub. Health (Feb. 7, 2007), *available at* http://www.cptech.org/blogs/ipdisputesinmedicine/2007/02/drmargaret-chan-to-dr-mongkoi-na.html.

asking the USTR (US Trade Representative) to not exercise trade pressure on Thailand.[64]

Moreover, Thailand's licenses are consistent with the practice of many nations. Over twenty-five countries have issued or considered issuing compulsory licenses.[65] Wealthy countries including the United States and European countries have issued compulsory licenses as well as engaged in government use of patents.[66] The United States routinely issues compulsory licenses as a remedy for anticompetitive actions and has threatened to do so in a variety of situations, including supplying generic ciprofloxacin and permitting government use of BlackBerrys.[67] If wealthy countries are permitted to grant such licenses, Thailand should not be denied such a right.

Thailand recognizes that its actions will subject it to political pressure and vilification. However, Thailand strongly believes it must stand up to the pressure from the pharmaceutical companies and the United States.[68] Thus, Thailand is a world leader in securing better treatment for its patients.

C. UBER-RIGHT (PhRMA) PERSPECTIVE

Here is the argument that would likely be made by those who view patents from an uber-right perspective, such as that adopted by PhRMA:

Thailand has overruled the international patent system by breaking patents on numerous medicines.[69] Thailand's military government is trampling the global patent system and expropriating intellectual property as its own, rather than paying its proper share.[70]

---

[64] Letter from Tom Allen et al., Members of Congress, to Susan Schwab, US Trade Rep. (Jan 10, 2007).

[65] JAMES PACKARD LOVE, KNOWLEDGE ECOLOGY INT'L, RECENT EXAMPLES OF THE USE OF COMPULSORY LICENSES ON PATENTS (2007), *available at* http://www.keionline.org/misc-docs/recent_cls_8mar07.pdf.

[66] WHITE PAPER, *supra* note 35, at 4 ("[D]eveloped countries including the USA, European countries, and other developing countries have previously attempted and implemented compulsory licensing and Government Use of patents."). *See also* Harvey E. Bale, Jr., Letter to the Editor, WSJ.COM, Mar. 14, 2007, http://online.wsj.com/article/SB117382283064835964.html (noting that Raymond Offenheiser, President of Oxfam America asserted "the United States has been a major user of compulsory licensing").

[67] LOVE, *supra* note 65, at 3–4.

[68] *See* Macan-Marker, *supra* note 62 (quoting Jiraporn Limpananont, associate professor in pharmaceutical science at Chulalongkorn University, as stating "[w]e have to stand up to the pressure form the pharmaceutical companies and the US").

[69] *See* Nirmal Ghosh, *Thailand Breaks Patents for Heart and AIDS Drugs*, STRAIT TIMES, Jan 31, 2007 (beginning news story with assertion that "Thailand has overruled the international patent system"); Darren Schuettler, *Angered US Firm Excludes Thailand from New Drugs*, REUTERS, Mar 14, 2007 (quoting an Abbott Spokesman for stating "Thailand has chosen to break patents on numerous medicines, ignoring the patent system").

[70] Cass, *supra* note 2 (characterizing Thailand as "trampling patents"); Editorial, *Theft in Thailand*, WALL ST. J., Feb. 10, 2007, at A8; Letter from Kenneth L. Adelman, Exec. Dir., USA for Innovation, to Members of Congress (May 9, 2007), *available at* http://lists.essential.org/pipermail/ip-health/2007-May/011154.html (characterizing licenses as "attacks on American intellectual property").

Thailand is threatening a well-established global system for innovation.[71] Everyone knows that without intellectual property, there is simply no innovation,[72] so by breaking patents, Thailand has decided not to support innovation.[73] Although strong patents benefit everyone, public debates improperly pit ignorant passion against profit.[74] Activists blindly believe the pipeline of drugs will continue even if there is less funding, seemingly oblivious to the fact each new drug costs about a billion dollars to develop.[75]

In addition to being shortsighted, Thailand's actions are selfish, and ironically will compromise the health of everyone, including its own citizens.[76] The costs of research and development must be borne by someone. If Thailand is not willing to contribute its share, the rest of the world (including countries at even lower stages of development) will be required to shoulder the burden unnecessarily and in contravention of a prior global accord to increase access to the poorest countries. Higher prices for everyone besides Thailand are simply unfair. As a middle-income country, Thailand should be respecting patents to promote innovation and investment opportunities. It is shocking that a country that ranked 34 out of 181 countries based on GDP and knowingly diverted funds from health care to defense spending can expect any public sympathy for stealing patents.[77] Moreover, Thailand's reckless action will also boomerang against any further economic growth. Why would any company be willing to invest in Thailand after it has shown such reckless disregard of property rights?

Compulsory licenses are antithetical to patent rights, and accordingly are granted rarely and only under very narrow circumstances. Compulsory licenses are rarely employed by any country, even if permitted under global rules.[78] Such extreme action is only permitted in extraordinary circumstances, such as when a critical patent is not being used to produce essential goods or the patent owner cannot provide adequate quantities of a drug in a health emergency.[79] However, such situations are very rare.[80]

---

[71] *See* Cass, *Patent Remedy, supra* note 2 (referring to licenses as a threat to the "world's system of protections for innovation").

[72] *A Gathering Storm*, ECONOMIST.COM, June 9, 2007, at 71 (quoting Fred Haasan of Schering–Plough).

[73] Zamiska, *supra* note 45 (quoting Jennifer Smoter, spokeswoman for Abbott).

[74] Ronald A. Cass, Op-Ed, *Drug Patent Piracy*, WALL ST. J. ASIA, May 7, 2007 [hereinafter Cass, *Drug Patent Piracy*] (stating "IP protection benefits everyone" and that "unfortunately, public debates on IP rights are skewed to pit passion against profits").

[75] *Id.* (asserting that those opposed to IPR "broadly assert that the drug companies won't stop inventing just because they will make less money").

[76] *See* Zamiska, *supra* note 45 (quoting Teera Chakajnarodom, president of PhRMA in Bangkok as stating that "what the government is doing is very shortsighted. The patients will need newer generations of antivirals for treating AIDS but won't have access to them").

[77] Cass, *Drug Patent Piracy supra* note 74.

[78] Adelman, *supra note* 19; *see also Bangkok's Drug War Goes Global, supra* note 2 (stating "no serious government has contemplated using compulsory licensing, even if its' [sic] allowed to do so under the WTO rules").

[79] Cass, *Patent Remedy supra* note 2.

[80] Attempts to characterize compulsory licenses as pervasive are simply wrong; a variety of activities are incorrectly characterized as compulsory licensing that have nothing to do with licensing of a patented drug. The

Moreover, such licenses should be very limited because they inevitably doom the patent owner to receive far less than a reasonable economic return.[81] Accordingly, global rules require that compulsory licenses be used only as a last resort.[82]

Any sympathizers to Thailand's position are simply opposed to protection of all property rights.[83] Thailand's action is not surprising as authoritarian governments and social activists take what is not theirs.[84] Thai sympathizers are just the usual crowd of anti-patent hooligans who fail to understand that without patents, there would be no drugs for anyone. These anti-property activists have tried hard to alter the meaning of TRIPS and were behind the 2001 Declaration that permits governments to deal with health emergencies posed by epidemics of HIV in Sub-Saharan Africa.[85] However, that Declaration does not support compulsory license of any drug patent for any reason at all—especially when a relatively developed country such as Thailand with no epidemic simply wants to save costs by stealing the property of another.[86]

Thailand's purported interest in fostering public health is suspect. After all, the recent compulsory licenses were issued after a military-imposed government suspiciously raised the

United States has no general provision to permit compulsory licenses; although permitted under the Clean Air Act—as repeatedly noted by activists—no license has ever issued under the CAA. *See* Posting of Siney Rosenzweig, *Bogus Reliance on the Clean Air Act's Compulsory Patent Licensing*, Progress & Freedom Foundation Blog, Mar. 17, 2009, at http://blog.pff.org/archives/2009/03/jamie_loves_bogus_reliance_on_the_clean_air_acts_c.html. Moreover, licenses issued as part of a remedy for actual anticompetitive conduct are similarly irrelevant to the health care context. Finally, a denial of a permanent injunction is not a compulsory license; although activists may try to equate the two as having a comparable effect, a compulsory license is something issued by a government—not a court—according to statute.

[81] Cass, *Drug Patent Piracy supra* note 74 ("Compulsory licenses—like any one sided deal that doesn't require consent from the person whose property is taken—almost always leave the rights holder with far less than a reasonable economic return.").

[82] *E.g.*, *A Gathering Storm*, *supra* note 72 (noting Jon Pender of British pharmaceutical giant GlaxoSmithKline insisted that compulsory licensing was only to be used "as a last resort."); *Drugs in Thailand*, Fin. Times, Jan. 31, 2007, *available at* http://www.ft.com/cms/s/0/37500f5c-b0cf-11db-8a62-0000779e2340.html?nclick_check=1 (stating "WTO rules say compulsory licenses should be a last resort").

[83] Cass, *Patent Remedy, supra* note 2 (stating that those who endorse compulsory licenses "oppose protection of property rights").

[84] *See* Ronald Cass, *Thai Patent Turmoil*, Wall St. J., Asia Mar. 3, 2007 (referring to "the free lunch crowd").

[85] Cass, *Patent Remedy, supra* note 2 ("While the system of IP protections has worked well to encourage investment in innovations, some groups oppose protection of property rights in general and IP rights in particular. Over the past decade, these groups have worked hard to alter the meaning of the TRIPS agreement and to encourage all governments to use compulsory licensing to break IP protections.").

[86] *Id.* (characterizing the Doha Declaration as a "small victory" for groups opposed to protection of all property rights and that "the activists used that declaration to argue that TRIPS authorizes any nation to impose compulsory licensing on any patent, especially any drug patent, for any reason. The military-backed Thai government's compulsory license initiative . . . is the first fruit of the activist' campaign—a relatively developed nation facing no epidemic, just seeking to shift its spending priorities by taking someone else's property").

defense budget by over thirty billion.[87] Any purported budget crisis should be viewed with skepticism considering that the military-imposed Thai government gave itself a huge pay increase together with a substantial increase in defense spending while simultaneously reducing the healthcare budget.[88] Why should pharmaceutical investment worldwide suffer just because Thailand has elected to fund its defense budget rather than pay for the drugs it uses? Why should Thailand's medical budget be underwritten by all other countries?

Thailand's asserted health commitment is a sham. Thailand only declared a universal access to health program as a populist measure.[89] Moreover, the licenses are issued to Thailand's Government Pharmaceutical Organization (GPO), which is not only historically corrupt, but also a state-owned monopoly for profit.[90] In addition, any drugs made by the GPO will be inferior to the actual patented product because the GPO facilities have never met WHO standards.[91] Poor products may lead to drug resistance or be ineffective, which would only further exacerbate costs. Any purported HIV resistance in Thailand is likely due to its reliance on inferior drugs manufactured by its own GPO.[92]

The alleged need for HIV treatment in Thailand is similarly a sham. There is no HIV epidemic in Thailand. Thailand has relatively low rates of HIV infection, with less than 1 percent of the population infected—a far cry from countries such as Sub-Saharan Africa where 20 percent of the population may be afflicted.[93] Thailand has contained its

---

[87] Roger Bate, *Thailand's Creaking Health System*, AMERICAN, June 9, 2008, *available at* http://www.american.com/archive/2008/april-04-08/thailand2019s-creakinghealth-system; Don Pathan, *Military Spending to Soar a Further 24%*, NATION, June 28, 2007, *available at* http://www.nationmultimedia.com/2007/06/28/politics/politics_30037960.php.

[88] Bate, *Thailand's Creaking Health System supra* note 87 (asserting that Thai health spending is now 3% of GDP, down from 3.5% a few years ago, and that if it simply spent an additional 1% of GDP, it would have more than $2 billion to spend on health care).

[89] *See generally* Jonathan Head, *Thailand Takes on Drug Giants*, BBC NEWS Apr. 26, 2007 ("The nationwide health scheme was first introduced in 2001 by then-Prime Minister Thaksin Shinawatra, who had a gift for coming up with populist policies that would keep getting him elected.").

[90] Roger Bate, *Thai-ing Pharma Down*, WALL. ST. J., Feb. 9, 2007, at 13 [hereinafter Bate, *Thai-ing Pharma Down*] ("The only winner will be Thailand's historically corrupt Government Pharmaceutical Organization, or GPO, the state-owned pharmaceutical monopoly."); *Patently Wrong*, WALL. ST. J. ASIA, Feb. 9, 2007 (stating that the licenses are a "boon" for the GPO).

[91] *E.g., Theft in Thailand, supra* note 70; Bate, *Thai-ing Pharma Down, supra* note 90 (stating that GPO drugs "are at best approximate copies" and that even with lax WHO standards, eighteen18 different "'WHO approved' HIV treatments have been withdrawn in 2005 due to lack of proof that the drugs actually worked.").

[92] *Theft in Thailand, supra* note 70 (HIV drug resistance in Thailand is high, most likely because of the reliance on faulty GPO copies); Bate, *Thai-ing Pharma Down, supra* note 90 (drugs that are not pre-qualified [WHO approved] may not directly kill people, but they could foster resistance to AIDS drugs.").

[93] *See, e.g.*, Christopher C. Horner, *Thailand Stealing out of WTO?*, WASH. TIMES, May 17, 2007, at A15 (suggesting Thailand cannot have a national emergency because it has a "comparatively low rate of AIDS infection" with less than one percent of the population infected, as opposed to other WTO members with rates as high as twenty percent); *Theft in Thailand, supra* note 70 ("it's hard to argue that Thailand has an AIDS epidemic, when its incidence is a little over one percent and countries such as South Africa are well over twenty percent"); Cass, *Thai Patent Turmoil supra* note 84 ("Thailand today has a relatively low rate of HIV/AIDS compared to much of the developing world, . . . , and has seen a dramatic decrease in its AIDS death rate since 2000.").

HIV numbers, such that there is no need for compulsory licenses of drugs. Moreover, to the extent HIV is a problem, it is one of Thailand's own doing that is fostered by its lucrative sex trade.[94]

The argument that there could be any emergency justifying breaking the patent on Plavix is completely laughable and highlights the extremely inappropriate action of Thailand. There is clearly no threat of an epidemic based on heart disease as that condition is neither infectious nor contagious.[95] Moreover, to the extent Thai citizens suffer from heart disease, their "problem" is simply that they are wealthy enough to be able to buy and consume large quantities of unhealthy foods.[96] Citizens with such wealth should be able to pay the fair price for medication. Alternatively, they should take better care of their health. Breaking a patent is completely unjustified for a condition that reflects an unhealthy lifestyle undertaken by affluent citizens.

In addition, patents are not the problem; poor health systems are a much bigger problem than the prices of drugs. Besides, drugs are already provided free, or at low cost when they are truly needed. Thailand's economy is in no way comparable to the countries that receive such low-cost drugs, such as Sub-Saharan Africa. Thailand's is cutting the wrong corners—instead of breaking patents, it should address corruption in the health services industry and provide better health training.[97]

Thailand's actions are not only suspicious, but also inconsistent with TRIPS. TRIPS only permits licenses to be imposed in limited circumstances, such as a national health emergency, and even then, only after legitimate and lengthy efforts to first negotiate with the patent owner. There is no specific provision in TRIPS that permits Thailand to issue licenses on drugs that treat nonepidemic conditions. Thailand has clearly violated the spirit, if not the letter of TRIPS by taking such extreme action in exploiting the vaguely worded agreement simply to balance its budget.[98] There is no mention in TRIPS that licenses can be used to get more drugs than a nation can legitimately pay for. After all, everyone knows that the goal of TRIPS was to ensure that patents were given increased, rather than reduced, protection. Otherwise, innovation will fail—and there will be no drugs to supply to the Thai market or any market in the world.

---

[94] *See, e.g.*, Roger Bate, *Thailand and the Drug Patent Wars*, HEALTH POL'Y OUTLOOK, Apr. 2007, at 2, *available at* http://www.aei.org/outlook/25890 (noting Thailand's AIDS epidemic is "fueled by its notorious sex industry").

[95] *See, e.g.*, Bate, *Thaiing Pharma Down, supra* note 90 (stating "it's hard for anyone to argue that heart disease meets" the criteria of a health emergency); Roger Bate, *Thailand's Patent Attack*, N.Y. SUN, Feb. 13, 2007, *available at* http://www.nysun.com/opinion/thailands-patent-attack/48499/ (asserting that heart disease and leukemia are not epidemics).

[96] Ghosh, *supra* note 69 (asserting that heart disease afflicts the affluent).

[97] Bate, *Thaiing Pharma Down supra*, note 90 (alleging other problems beyond drug costs are a problem, including poorly paid medical workers and corruption in the healthcare industry that are exacerbated by limited government funding); Bate, *Thailand's Creaking Health System, supra* note 90 (suggesting "the real risk to the poorest of the ill and HIV sufferers in particular, is not drug prices, but bad health systems and poor training of medical professionals").

[98] *Theft in Thailand, supra* note 70; *Lonely Thailand*, WALL ST. J. ASIA, May 23, 2007, at 11.

Thailand's extraordinary bad faith in issuing the compulsory licenses is evident at multiple junctures. First of all, Thailand did not engage in any serious negotiations with patent owners. Although Thailand repeatedly invokes TRIPS, it seems to gloss over the fact that TRIPS requires patent owners to be consulted prior to the imposition of a compulsory license. In fact, patent owners were shocked to have their patents broken without there first being an attempt to negotiate a reasonable price.[99]

The inappropriateness of Thailand's action is broadly recognized in the international community. As stated by Switzerland, "patents are part of the *solution* to long term access to innovative medicines in Thailand," such that compulsory licenses must be used "only in emergencies and other exceptional cases" or else the entire patent system will be undermined, and with it, the "incentive to invest in research and development of new and more effective medicines."[100] Similarly, the EU Commissioner of Trade has repeatedly written to Thailand to clarify that compulsory licenses are inappropriate to use as a standard business practice as this will inevitably be detrimental to the development of new medicines, and to insist that Thailand stop such inappropriate use.[101] Even the Director-General of the WHO has recognized that patent protection is critical and that compulsory licenses must be pursued cautiously.[102]

Thailand's repeated stealing of drugs must be condemned. Abbott's decision to withdraw new drugs from the Thai market is very reasonable as Thailand has indicated that if given the opportunity, it will take property of others.[103] It is completely inappropriate for a middle-income country to steal from companies.

Thailand's actions also destroy a fragile global framework. Companies currently use a tiered pricing system. They already sell drugs at higher costs to wealthy countries so that they can provide drugs at low or no cost in the poorest markets. However, this is only sustainable if the companies can enjoy profits in middle-income countries that can afford to pay. If other middle-income countries follow Thailand's bad example, the entire structure of subsidizing drugs to poor countries will collapse. Not only will this result in less

---

[99] *E.g.,* Darren Schuettler, *Angered U.S. Firm Excludes Thailand from New Drugs*, Reuters, Mar. 14, 2007, *available at* http://www.reuters.com/article/companyNewsAndPR/idUSBKK26805820070314 (noting that the drug makers were stunned they had to received no prior warning); Harvey E. Bale, Jr., Letter to the Editor, WSJ.Com, Mar. 14, 2007, http://online.wsj.com/article/SB117382283064835964.html (quoting Dr. Harvey Bale, Director General of IFPMA as stating that Thailand had "no serious contacts" with patent owners prior to the imposition of the licenses).

[100] Swiss Aide Memoire, *supra* note 44.

[101] Letter from Peter Mandelson, EU Trade Comm'r to Krirkkrai Jirapaet, Thai Minister of Commerce (Sept. 10, 2007), *available at* http://www.vittorioagnoletto.it/uploads/file/Mandelson4.pdf.

[102] *Theft in Thailand*, *supra* note 70 (quoting Chan as stating "I'd like to underline that we have to find a right balance for compulsory licensing.... We can't be naïve about this. There is no perfect solution for accessing drugs in both quality and quantity.").

[103] Zamiska, *supra* note 45 (noting "pharmaceutical executives say the Thai government's decision, which they say effectively steals the drugs from companies that own them, has left the industry with little choice").

access for poor countries, but it will reduce incentives for all drug development, including drugs that address the needs of developing countries.[104]

IV. Impact of Perspectives on Compulsory Licensing Controversy

This section focuses on how perspectives have played a key role in the discussion of whether Thailand's actions are proper under TRIPS. The impact of perspectives was slightly foreshadowed in the earlier vignettes. For example, whereas holders of the privilege perspective emphasize TRIPS, holders of an uber-right perspective often give short shift to discussing TRIPS and instead argue that the stealing of patent rights will doom innovation to the detriment of all.[105] This section provides a deeper examination of how patent perspectives have introduced confusion into clear text and raised red herrings that detract from issues in need of actual attention. Although patent-owning pharmaceutical companies are frequently accused of ill intent or obfuscation, that is not the goal of this section. Rather, this section suggests that the positions of patent owners as well as consumer advocates may be better explained as reflecting the schemas of patent perspectives.

A. TRIPS Article 31

This section provides a brief recap of TRIPS article 31 to lay the foundation for examining how perspective impact legal interpretations. As discussed in greater length in Chapter 5, member states can issue compulsory licenses so long as they comply with the procedures in TRIPS article 31. Any country can issue a compulsory licenses under this provision; it is not limited to the poorest countries. In addition, TRIPS permits nations to decide for themselves what subject matter may be subject to compulsory licensing. This is implicit in TRIPS and also reinforced by the unanimously concluded Doha Public Health Declaration.[106] TRIPS focuses on ensuring that proper procedures are followed, such as consideration of the "individual merits" of each compulsory license rather than

---

[104] *See, e.g.*, Bate, *Thai-ing Pharma Down*, *supra* note 91 (suggesting that if other countries demand no-profit African pricing, the incentives for drug development are weakened "especially for diseases uniquely affecting the Asian region, such as dengue fever and leishmaniasis").

[105] *E.g.*, Benjamin Krohmal, Knowledge Ecology International, Notes from March 16th 2007 U.S. Capitol Briefing on Thailand's Compulsory Licenses, Statement of Richard Kjeldgaard, of PhRMA (Mar. 16, 2007) http://www.keionline.org/index.php?option=com_content&task=view&id=37 (avoiding discussion of whether Thailand's licenses were consistent with TRIPS by characterizing such issues as "legal technicalities" and instead generally arguing that compulsory licenses are destructive to drug development).

[106] Doha Public Health Declaration, ¶ 5(b). While some patent owners (and their sympathizers) tend to discount the Declaration as a political statement of no consequence, most scholars consider it relevant to interpreting TRIPS, as noted in Chapter 5. In addition, even if the Declaration were not utilized, the interpretation of TRIPS article 31 should be the same as the Declaration simply makes explicit the principles already set forth under TRIPS.

giving blanket licenses for entire categories of subject matter.[107] TRIPS also requires the patent owner be given "adequate remuneration" for use of the patent, and generally requires prior negotiation with the patent owner before a compulsory license is issued (although allowing for waiver of this requirement in three situations: a national emergency, situation of extreme urgency, or public noncommercial use).[108] In addition, although an early draft suggested that compulsory licenses should be generally minimized,[109] the final text contains no such limitation. The only restriction to the scope of compulsory licenses is on the level of individual patents; the scope and duration of individual licenses are to be limited to the authorized purpose.[110] However, as discussed in Chapter 5, this is not a substantial limitation.

B. DISTORTED TRIPS INTERPRETATIONS FOSTERED BY AN UBER-RIGHT PERSPECTIVE

This section highlights the most extreme impact of patent perspectives on a proper understanding of TRIPS. It is important to first recall that the prior section noted that TRIPS provides no limits to what patents may be licensed, as well as no restrictions on what countries can use compulsory licenses. Although the Doha Public Health Declaration has reinforced these rules, they have been repeatedly contradicted and distorted by patent owners and others holding an uber-right view. This section is not intended to demonize those perspectives. Rather, the goal is to show the power of perspectives in influencing interpretations that misrepresent TRIPS.

1. Mythical Emergency Limitation

Two related myths are propagated by those with an uber-right perspective concerning the extent to which an emergency is relevant to a compulsory license. The first myth is that under TRIPS, a compulsory license is only appropriate if a national emergency or health crisis exists. This is false. As noted above, TRIPS does not restrict what subject matter may be licensed, and the Doha Public Health Declaration affirms this. A national emergency, including a health crisis is only relevant in one type of situation—if a compulsory license is issued without prior negotiation with the patent owner. The second myth acknowledges that a national emergency is not always required, but still misstate TRIPS

[107] *Compare* TRIPS art. 31(a) (requiring licenses be considered on "individual merits") *with* Patent Act, S.C., ch.23, § 187 (1923) (Can.).

[108] TRIPS art. 31(b).

[109] For example, a 1990 draft suggested that "parties shall minimize the grant of compulsory licenses in order not to impede adequate protection of patent rights." Draft of July 23, 1990, W/76, reprinted in DANIEL GERVAIS, THE TRIPS AGREEMENT: DRAFTING HISTORY AND ANALYSIS 387 (3d ed. 2008).

[110] TRIPS art. 31(c).

rules by omitting public noncommercial use as a ground for issuing a compulsory license with prior negotiation. Examples of each of these are shown below.

a. Mythical Emergency Limitation

Those holding an uber-right view repeatedly suggest that compulsory licenses must be limited and suggest that an emergency is required. For example, a US report on foreign trade barriers discusses "Thailand's ability to issue compulsory licenses to address *public health emergencies*."[111] Similarly, a letter from several congressmen to US Trade Representative Susan C. Schwab declares the congressmen "strongly support WTO rules that recognize the rights of countries to consider actions, including compulsory licensing, to address *urgent* public health needs, such as those resulting from HIV/AIDS, tuberculosis, malaria and other pandemics. But, *we do not believe* that WTO members intended those rules to be used to allow compulsory licenses on any medicine whatsoever."[112] In addition, Juliano Froehner, an assistant professor at the University of Sao Paulo in Brazil, states that although "negotiators did authorize signatory nations to decide what constitutes a public health emergency that would warrant a CL,. . . it was and is understood that compulsory licenses are intended to be used rarely and in response to genuine emergencies."[113] However, as noted above, nothing in TRIPS indicates that compulsory licensing is limited to emergencies.

b. Mistake by Omission–An Emergency is Not the Only Time When Prior Negotiation with the Patent Owner May be Waived

A less distorted, but nonetheless incomplete characterization of TRIPS is the suggestion that a national emergency is the only condition that may waive the usual prior negotiation requirement. For example, Roger Bate, a fellow of the conservative think tank American Enterprise International (AEI), stated that compulsory licenses are permissible if "efforts to obtain authorization from the right holder on reasonable commercial terms and conditions have failed, or in cases of national emergency."[114] Similarly, *The Economist*

---

[111] Office of the U.S. Trade Representative, 2008 National Trade Estimate Report on Foreign Trade Barriers: Thailand (2008), *available at* http://www.ustr.gov/sites/default/files/uploads/reports/2008/NTE/asset_upload_file823_14611.pdf (emphasis added).

[112] Letter from Henry A. Waxman et al., Members of Congress, to Susan Schwab, U.S. Trade Rep. (Apr. 9, 2008), *available at* http://waxman.house.gov/UploadedFiles/Letter_to_US_Trade_Representative_Susan_Schwab_Urging_Focus_on_Access_to_Medicines.pdf (emphasis added).

[113] Juliano Froehner, Compulsory Licensing: Inevitable Failures of the Thailand Strategy in the Global Fight Against HIV/AIDS 5 (July 20, 2007) (unpublished manuscript) (on file with author). Whether Professor Froehner continues to hold these views is unclear since the quoted material has since been removed from its earlier location on the web.

[114] Roger Bate, *Thailand's Drug Wars*, American, Mar. 12, 2008, [hereinafter Bate, *Thailand's Drug War*] http://www.american.com/archive/2008/march-02-08/thailand2019s-drug-wars (internal quotation marks omitted).

quotes Jon Pender of GlaxoSmithKline that "although compulsory licensing is legal, TRIPS rules allow it only under limited circumstances, such as national health emergencies, and only after lengthy efforts to negotiate with firms."[115] Although the article clearly notes the source and that GlaxoSmithKline is a "British drugs giant," thus suggesting some self-interest, there is nothing in the article to indicate this position is possibly inaccurate. An uber-right view may also be at play in an article that correctly quotes the complete text of TRIPS that includes all situations where prior negotiation can be waived (including public non-commercial use)[116] but in the very next sentence states that a compulsory license without initial negotiation is permissible "*only in situations of national emergency or of extreme urgency.*"[117] This omission may reflect the uber-right view that compulsory licenses should be granted only in emergencies—regardless of the actual law.

In addition, some academics also support an uber-right view. For example, Professors Richard Epstein and Scott Kieff state that TRIPS permits compulsory license in times of national emergency, but that the Thai licenses do not qualify because they are directed towards chronic and long-term conditions.[118] Similarly, Stanford Law Professor Alan Sykes states that TRIPS "limits compulsory licensing without prior negotiation to genuinely extreme circumstances," but mentions only a "national emergency."[119] He seems to entirely overlook the possibility that a country could issue a compulsory license without prior negotiation in a situation other than a national emergency and instead tries to characterize what would likely be a public noncommercial use as being a twisted situation of national emergency.

Interestingly, Professor Sykes seems to be concerned about the waiver of prior negotiation even in the limited circumstances he describes. In particular, he is skeptical that a developing country should be able to "unilaterally determine that they are unable to afford pharmaceuticals at current prices, declare that a national emergency results, and then implement policies that leave patent holders with rents near zero."[120] This statement may reflect an uber-right view of patents that leads him to assume a contorted interpretation of

---

[115] *A Gathering Storm*, *supra* note 72.

[116] Christopher E. Chalsen & Lawrence T. Welch, *Growing Hostility Toward Intellectual Property?* 8, Paper Presented at 1st IPO-JIPA Asian Practice Int'l Congress (Sept. 14, 2005), *available at* http://www.ipo.org/AM/Template.cfm?Section=Home&Template=/CM/HTMLDisplay.cfm&ContentID=14899.

[117] *Id.* (emphasis added).

[118] Richard A. Epstein & A. Scott Kieff, *Questioning the Frequency and Wisdom of Compulsory Licensing for Pharmaceutical Patents*, John M Olin Law & Economics Working Paper No. 527, at 4, available at ssrn.com/abstract=1660702. *See also* Kristina M. Lybecker & Elisabeth Fowler, *Compulsory Licensing in Canada and Thailand: Comparing Regimes to Ensure Legitimate Use of the WTO Rules*, 37 J.L. MED. & ETHICS 222, 233 (2009) (criticizing Thai license of heart medication as not a health emergency on the assumption that such an emergency is required).

[119] Alan O. Sykes, *TRIPS, Pharmaceuticals, Developing Countries, and the Doha Solution*, 3 CHI. J. INT'L L. 47, 56 (2002).

[120] *Id.* at 56.

what constitutes a national emergency and overlook that TRIPS plainly permits public noncommercial use without prior negotiation. Alternatively, perhaps the idea that any patent could be licensed for public noncommercial use is so terrifying that framing the issue as a clearly improper emergency seems more appropriate—even if this happens at an unconscious level.[121] All of the above statements misrepresent the actual text of TRIPS, which permits waiver of prior negotiation with a patent owner not only for emergencies, but also in the case of extreme urgency or public noncommercial use.[122] Nonetheless, the misconception is repeatedly emphasized by those with an uber-right perspective as well as the mainstream media[123]—although they could potentially be simply parroting the vocal uber-right.[124]

2. Mythical Exclusion of Some Countries

Although discussion of the perceived lack of emergency in Thailand seems to overshadow most other topics, there are a couple of other areas where the influence of the uber-right perspective seems to result in new requirements not within the literal text of TRIPS. Two examples exist with respect to what countries may qualify to use compulsory licensing. As noted above, TRIPS permits any member state to issue a compulsory license. Nonetheless, there have been suggestions that licenses be denied to middle-income or military-governed countries. For example, an article in an academic publication recently asserted "compulsory licensing is . . . intended . . . for developing countries."[125] There is no citation to TRIPS for this statement—nor can there be because TRIPS is not so limited.[126]

---

[121] Similarly, an article coauthored by Professor Lybecker suggests that although "it *may* be reasonable to utilize compulsory licensing for national emergencies," there is a need to distinguish "legitimate compulsory licensing regimes from abusive ones." Lybecker & Fowler, *supra* note 118, at 225.

[122] TRIPS art. 31(b).

[123] Roger Bate & Kathryn Boateng, *Drug Pricing and Its Discontents: At Home and Abroad*, HEALTH POL'Y OUTLOOK, Aug. 2007, at 4, http://www.aei.org/outlook/26622; PHARM. RESEARCH & MFRS. ASSN., PARTNERING FOR BETTER HEALTH: AN INDUSTRY PERSPECTIVE 26 (2007) [hereinafter PARTNERING FOR BETTER HEALTH], *available at* http://www.qplushost.com/portfolio/prema/upload/publications/PReMAWhitePaper_Thai_English.pdf (suggesting by omission that the Doha Public Health Declaration only affirms the right to use flexibilities for health crises and emergencies); Bate, *Thailand's Drug Wars, supra* note 114 (neglecting to mention public noncommercial use as a possible ground for issuing a license without prior negotiations); Nicholas Zamiska, *Thai Move to Trim Drug Costs Highlights Growing Patent Rift*, WALL ST. J., Jan. 30, 2007, at A8 (stating "WTO rules allow a government to unilaterally declare an emergency and make or sell patented drugs without the permission of the drug companies").

[124] Harish Mehta, *Cheap Life-Saving Drugs: Thailand Shows the Way*, BUS. TIMES SINGAPORE, Oct. 18, 2007, *available at* http://www.biotechsingapore.com/ttd_Bizenterprise/Singlenews.aspx?DirID=78&rec_code=107882 (asserting a national emergency is the only ground for issuing a compulsory license without prior consultation with the patent owner); *Bangkok's Drug War, Round Two, supra* note 49 (mentioning only a national emergency as a possible exception to prior negotiation).

[125] Lybecker & Fowler, *supra* note 118, at 223.

[126] *Id.* at 225 (emphasis added).

Critics repeatedly suggest that Thailand should not have been permitted to issue compulsory licenses because it is a middle-income country.[127] However, nowhere in TRIPS article 31 is there a limitation on compulsory licenses based on the economic status of the country. Many of the criticisms are also tied to the general premise that middle-income countries can afford to pay full price. This factual premise may be disputed—at least 25 percent of the Thai population lives on less than $2/day.[128] However, more importantly, there is nothing in TRIPS that requires a country to demonstrate financial incapacity or to be below a certain income level to utilize compulsory licenses. Any suggestions otherwise may reflect an uber-right perspective that reads in additional limitations to justify the position that compulsory licenses must be limited.

Another issue that is sometimes raised with respect to Thailand is that the licenses were imposed by a military-installed government, as if this were a relevant factor. However, just as with the economic status of a country, the type of government is not a relevant issue under TRIPS. There is nothing under the terms of TRIPS article 31 referring to the type of government entitled to use a compulsory license, let alone any suggestion that use of licenses by a military-based government should be subject to increased scrutiny.[129]

C. NON-ISSUES RAISED VERSUS REAL ISSUES IN NEED OF RESOLUTION

This section focuses yet again on the impact of patent perspectives on compulsory licenses, but from a different angle. Whereas the prior section provided examples of the power of perspectives in distorting the clear language of TRIPS, this section highlights a more subtle, yet nonetheless significant impact of perspectives. In particular, this section focuses on the fact that perspectives may result in the propagation of nonissues under TRIPS that mask fundamental questions about TRIPS terms that are in actual need of interpretation. The competing perspectives are posited as being responsible for the misplaced discussion because each perspective may yield differing, yet valid interpretations. For each non-issue raised, this section aims to suggest the real issue under TRIPS that in fact needs attention.

1. No Emergency versus Public Noncommercial Use

Critics have spent a great deal of time discussing whether there is an appropriate epidemic that would justify any of Thailand's licenses. Some have criticized Thailand's licenses as

---

[127] *E.g.*, Cass, *Patent Remedy supra* note 2; Kazmin & Jack, *supra* note 37.

[128] United Nations Dev. Programme, Human Development Report 2007/2008—Thailand, *available at* http://hdrstats.undp.org/countries/data_sheets/cty_ds_THA.html.

[129] In fact, other provisions of TRIPS suggest that discretion is given to the national authority without regard to how it is organized. For example, the decision of what constitutes permissible subject matter is one that is within the province of the national government.

improper for lack of any public emergency regarding AIDS, let alone heart disease or cancer.[130] Moreover, the comments seems to suggest that what constitutes an emergency should be qualified—in contrast to the plain language of the Doha Public Health Declaration, which states that each nation is entitled to decide for itself what constitutes a national emergency, *including HIV.*

These criticisms may reflect an uber-right view that wants to limit exceptions. In particular, perhaps an uber-right view is willing to permit compulsory licenses in the case of an emergency, but *not* for public noncommercial use, such that all the arguments are framed within the emergency context. Moreover, the attempt to further limit what counts as an emergency when the Doha Public Health Declaration states that this is up to national authorities also suggests the imprint of the uber-right view.

The relevant TRIPS issue that may be masked by the unnecessary emergency discussion is what constitutes "public noncommercial use." As noted above, this is one of several grounds upon which a nation may issue a compulsory license without prior negotiation with the patent owner. Thailand has assumed that it clearly qualifies and although some scholars support Thailand's opposition, TRIPS is ambiguous on what this term means. However, what constitutes an adequate public non-commercial use has important implications since it could be a much more common ground than an emergency.

It may be more appealing to paint Thailand's actions as failing to meet the requirement of an emergency—even if this is not an issue—rather than address the more difficult question of what constitutes public noncommercial use.[131] However, some of the comments that attempt to fit criticism concerning the license of Plavix as an inappropriate emergency could be easily replaced as criticism against an overbroad interpretation of public noncommercial use. For example, if heart disease is considered an emergency, the Plavix license has been sharply criticized as the first step down a slippery slope toward licensing any and all patents.[132] However, it would seem equally apropos for the pharmaceutical industry to say that if public noncommercial use were a permissible ground for a compulsory license, it would be a slippery slope toward licensing all patents. Perhaps the public noncommercial use is the true slippery slope that terrifies patent owners such that they have displaced their concerns onto the emergency criterion instead.

A few critical articles address the question of what constitutes public noncommercial use, but they generally take a very dismissive and conclusive approach. For example, some state that because the licenses were issued to a government agency that is for-profit, the

---

[130] *See, e.g.*, Lybecker & Fowler, *supra* note 118 at 233 (suggesting heart disease may not be an emergency because of only 300,000 patients being implicated, and that the 15,000 afflicted with cancer may similarly not constitute an adequate emergency).

[131] This is a challenging question under TRIPS as what is considered noncommercial as well as public use have been thorny questions about national laws with respect to what should be a legitimate research exception in patent law, as well as a legitimate fair use in copyright law.

[132] Cass, *Thai Patent Turmoil*, *supra* note 84.

licenses presumably must also be for-profit. Others declare that for Thailand to assert the application of public noncommercial use would be simply "taking advantage of vague language."[133] However, it is unclear whether the term "public noncommercial use" is really any more vague than other terms that lack definition in TRIPS; after all, when other terms are undefined, countries have been given freedom to define these terms themselves. In addition, to the extent this term should have a uniform meaning, these dismissive comments fail to foster a productive conversation. Although the existing comments about how Plavix is not an emergency can be easily dismissed as legally irrelevant, there is an outstanding issue of whether Thailand—or any other country—could broadly license any and all drug patents without any negotiation with the patent owner based on the ground of public noncommercial use.

2. No Research versus Adequate Remuneration

A disproportionate amount of commentary suggests that compulsory licenses are wrong or improperly issued because they will negatively impact research. Out of the many articles criticizing the Thai licenses, almost none mention the amount of remuneration, whereas most suggest the licenses would negatively impede innovation.[134] Although innovation is relevant, TRIPS does actually address this issue by mandating the patent owner be provided "adequate remuneration." This term is not defined under TRIPS. However, if the remuneration is truly adequate, the pipeline of new drugs should not suffer. Moreover, in addition to the uneven discussion of adequate remuneration, none of the patent owners brought a legal challenge to the amount of remuneration provided by Thailand, although that avenue was available to them.[135]

The uber-right may believe that compulsory licenses are inherently inconsistent with the principle of adequate remuneration such that it is not worth even discussing that term. In the few instances where remuneration is mentioned, it seems clear an uber-right perspective would deem any amount below the usual market value to be inadequate. For example, one editorial states that royalties paid for compulsory licenses "almost always leave the rights holder with far less than a reasonable economic return," thus suggesting

[133] Editorial, *The Thai Flu*, Wall St. J., Mar. 14, 2007, at A14.

[134] *E.g.*, Swiss Aide Memorie, *supra* note 44 ("Switzerland is convinced that patents are part of the solution to long term access to innovative medicines in Thailand"); *US Drugmaker Abbott, Thailand Face Off in AIDS Drug Patent Stalemate*, Int'l Herald Trib, June 6, 2007 (quoting PhRMA president as stating that Thailand's "misguided focus" would have "longer term cost, potentially limiting important incentives"). Even some academics focus solely on arguing that incentives will be reduced and provide no discussion of whether the Thai remuneration complied with TRIPS; remuneration is only discussed with respect to the fact that compulsory licenses fail to provide adequate remuneration to patent owners. *E.g.*, Lybecker & Fowler, *supra* note 118, at 236; Epstein & Kieff, *supra* note 118, at 6.

[135] Of course, an official challenge to remuneration may be viewed as an unpalatable concession to compulsory license.

that *any* compulsory license would be unacceptable.[136] No different amount of remuneration is proposed—the only alternative discussed is simply not imposing a license.[137]

Despite the lack of discussion of adequate remuneration thus far, this could be a very important issue. After all, if a compulsory license could be imposed with a remuneration amount deemed adequate to patent owners, those opposing compulsory licenses would theoretically lose a major objection.[138] At least one academic suggests that it is a myth to assume that compulsory licenses should result in cost savings; instead, he suggests that compulsory licenses should ensure patent owners receive appropriate value.[139] If that were the case, a compulsory license would be of little value to countries attempting to use "TRIPS flexibilities" to enhance access to medicine as the Doha Public Health Declaration suggests is possible.

3. Factual Dispute on Prior Negotiation versus Defining Prior Negotiation

Another nonissue is the seemingly factual question of whether Thailand negotiated with patent owners before issuing compulsory licenses. Although this may seem to be a relevant factual dispute with a single correct answer, the real problem is that the "facts" seen are a function of how holders of differing patent perspectives define "prior negotiation"—another important, yet undefined term under TRIPS. As discussed earlier, whereas Thailand asserts it is not required to negotiate with patent owners prior to exercising a compulsory license for public noncommercial use, but in fact did attempt to negotiate for years, critics of its licenses generally contend that Thailand failed to negotiate and that patent owners were shocked to find out about the licenses.

The prior negotiation controversy highlights the impact of patent perspectives on undefined TRIPS terms. The privilege view of patents may view any negotiation—even years before the compulsory license—as prior negotiation. After all, there was some negotiation, and it was prior to the license. If patents are viewed as a privilege, the extent of negotiation would not seem important, nor would giving notice to the patent owner of an imminent imposition of a compulsory license because a patent is only a privilege, not a right. However, judging from the response of patent owners that they were shocked the licenses were imposed with *no negotiation* at all, it seems the holder of an uber-right view may take a different perspective on what constitutes prior negotiation. Perhaps the uber-right view requires prior negotiation to occur immediately before the license is

[136] Cass, *Drug Patent Piracy*, *supra* note 74.

[137] Bate & Boateng, *supra* note 123, at 7; Cass, *supra* note 2; Bate, *Thailand and the Drug Patent Wars*, *supra* note 94.

[138] There may still be fears of parallel imports, although some suggest these fears are disproportionate to actual evidence. *E.g.*, Kevin Outterson, *Pharmaceutical Arbitrage: Balancing Access and Innovation in International Prescription Drug Markets*, 5 YALE J. HEALTH POLICY 193, 261-62, & n.304 (2005).

[139] Daniel Cahoy, *Confronting Myths and Myopia on the Road from Doha*, 42 GA. L. REV. 131, 155–56 (2007).

imposed and perhaps with explicit mention of an impending compulsory license, rather than general negotiations years in advance of such a license.

The importance of prior negotiation to the uber-right view may have resulted in at least one misstatement of the TRIPS rule. TRIPS plainly provides an exception to prior negotiation in the case of an emergency, public urgency, or public noncommercial use.[140] Nonetheless, Professor Froehner alleged that "[e]ven in a legitimate emergency . . . the owner of the IP rights that are confiscated is entitled under the WTO rules to consultation before the decision is made."[141] Although this statement is legally inaccurate, it may accurately reflect what holders of the uber-right view consider to be essential such that it becomes their reality.

4. Factual Dispute Regarding Frequency of Compulsory Licenses

The final nonissue shows some exaggeration by holders of both the uber-right and privilege views.[142] In particular, whereas the privilege perspective asserts that compulsory licenses occur frequently in a variety of countries,[143] the uber-right perspective asserts that compulsory licenses are rare.[144] This seems like another situation where there must be only one correct answer. However, the differing perspectives may once again play a role. The dispute here focuses not only on how often compulsory licenses are issued, but on what constitutes a compulsory license in the first place. Those who hold a view of patents as a privilege take a broad view of compulsory licenses that includes not only situations such as the Thai licenses, but also compulsory licenses issued as a penalty for anticompetitive practices, government uses, and even court cases that elect to deny a permanent injunction. The pertinent TRIPS provision is broad enough to include all but the court cases; after all, the key term under TRIPS is "use not authorized by the patent owner."

[140] TRIPS art. 31(b) ("This requirement [of prior negotiation with the patent owner] may be waived by a member in the case of a national emergency or other circumstance of extreme urgency or in cases of public non-commercial use.").

[141] Froehner, *supra* note 113, at 5.

[142] Although this is the final "nonissue" discussed in this article, there are other nonissues that could be addressed. For example, as shown in the contrasting narratives, holders of the uber-right view seem to second-guess whether Thailand is acting in the interest of public health, although this is not technically required by TRIPS. This could be another situation where holders of the uber-right perspective would *prefer* compulsory licenses to be limited to health crises, such that they may seize any opportunity to suggest the licensor country does not have a legitimate health interest.

[143] *E.g.*, Raymond C. Offenheiser, Letter to the Editor, WSJ.COM, Mar. 14, 2007, http://online.wsj.com/article/SB117382283064835964.html; LOVE, *supra* note 65.

[144] *E.g.*, *Bangkok's Drug War Goes Global*, *supra* note 2 (asserting that "no serious government has contemplated using compulsory licensing, even if it is allowed under the WTO rules"). *See also* Sidney A. Rosenzweig, *The False Connection between Strong Patent Rights and Global Inequity: A Response to Professor Stiglitz*, PROGRESS ON POINT, Mar. 2009, at 1, 19.20, http://www.pff.org/issues-pubs/pops/2009/pop16.8patentstiglitz.pdf (suggesting compulsory licenses are rarely used as remedies for anticompetitive conduct, and moreover that "it is hard to see how antitrust-related licensing is relevant to public health licensing").

This term is considered to encompass not only the type of compulsory license issued by Thailand, but also the "government use" of patents commonly practiced by the United States whereby use of patented inventions by government contractors are subject only to remuneration, but never injunctions. The uber-right view somewhat predictably attacks assertions that denials of injunctive relief as proof that defenders of compulsory licenses are inaccurate in overstating the frequency of compulsory license – although they seem to ignore the many other situations that do indicate worldwide use of compulsory licenses.[145]

In this case, although the frequency of compulsory licenses being issued may on its face seem pertinent to whether they should be permitted or granted, that is a nonissue in the context of what is permissible under TRIPS. Granted, the frequency of compulsory licenses may seem like a relevant policy consideration, but to the extent TRIPS states the rule of law, that would seem the relevant metric. So, what does each side have to gain by focusing on this nonissue?

One hypothesis for the posturing concerning what counts as compulsory licensing and how frequently it occurs is that the accusations are intended to mask other issues. For example, if holders of the uber-right perspective can show the privilege view on the number of compulsory licenses granted is an exaggeration, they can perhaps attempt to suggest that all statements concerning the privilege view are suspect. However, holders of the privilege view may be attempting to indirectly tackle the nonissue of whether compulsory licenses impact innovation by suggesting they occur frequently. This could be an admission by the privilege view that policy is relevant on some level even if not explicitly mentioned in TRIPS.

V. Examining Alternate Explanations

The competing patent perspectives theory is important both because it is real and because it has not been previously recognized. Like all new theories and observations, challenges are anticipated. Accordingly, this section attempts to anticipate and address possible challenges to the competing perspectives theory set forth above.

A. A FAKE PHENOMENON?

One potential objection to the competing views presented here may be that the two poles of patent perspectives articulated do not accurately represent true views, or at most, represent mere rhetoric. However, the intensity of debates concerning compulsory licenses

---

[145] *E.g.*, Epstein & Kieff, *supra* note 118, at 9-13 (suggesting that broadcast licensing of copyrighted songs and judicial denials of injunctive relief do not constitute evidence of frequent compulsory licenses in the US, when none of the cited support for defenders of compulsory licenses suggests broadcast licenses are relevant and most of the other cited material relates to global use of compulsory licenses issued on drugs).

(including name-calling and full-page advertisements to support differing positions) suggests there is in fact something real to discuss.[146] Moreover, even if some statements are rhetoric, when similar positions are evidenced by academics, it seems less likely everyone is engaged in simple posturing. Also, patent perspectives may be the source of rhetoric as rhetoric does not arise in a vacuum.

This is not to suggest the competing perspectives provide the only reason for discord. There are also other legitimate issues that may be at play, including but not limited to the fear that compulsory licenses will lead to other problems for patent profits even in countries where no such licenses issue. This could happen, for example, if low-cost drugs made under compulsory license become parallel imports in another country where they undercut the price of the patent owner.[147] Alternatively, the price of drugs made under a compulsory license may result in pressure to reduce prices in other markets, either because some nations use reference pricing based on what other countries pay or because of public pressure to offer discounted prices. However, the existence of these other factors does not mean that perspectives do not exist.

B. ALL ABOUT PROPERTY

Another possible objection to the competing views presented is that they simply reflect different views of property, with the observations not being unique to patent law. However, there should be some similarity between patents and other property in general as by law, patents are often considered property. In addition, even if the perspectives reflect different views of property, this may simply suggest they have broader application.

Alternatively, perhaps controversy over any property—whether real or intellectual—is more a function of a disconnect between expected and actual rights. In particular, to the extent compulsory licenses are not expected, patent owners may be rightfully surprised about their rights being taken away. Similarly, when the rights of private property are seemingly disturbed, as happened when the US Supreme Court permitted a broad

[146] Full-page advertisements criticizing Thailand's actions were taken out in the *Wall Street Journal* and in the Thai media by a group called USA for Innovation. *See, e.g.*, http://www.2bangkok.com/07/news07apr.shtml; http://www.thaipr.net/nc/readnews.aspx?newsid=2FF823462E1DE002F0DBA754EE0ABCC3; http://blog.light-of-reason.com/archive/usa-for-innovation-ad-campaign-targets-thailands-health-ministers-web-of-deceit/. However, competing ads supported Thailand's actions. *See, e.g.*, http://www.atimes.com/atimes/Southeast_Asia/IE17Ae02.html;http://www.natap.org/2007/newsUpdates/052407_05.htm

[147] The opposition to compulsory licenses may be very closely tied to opposition to parallel imports (and international exhaustion of patent rights). Indeed, some who object to compulsory licenses suggest the same types of health and safety problems that have been raised as support for denying anyone but the original patent owner the right to distribute goods. *E.g.*, Epstein & Kieff, *supra* note 118, at 7 (suggesting inferior distribution systems would lead to a "failure to maintain sensitive pharmaceutical compounds in proper condition.").

reading of what property could be taken, rights holders may react strongly.[148] Although there may be a similar reaction in both cases, they can also each be alternatively characterized as a disagreement between rights (whether property or patent) as being an uber-right or as a privilege that is subject to exception.

However, patents have unique characteristics that make them distinguishable from real property. One very important issue is that patents can impede access to medicine—something real property normally does not do. In fact, whereas patents routinely impede access to medicine (at least with respect to immediate access to patented medicine), real property does not generally stand in the way of health and safety. Another special characteristic of patents is that they are limited in time, such that some may suggest the patent rights should be more exclusive during that period. However, the privilege perspective would seem to expect additional exceptions to patent rights despite this factor.

C. GENUINE CONFUSION

An alternative explanation to patent perspectives may be that they simply represent confusion. Perhaps in recognition of this confusion, the WTO has aimed to clarify via a list of frequently asked questions on its website that a national emergency is not required for a compulsory license.[149] However, confusion still abounds. Even among scholars, there is frequently an assumption that an emergency is required, or at least that only an emergency or other extreme urgency could be grounds for waiving prior negotiation despite TRIPS listing public noncommercial use in addition to these two grounds.[150] Some of this confusion may stem from the first global controversy involving whether a nation's use of a compulsory license complied with TRIPS. As will be discussed in Chapter 11, shortly after TRIPS was enacted, South Africa faced a major HIV epidemic and enacted a law that promoted broad use of compulsory licenses to help address the health crisis. Although South Africa's law was controversial, there was no question it was facing an HIV epidemic. Thus, news reports justifying South Africa's actions as being proper in the face of

---

[148] Kelo v. City of New London, 545 U.S. 469 (2005)(upholding the government right to transfer property from one private owner to another in the interest of economic development); David Fagundes, *Property Rhetoric and the Public Domain*, 94 MINN. L. REV. 652 (2010).

[149] WTO, TRIPS and Health: Frequently Asked Questions, Compulsory Licensing (Sept. 2006), http://www.wto.org/english/tratop_e/trips_e/public_health_faq_e.htm.

[150] For example, in an article discussing the history of TRIPS, as well as the discussion of access to medicine that is sympathetic to the interests of developing countries, article 31 of TRIPS is stated as requiring prior negotiation except "in the event of a national emergency or other circumstances of extreme urgency," without any mention of public noncommercial use. Duncan Matthews, *Is History Repeating Itself? The Outcome of Negotiations on Access to Medicines, the HIV/ADIS Pandemic and IPR in the WTO*, L. SOC. JUSTICE & GLOBAL DEV., June 4, 2004, at 5.

an emergency may have left the impression that an emergency was required.[151] It is likely holders of the uber-right view would not discourage any confusion that an emergency is required for a compulsory license.[152]

Confusion regarding compulsory licenses is also exacerbated by the fact that the requirements of compulsory licenses relevant to Thailand are not the ones that have been most discussed since the South African controversy, both in the popular press and the scholarly literature. Rather, there is a complicated procedure for waiver of one of the usual TRIPS requirements *not* at issue in the Thai licenses—that a compulsory license be predominantly for domestic use (as further discussed in the next chapter).[153] This Waiver involves many additional requirements, some of which seem to have been incorporated into the uber-right view of compulsory licenses in general. For example, the Waiver is stated to apply to any LDC member, perhaps giving credibility to the assumptions of some that all compulsory licenses should be limited to such countries.[154] In addition, a number of countries agreed to only use the procedure in case of national emergency or other circumstances of extreme urgency, perhaps giving rise to confusion about whether an emergency is required. Also, the confusion about whether only patents for some subject matter may be licensed could also derive from discussions of whether the Waiver should be limited to the conditions listed in the original Doha Declaration (such as HIV and malaria); although the Waiver itself does not explicitly state it is limited to certain diseases, some have nonetheless made that contention.

Another possible source of confusion that promotes the uber-right view that compulsory licenses should be limited to national emergencies could be based on an assumption that TRIPS reflects what was originally proposed by patent-owning companies. After all, TRIPS was the brainchild of US companies seeking to protect their global interest in intellectual property, and most agree the final language in TRIPS often reflects the desires of developed countries, such as the United States.[155] In the context of compulsory licenses, the

---

[151] *E.g.*, Merrill Goozner, *Third World Battles for AIDS Drugs*, CHI. TRIB., Apr. 28, 1999 ("global rules of trade do allow for what is known as compulsory licensing if it is done to combat a national emergency"); Karl Vick, *African AIDS Victims Losers of a Drug War*, WASH. POST, Dec. 4, 1999 (stating TRIPS provided a "loophole" for countries facing a health emergency to make patented drugs).

[152] Media reports read by the public (including academics) frequently reflect an uber-right perspective and seem to give it credence. For example, Ron Cass has written a number of editorials in the *Wall Street Journal* criticizing Thailand's licenses; his editorials definitely fall within the uber-right mold, but are given the imprint of legitimacy because he is often referred to as a former dean and professor of Boston University, as well as being associated with the authoritative sounding "Center for the Rule of Law," which is in fact a conservative group.

[153] WTO, Decision of the General Council, *Implementation of Paragraph 6 of the Doha Declaration on the TRIPS Agreement and Public Health*, WT/L/540 (Aug. 30, 2003); WTO, Decision of the General Council, *Amendment of the TRIPS Agreement*, Annex to the Protocol Amending the TRIPS Agreement (Art. 31bis), WT/L/641 (Dec. 8, 2005).

[154] Art. 31bis, Annex 1(b) (defining "eligible importing member").

[155] *See, e.g.*, Matthews, *supra* note 150, at 4 (stating the final version of TRIPS "was perceived to closely reflect the demands of developed countries and their industries").

United States did suggest that licenses be limited to national emergencies.[156] This prior proposal could account for continued suggestions from the uber-right that compulsory licenses are only appropriate in limited situations—despite TRIPS clearly stating otherwise.

Although there are many sources of possible confusion for the distorted interpretations of TRIPS beyond the competing patent perspectives, confusion does not necessarily mean the perspectives do not exist. Rather, perspectives may result in more confusion and distortion. In addition, although the general public could understandably be confused by compulsory license requirements, confusion alone does not seem an adequate explanation for inappropriate interpretations by lawyers and scholars. At a minimum, competing patent perspectives may help provide a more nuanced explanation as to why compulsory licenses are so controversial despite their clearly being permissible under TRIPS.

VI. Conclusion

Although this chapter has focused on documenting how perspectives can have a dramatic impact on the way a single article of TRIPS is viewed, the existence of these perspectives has broader implications. For example, the conclusion of TRIPS (as well as subsequent disagreements) can be explained by the competing perspectives. Although many believe TRIPS was ultimately accepted by developing countries as part of a package deal involving broader access to markets, an equally important factor may have been the skeletal final language that permitted acceptable interpretations to all perspectives. For example, when TRIPS was concluded, the important term *invention* was undefined; holders of an uber-right view could assume that TRIPS would permit a broad range of subject matter to be patentable, whereas holders of an privilege view could have an opposing assumption. The different views may collide when a country implements a law that more clearly reflects a single perspective. For example, when India revised its patent law to limit patentability of compounds that were similar to known compounds, its law could be seen as consistent with a privilege view. However, to those with an uber-right view, the law would be deemed improper.

Other chapters will revisit these perspectives to provide a more nuanced analysis of other controversies. For example, Chapter 10 examines whether an EU regulation that permits suspension of in-transit goods as possible patent infringement complies with TRIPS while also considering how competing perspectives have played a role in the controversy. Chapter 11 considers how competing perspectives have played a role in the development of international law and policy since the conclusion of TRIPS. Finally, Chapter 12 considers how competing perspectives implicate proposals to address access to medicine.

---

[156] Draft Agreement on the Trade-Related Aspects of Intellectual Property Rights - Communication from the United States, art. 27, MTN.GNG/NG11/W/70 (May 11, 1990).

7 Complicated Compulsory Licenses: The Waiver/Article 31bis "Solution"

This chapter explains an important, yet often misunderstood exception to the TRIPS requirement that a compulsory license must be issued predominantly for domestic use (except where issued to address anticompetitive conduct). It assumes an understanding of the usual compulsory licensing requirements addressed in Chapter 5 and builds upon it. As is further explained in this chapter, the WTO Council has created an official waiver ("the Waiver") of the usual domestic use requirement to help countries export low-cost generic drugs made under a compulsory license to predominantly poor countries.

This chapter explains when the Waiver is relevant, as well as the additional requirements necessary to use it beyond those that apply to all compulsory licenses. It ends with an evaluation of the effectiveness of the Waiver in achieving its intended purpose. By the conclusion of the chapter, a reader should understand the following issues:

- Why is the Waiver necessary?
- Does the Waiver apply to every compulsory license?

- What countries may import drugs made pursuant to the Waiver?
- What steps must take place pursuant to the Waiver for drugs to reach the intended country?

I. Introduction 197
  A. History 197
    1. Impetus of the Waiver 197
    2. Creation of the Waiver 198
  B. Overview of the Waiver 200
    1. Waiver Not Applicable 200
    2. Waiver Applicable 201
II. The Waiver Proposed article 31bis 202
  A. Coverage 202
    1. Countries 202
    2. What Patented Products May be Licensed 204
    3. Trade Diversion Issues 204
    4. Other Restrictions 205
  B. Overview of Process 206
    1. Importing Country 206
    2. Exporting Country 207
III. National Implementation 209
  A. Required Changes 209
    1. Importing Country 209
    2. Exporting Country 210
  B. Possible Additional Restrictions under National Law 210
    1. Mandatory Prior Negotiation with Patent Owner 211
    2. Limited Types of Drugs 211
    3. Eligible Countries—and Additional Requirements 212
    4. Eligible Importer 212
    5. Limited Scope of License 212
    6. Additional Regulatory Approval 213
    7. Remuneration 213
  C. Case Study: Canada/Rwanda 214
IV. Evaluation 217
  A. Current Criticisms 217
  B. Next Steps? 219

I. Introduction

A. HISTORY

1. Impetus of the Waiver

In 2001, when countries concluded the Doha Public Health Declaration, there was unanimous agreement that the usual compulsory license rules presented a problem for poor countries. In particular, although a TRIPS-compliant compulsory license should be a legitimate tool to promote access to cheaper medicines, it is of no utility to a country that is too poor to have adequate manufacturing capacity to make its own generic drugs. Moreover, such a country is unable to import drugs made under a compulsory license elsewhere because TRIPS mandates nations to use compulsory licenses "predominantly" for domestic use.[1]

Consider a typical LDC such as Ethiopia. As a LDC, Ethiopia need not provide patents on drugs until at least 2016. However, although it can lawfully make lower-cost generics, it cannot realistically do so because it lacks adequate manufacturing capacity. Moreover, although a country such as Brazil may issue a compulsory license to make lower-cost generic drugs that are otherwise patented, Brazil cannot export the licensed drugs to Ethiopia because of the usual rule that requires licensed drugs be predominantly for domestic use. The usual TRIPS requirement that compulsory licenses be granted predominantly for domestic use results in the poorest countries (such as Ethiopia) having less access to low-cost medicines than middle-income countries (such as Brazil) that are capable of manufacturing drugs under a compulsory license.

Developing countries raised this issue before and during the Doha Ministerial Conference[2] and suggested a number of solutions to address the problem.[3] Although countries could not reach a consensus on a specific solution at the Conference, they did

---

[1] TRIPS: Agreement on Trade-Related Aspects of Intellectual Property Rights, Apr. 15, 1994, Marrakesh Agreement Establishing the World Trade Organization, Annex 1C, art. 31(f), The Legal Texts: The Results of the Uruguay Round of Multilateral Trade Negotiations 320 (1999), 1869 U.N.T.S. 299 [hereinafter TRIPS].

[2] WTO Council for Trade-Related Aspects of Intellectual Property Rights, Submission by the African Group, Barbados, Bolivia, Brazil, Cuba, Dominican Republic, Ecuador, Honduras, India, Indonesia, Jamaica, Pakistan, Paraguay, Philippines, Peru, Sri Lanka, Thailand and Venezuela, ¶ 34, IP/C/W/296 (June 29, 2001) [hereinafter Submission by African Group et al., IP/C/W/296]; WTO Council for Trade-Related Aspects of Intellectual Property Rights, Proposal from a Group of Developing Countries, *Draft Ministerial Declaration on the TRIPS Agreement and Public Health*, ¶¶ 5, 9, IP/C/W/312, WT/GC/W/450 (Oct. 4, 2001) [hereinafter Developing Country Draft Ministerial Declaration, IP/C/W/312].

[3] For example, developing countries suggested that export of drugs made under compulsory licenses necessarily be a "limited exception" under article 30. Developing Country Draft Ministerial Declaration, IP/C/W/312, ¶ 9; *see also* Submission by African Group et al., IP/C/W/296, ¶ 34.

unanimously agree a problem existed; as noted in the Doha Public Health Declaration, the TRIPS Council[4] was instructed to find an expeditious solution.[5]

2. Creation of the Waiver

Following substantial international discussion, the TRIPS Council provided a solution in 2003 in the form of a formal waiver of the TRIPS requirement of domestic use for compulsory licensing.[6] Under the Waiver, a country can issue a compulsory license predominantly for *export* to a country in need so long as additional requirements are met. However, to effectively utilize the Waiver, national laws needed to be amended to permit compulsory licenses for export. As discussed in the next section, national laws implementing the Waiver sometimes impose additional obligations beyond those in the Waiver itself.

Although the Waiver may remain in effect indefinitely, the goal is to formally amend TRIPS to make it a permanent part of TRIPS as article 31bis. This provision was proposed as an amendment in 2005 with the initial hope of securing member ratification (requiring agreement by two-thirds of member countries) within six months.[7] However, after four years, not even one-third of the members have ratified the amendment.[8] The lukewarm international response may in part reflect a lack of enthusiasm for the original

[4] A body created by TRIPS that is intended to ensure that countries comply with TRIPS and that provides assistance to members with resolving issues. TRIPS art. 68.

[5] World Trade Organization, Ministerial Declaration on the TRIPS Agreement and Public Health of November 14, 2001, ¶ 6, WT/MIN(01)/DEC/W/2, 41 I.L.M. 755 (2002) [hereinafter Doha Public Health Declaration] ("We recognize that WTO Members with insufficient or no manufacturing capacities in the pharmaceutical sector could face difficulties in making effective use of compulsory licensing under the TRIPS Agreement. We instruct the Council for TRIPS to find an expeditious solution to this problem and to report to the General Council before the end of 2002.").

[6] WTO, Decision of the General Council, *Implementation of Paragraph 6 of the Doha Declaration on the TRIPS Agreement and Public Health*, WT/L/540/Corr. 1 (Aug. 30, 2003) [hereinafter 2003 Waiver]. This Waiver is permissible because any requirement of a WTO agreement, including TRIPS, may be waived. *See* WTO art. IX(3–4).

[7] WTO Decision of the General Council, *Amendment of the TRIPS Agreement*, Annex to the Protocol Amending the TRIPS Agreement (art. 31bis), WT/L/641 (Dec. 8, 2005) [hereinafter TRIPS art. 31bis]; WTO art. X:3 ("amendments shall take effect for the Members that have accepted them upon acceptance by two thirds of the Members and thereafter for each other Member upon acceptance by it.").

[8] WTO, Members Accepting Amendment of the TRIPS Agreement, http://www.wto.org/english/tratop_E/trips_e/amendment_e.htm (listing 28 of the over 150 member countries that have ratified the amendment) (last visited on Nov. 10, 2010); *see also* Matthew Kennedy, *When Will the Protocol Amending the TRIPS Agreement Enter into Force*, 13 J. Int'l Econ. L. 459, 461 (2010).

2003 Waiver,[9] which was criticized, but proposed as a formal amendment because the parties could not agree on a better alternative.[10]

Even if article 31bis becomes part of TRIPS, there may be a remaining legal ambiguity. When article 31bis was formally adopted by the WTO Council as an amendment ready for ratification by member states, a contemporaneous oral statement was read by the General Council Chairperson suggesting additional limitations on the use of the Waiver.[11] For example, according to this statement, certain developing countries agreed to only use the Waiver in circumstances of extreme urgency.[12] If this statement is not binding, those developing countries would have no limitations on using the Waiver because the formal text of proposed article 31bis makes no mention of restrictions for these states.[13] The legal status of the General Council Statement is unclear. However, it will be incorporated into

---

[9] Alternatively, the lack of enthusiasm could simply reflect a lack of urgency as the Waiver continues to exist until article 31bis is ratified. Another possible reason is that some countries are wary of amending a WTO agreement—and this would be the first such amendment.

[10] *E.g.*, TRIPS Council Remains Divided on Public Health Amendment, BRIDGES Wkly Trade News Dig., Oct. 26, 2005, *available at* http://ictsd.org/i/news/bridgesweekly/6210/ (discussing lack of agreement regarding alternatives to 2003 Waiver). The 2003 Waiver was criticized as overly complex and unworkable—criticisms that similarly apply to article 31bis as the two are identical. *E.g.*, Frederick M. Abbott & Jerome H. Reichman, *The Doha Round's Public Health Legacy*, 10 J. INT'L ECON. LAW 921, 956 (2007); Frederick M. Abbott, *The WTO Medicines Decision: World Pharmaceutical Trade and the Protection of Public Health*, 99 AM. J. INT'L L. 317 (2005); Brook K. Baker, *Arthritic Flexibilities for Accessing Medicines: Analysis of WTO Action Regarding Paragraph 6 of the Doha Declaration on the TRIPS Agreement and Public Health*, 14 IND. INT'L. & COMP. L. REV. 613 (2004); Matthew Rimmer, *Race against Time: The Export of Essential Medicines to Rwanda*, 1 PUB. HEALTH ETHICS 73 (2008), draft available at http://works.bepress.com/matthew_rimmer/53/; Letter from CPTech, Oxfam, MSF and HAI to WTO Delegates Regarding December 16, 2002 Chairman's Text for "Solution" to Paragraph 6 of the Doha Declaration on TRIPS and Public Health, *available at* http://www.cptech.org/ip/wto/p6/ngos12192002.html.

[11] WTO General Council, General Council Chairperson's Statement, WT/GC/M/82 (Nov. 13, 2003), *available at* http://www.wto.org/english/tratop_e/trips_e/gc_stat_30aug03_e.htm [hereinafter 2003 Chairperson's Statement]; WTO, Statement by the Chairperson to the General Council (Dec. 6, 2005), *available at* http://www.wto.org/english/news_e/news05_e/trips_319_e.htm [hereinafter 2005 Chairperson's Statement].

[12] 2005 Chairperson's Statement, *supra* note 11 ("As we have heard today, and as the Secretariat has been informed in certain communications, some other Members have agreed that they would only use the system as importers in situations of national emergency or other circumstances of extreme urgency. These are the following: Hong Kong, China; Israel; Korea; Kuwait; Macao, China; Mexico; Qatar; Singapore; the Separate Customs Territory of Taiwan, Penghu, Kinmen and Matsu; Turkey and the United Arab Emirates.").

[13] In contrast, the Annex to article 31bis lists countries who promise to never use the Waiver. TRIPS art. 31bis, Annex to the TRIPS Agreement, ¶ 1, note 7 (noting that Australia, Canada, the EU Member States, Iceland, Japan, New Zealand, Norway, Switzerland, and the United States will not use the Waiver for importation). However, no agreement could be reached on incorporating their agreement into article 31bis, such that the same statement reached when the 2003 Waiver was adopted was read again in 2005 when article 31bis was adopted by the WTO Council.

this chapter because it may be a supplementary text that helps to clarify the Waiver and article 31bis.[14]

B. OVERVIEW OF THE WAIVER

This section provides a brief overview of when the Waiver is appropriate. Although the Waiver was created to help provide low-cost drugs to poor countries, it is not necessarily involved in every instance where drugs are provided to a poor country. Determining when the Waiver is applicable has spawned great confusion. Given the existing confusion of basic compulsory licenses (as discussed in Chapter 6), the additional layer of restrictions set forth in the multipage Waiver seems to have compounded confusion. Before attempting to discuss the Waiver requirements in detail, This section first clarifies when the Waiver is *not* necessary, followed by when it is.

1. Waiver not Applicable

The Waiver is not needed if:

(1) A country is able to secure adequate amounts of patented drugs at a price it can afford. There are a number of ways a country could secure such drugs, including:

- a. A country could purchase the necessary drugs directly from the patent owner at a discounted price.
- b. A country could receive desired drugs as a gift from a humanitarian organization.[15]
- c. A country could purchase drugs made as generics in another country where they are not patented.[16]

OR

[14] Alternatively, it can be argued that the statement should not have interpretive force because a "corrigendum" was added to it that indicates the statement is "for information purposes only and without prejudice to Members' legal rights and obligations." 2003 Waiver, *supra* note 6; *see also* Andrew D. Mitchell & Tania Voon, *The TRIPS Waiver as a Recognition of Public Health Concerns in the WTO*, *in* INCENTIVES FOR GLOBAL PUBLIC HEALTH: PATENT LAW AND ACCESS TO ESSENTIAL MEDICINES 66 (Thomas Pogge et al. eds., 2010).

[15] If the organization bought drugs that are patented in the importing country, the importing country must issue a compulsory license, to avoid infringement but the Waiver procedure is not applicable.

[16] If the importing country has a patent on the desired drug, a compulsory license is necessary for the import, but the more complex Waiver procedure is not applicable.

(2) A country has adequate manufacturing to make its own generic versions of desired patented drug under a routine compulsory license.[17] Typically, such a country would be a middle-income country that has a established capacity for making generic drugs, such as India, Brazil, or China. A country in this position could issue a compulsory license consistent with TRIPS article 31.

As noted above, there are an array of conditions under which a country could secure cheaper versions of patented drugs. However, there are some situations where the above scenarios are not adequate to supply all the drugs desired.

A key issue to determining whether use of the Waiver is necessary is whether a country wants to manufacture drugs under a compulsory license *for export* to a country that has indicated its intention to import such drugs. The reason the Waiver is necessary is that TRIPS usually requires that drugs made under a compulsory license be predominantly for domestic use. However, the domestic use requirement may be waived if a number of additional procedures are followed. Notably, whether the importing country provides patent protection is irrelevant to whether the Waiver is necessary because even if the importing country needed to issue a compulsory license to import patented drugs, that license would comply with the requirement that the license be for domestic use.

2. Waiver Applicable

The Waiver is necessary if all three of the following conditions exist:

(1) the country needs more drugs at a lower price than they can currently obtain; *and*

(2) the country has inadequate manufacturing capacity to make its own low-cost generic version (regardless of whether it provides patent protection); *and*

(3) the country is seeking to import drugs that would be made under compulsory license *for export* from another country where they are currently patented.

Notably, the use of the Waiver is not necessary to import drugs that are made as legitimate generics in a country where they are not patented. For example, if Ethiopia wanted to import a drug made as a generic in India, the Waiver would not be applicable. This does not necessarily mean that Ethiopia does not need a compulsory license. If the desired drug is patented in Ethiopia, it will need to issue a compulsory license to import the drug. However, following the usual requirements for compulsory licenses to address imports is

[17] However, as noted in Chapter 5, although compulsory licenses are permissible under TRIPS, countries that impose them may nonetheless be subject to retaliation and pressure by drug companies, as well as national governments.

much simpler than utilizing the Waiver. Accordingly, it is important to understand when a basic compulsory license is required versus the Waiver. Essentially, the Waiver only comes into play when a country wants to make generic versions of drugs under a compulsory license *for export*, such that the usual domestic use requirement for such licenses must be waived. In addition, the drugs can only be exported to a country that meets the Waiver requirements.

Now that the broad outlines of the problem prompting the Waiver as well as when it might apply have been addressed, this chapter will turn to the many additional requirements of the Waiver. In other words, although countries can waive the usual requirement to issue a compulsory license for domestic use only, this Waiver is contingent on compliance with additional obligations that do not exist for other compulsory licenses (for products that are made and used on a domestic basis). Notably, in addition to the requirements imposed by the Waiver, requirements may be imposed by national implementing legislation. The next two sections address each of these in turn. The last section concludes with an evaluation of the Waiver.

II. The Waiver (Proposed article 31bis)

This section addresses the procedural requirements imposed by the Waiver. In particular, it explains what countries may use it (whether as importing or exporting countries), as well as what drugs may be licensed. This section also explains the many procedural requirements that must be established to effectively use the Waiver.

A. COVERAGE

1. Countries

a. What Countries Can Import under the Waiver?

Determining what countries can use the Waiver (as an importing country) is a tricky issue. All LDCs can use the Waiver. Other countries may also be able to use it *if* they have not previously promised not to and lack adequate manufacturing capacity for the desired product. Notably, it is up to member states to self-identify whether they lack adequate manufacturing capacity. Although non-LDCs must notify the TRIPS Council of lack of manufacturing capacity, there is nothing in the procedures that permits the TRIPS Council to either approve or second-guess a national decision. In addition, the lack of adequate manufacturing capacity need not be for all drugs—it can apply only with respect to a single desired product. An LDC is likely to have general inadequate manufacturing capacity because of inadequate resources. However, the possibility of importing in cases where there is a more limited capacity for a desired product opens up the possibility that other countries could use the Waiver—subject to other conditions.

Some members have agreed they will not use the system as importers, or will only use it as importers in limited situations. Countries that have promised to *never* use the Waiver procedure as importing members are predominantly wealthy industrialized countries: Australia, Canada, EU Member States, Iceland, Japan, New Zealand, Norway, Switzerland, and the United States. Countries that have promised to only use the Waiver as an importing member in the case of national emergency or other circumstance of extreme urgency are: Hong Kong, China, Israel, Korea, Kuwait, Maco Mexico, Qatar, Singapore, Taiwan, Turkey, and UAE. Whether a national emergency exists is subject to national determination and there is no mechanism for such a decision to be independently evaluated.

b. Recap—What Countries Can Utilize the Waiver to Import Cheaper Drugs under License?

(1) Any LDC

(2) Any country besides a LDC *if*

   a. It has not previously promised not to use the system as an importer (absolutely, or only in the case of an emergency or situation of extreme urgency)

   b. It has inadequate manufacturing capacity for the desired patented product—as certified by the member state and not subject to independent approval

c. What Countries Can Export Drugs under the Waiver?

Any WTO member can theoretically export drugs pursuant to the Waiver. However, to actually export drugs, a country must have national laws in place to permit issuance of a compulsory license *for export*—something most countries did not have prior to the creation of the Waiver. Thus far, only a handful of countries have amended their laws to permit issuance of a license for export; they include Canada, India, EU Member States, Norway, Switzerland, China, and South Korea. Even for these countries, a careful examination of their national laws is required to assess whether additional requirements are imposed; this issue is discussed in more detail in the section on National Implementation.

d. Can Drugs Imported Pursuant to this Procedure be Exported to Another Country?

Although the Waiver procedures predominantly address how to ship goods to a qualifying importing country, there is a narrow class of cases where the importing country would be permitted to export the goods. If an importing country is a party to a regional trade agreement where more than half of the members are LDCs, the importing country is permitted to export some of the goods to another member whose citizens have the same health problem.[18] In such a situation, there is no need to even notify the TRIPS Council of export to a party to the regional trade agreement. But this provision is likely to have only limited impact as there are only a few regional trade agreements in Africa that would meet these requirements. In addition, if the original license provides the good for a single country, it is questionable whether there would be any surplus available to export to another country.

2. What Patented Products May be Licensed?

What products are subject to license is an important and contentious issue. Contrary to the desires of the pharmaceutical industry, there are no limitations on the types of diseases or conditions for which products may be licensed. Limitations were proposed but *rejected* during negotiations, such that a proper interpretation would not impose any.

Nonetheless, some confusion concerning the scope is understandable because the Waiver does not explicitly say that any patented product may be licensed. Rather, it refers to pharmaceutical products "needed to address the public health problems as recognized in paragraph 1 of the Doha Public Health Declaration."[19] That paragraph, in turn, does not list any specific drugs. Instead, it states that members recognize "the gravity of the public health problems afflicting many developing and least-developed countries, especially those resulting from HIV/AIDS, tuberculosis, malaria and other epidemics."[20] Although some countries have tried to suggest this means that only drugs for the listed diseases are covered, that is not consistent with the actual wording, which recognizes that developing countries have public health problems in general. Although the clause does single out a few diseases that are *especially* problematic, it clearly does not limit consideration to those diseases.

One nuanced issue that is often overlooked is that the Waiver permits countries to license any "product" of the "pharmaceutical sector," and specifically indicates that it includes active ingredients and not simply final products, such as pills. Although this may at first glance seem to be a distinction without a difference, it may have a significant impact on a poor country as it is easier to manufacture final products once provided the active ingredient instead of having to manufacture the active ingredient from scratch. Accordingly, a country could import an active ingredient pursuant to the Waiver and use that to make a desired drug.

3. Trade Diversion Issues

One important distinction for drugs made under the Waiver is that they must be clearly identified to help prevent trade diversion. In particular, Proposed Article 31bis states that

[18] TRIPS art. 31bis, ¶ 3.

[19] 2003 Waiver, pmbl; TRIPS art. 31bis, Annex to the TRIPS Agreement, ¶ 1.

[20] Doha Public Health Declaration, ¶ 1.

products made under the license must be "clearly identified" through special packaging, or of special color or shape of the product themselves (although such efforts must not have a "significant impact on price"). This marking can help prevent trade diversion by making the licensed products have a different appearance than the trademark of the brand name drug sold elsewhere. As discussed in Chapters 1 and 2, drug companies may use trademarks in addition to patents to help promote the sales of their products, as well as to distinguish products sold in different markets and control the sales of their products. The special labelling requirements were important to pharmaceutical companies, as well as the US and EU to help limit re-sale of the licensed goods in wealthier markets. In other words, the special marking seems consistent with the desire of many companies to minimize global trade of drugs even where wealthy markets already consider imports to be patent infringement. Drugs that are marked differently could constitute trademark in addition to patent infringement and be much easier to police at the borders.

In addition to distinctive labelling, all members have a duty to help police trade diversion. For importing countries, they are to take reasonable measures within their means. In addition, the Waiver contemplates that developed countries may provide assistance for this goal. For all other countries, the Waiver specifically requires they have "effective legal means" to prevent importation of goods intended for poor countries. While this does not explicitly state that members should decline to adopt international exhaustion, that could be one method to help prevent imports (as previously disussed in Chapter 2). In addition, members explicitly have a right to request review by the TRIPS Council if the legal means are considered inadequate.

4. Other Restrictions

Although the text of the Waiver does not set forth any additional restriction, the contemporaneous statement read by the chairperson ("Chairperson Statement") could potentially be read to limit the use of the procedure. In particular, the Chairperson Statement indicates the system "should be used in good faith to protect public health . . . not be an instrument to pursue industrial or commercial policy objectives."[21] It is unclear what it means that the Waiver not be used for industrial or commercial policy objectives—or why such objectives would necessarily be inconsistent with protecting public health. However, given the many cumbersome steps of this procedure, it is unlikely a country would be relying on the Waiver unless public health were at issue. In addition, it is not necessary that protection of public health and industrial objectives be mutually exclusive. After all, the Waiver itself refers to a desired goal of improving manufacturing capacity in some member countries.

---

[21] 2005 Chairperson's Statement, *supra* note 11, at ¶ 1; 2003 Chairperson's Statement, *supra* note 11, at ¶ 1.

Although this text is ambiguous, one thing it does *not* require is that licenses be limited to nonprofit situations - contrary to an earlier proposal (at one point suggested by the United States). Thus, companies should be able to make licensed drugs for export at a profit. However, the profit will on a practical level likely be negligible as the Waiver procedure allows firms to make only limited quantities for a specific country. In addition, given that many developing countries only purchase drugs through a competitive bidding process, the actual price will likely be low.

B. OVERVIEW OF PROCESS

This section addresses the crux of complexity surrounding article 31bis—that is, the myriad procedures and conditions that must be complied with to provide low-cost drugs made under compulsory license in one country for export to another country. Essentially three components are required: an eligible importing country willing to use the Waiver process, a country with laws permitting compulsory license for export, and an entity within the country with the export laws willing to manufacture a drug for export pursuant to the Waiver.

There must always be an eligible WTO member state that seeks to import a product from another country where it is patented. The eligible member must make a notification to the TRIPS Council with details concerning the desired product, lack of manufacturing capacity for the desired product (waived for LDCs), and assurance that a compulsory license is or will be granted to import the product if necessary. Before clarifying the details of these three requirements, it is important to point out that this notification alone is not adequate to ensure the arrival of the desired drugs. Rather, the eligible importing country must rely on other factors outside its control, that there will be an entity in an exporting country (with laws that permit licensing for export) willing to seek a compulsory license to export the desired drug in accordance with all the conditions of the Waiver, as well as any additional conditions imposed by national law. In other words, there are procedures to be satisfied by both the importing and exporting countries, as will be further detailed below in separate sections.

1. Importing Country

a. Official Notification

A potential importing country must decide what product is desired, then make a formal notification. The eligible importing country must officially notify the TRIPS Council of the following:

(1) the name and expected quantity of the desired product
(2) that the country has a self-identified insufficient manufacturing capacity to make the desired product (*not applicable to LDCs).*

(3) that a compulsory license consistent with TRIPS is granted or will be granted if the desired product is patented in the importing country.

b. Expected Quantity of Desired Product

It is important to clarify the requirement that an importing country specify the name and expected quantity of the desired product. Importantly, the language is "expected quantity" rather than a specific quantity. Accordingly, the expected quantity need not match the eventual quantity shipped. What this means is that the expected quantity can be an estimate, although a firm number will eventually be needed for a compulsory license to be granted. In addition, because there are no specific requirements on how to articulate the expected quantity, this requirement can be satisfied by reference to the number of doses, number of patients to be treated over a given period, or any other method that indicates a quantity.

c. Insufficient Manufacturing Capacity (non-LDCs Only)

All countries besides LDCs must certify that they have insufficient manufacturing capacity to make the desired product. As noted earlier, this certification is not subject to review by the WTO. In addition, the lack of manufacturing capacity of relevance here is *with respect to the specific product* and not general capacity to make some drugs. Accordingly, a country may be able to certify lack of manufacturing capacity for a specific patented product even though it has some general capacity to manufacture pharmaceuticals. In addition, this inquiry should exclude any production facilities owned or controlled by the patent owner.

d. Compulsory License—If Relevant

If the desired product is patented in the importing country, the notification should also indicate that a compulsory license has been or will be granted. Notably, no compulsory license must be issued at the time of notification. The Waiver does not set forth any time line for the compulsory license to be issued. As a practical matter, however, to avoid infringement, the license must be issued before the desired product may be imported.

If the importing country does provide patent protection for the desired drug, all the requirements of TRIPS article 31 must be followed in issuing the compulsory license except for one—the usual requirement of remuneration. The Waiver explicitly notes the remuneration need only be paid by the exporting country.[22] However, other elements of article 31 still apply. For example, the default rule is that the importing country must attempt voluntary negotiations with the patent owner (unless the license is issued pursuant to a national emergency, situation of extreme urgency, or public

[22] TRIPS art. 31bis(2).

noncommercial use). In addition, the compulsory license of the importing country must state the scope and duration of the license. For the actual license—as opposed to the notification to the TRIPS Council—a correct determination of the desired quantity of the drugs is important to avoid having another license to be unnecessarily issued.

2. Exporting Country

Three things must occur before licensed drugs may be exported to the country needing them. First, the exporting country must have laws in place to permit compulsory licensing for export. Then, a company within the exporting country must be willing to make and export a drug pursuant to article 31bis. That company must also secure a compulsory license that complies with article 31, article 31bis, and any additional conditions imposed by national law.

a. Basic Article 31 Requirements

Although the complete TRIPS requirements of basic compulsory licenses are discussed in more detail in Chapter 5, a few points bear noting here. First, although article 31bis waives the requirement of predominantly national use, it does not waive any other requirement of article 31. This means that before an exporting country can impose a compulsory license, the country (or the entity seeking the license) must engage in prior negotiation with the patent owner on reasonable terms and conditions unless one of the exceptions to prior negotiation exists (a national emergency, situation of extreme urgency, or public noncommercial use). Moreover, even if a license is issued, article 31 still requires a nation provide procedures for the patent owner to contest both the license and the amount of remuneration. Accordingly, even when a license is issued with the goal of exporting to a poor country, the patent owner may legally be permitted to contest it. If the license is contested, this could delay the expected shipment of goods.

b. Additional Compulsory License Requirements

Beyond the usual requirements of article 31, there are additional ones for licensing patented products for export. In particular, the scope of the license must be limited to the amount necessary to meet the needs of the importing member and must state that the entirety of the goods will be exported to the stated member. In addition, the licensed product must be identified with "specific labeling or marking" to distinguish it from others in the normal supply chain. For example, the product could be given a different brand name or packaging.

Assuming that an entity within the exporting country secures the license and it is not successfully challenged by the patent owner, there is one last step to complete before drugs are exported: before shipment of the goods, the exporting country must make its own notification to the TRIPS Council and post information to the Internet. The posting must specifically state the quantities of the licensed drug being supplied and its distinguishing features along with the name and address of the licensee, the product being licensed, the quantities of the product, the country(ies) for which the product will be supplied, and the duration of the license.

c. Summary of the Article 31bis Process

The following is a summary of the requirements as to both importing and exporting countries under article 31bis. The italicized items indicate requirements that do not exist in every case. In particular, an LDC that currently provides no patent protection for drugs would not need to comply with any of these italicized items.

1. Importing Country
   a. Eligible Importing Member
   b. Certification to TRIPS Council
      i. Product and expected quantities
      ii. *Self-identified lack of manufacturing capacity [waived if LDC]*
      iii. *Intent to issue TRIPS-consistent compulsory license (if applicable)*
   c. *Compulsory License (if applicable)*

2. Exporting Country
   a. Laws in place to permit compulsory license for *export*
   b. Willing domestic entity to make product under license for export
   c. Compulsory license of patented product—possibly with requirements beyond TRIPS imposed by national law
   d. Certification to TRIPS Council

3. Shipment of Goods Consistent with Official Certification

III. National Implementation

A. REQUIRED CHANGES

1. Importing Country

Countries that wish to import drugs pursuant to article 31bis may need to modify their compulsory license laws and possibly their regulatory laws—especially if a country has enacted TRIPS-Plus agreements. Thus far, very few countries have actually modified their domestic laws to permit compulsory license of drugs for export.[23]

Even if a country already has laws in place to permit compulsory licenses consistent with TRIPS article 31, additional modifications are likely necessary to comply with article 31bis. First, countries may need to amend their laws to ensure they permit use of a patented product *by importation* rather than by local manufacture.

In addition, importing countries need to amend their compulsory license laws to waive the usual requirement of remuneration; while most such licenses must provide remuneration, the Waiver explicitly states that this is not required since the manufacturer of the drug would otherwise receive remuneration in both the exporting and importing country. The Waiver requires remuneration to be paid only in the exporting country

A country that currently recognizes either data exclusivity or patent linkage may also need to modify those laws to ensure they do not pose a barrier to sale of the licensed drug. As noted in Chapter 9, regulatory laws may, through data exclusivity or patent linkage, impose additional limitations that impede the existence of lower-cost generic versions. Because these laws operate separately from the patent system, an exception to patent rules does not automatically result in an exception from these regulatory laws. Accordingly, to ensure that a compulsory license of a patent is effective, a corollary exception from the regulatory laws must be in place.

2. Exporting Country

All countries that wish to implement article 31bis as exporting countries need to amend their laws to permit compulsory licenses of patented drugs *for export*, as well as meet the various other procedural requirements under the article. The exporting country may similarly need to amend its regulatory laws to ensure neither data exclusivity nor patent linkage bar effective use of a compulsory license for export.

B. POSSIBLE ADDITIONAL RESTRICTIONS UNDER NATIONAL LAW

Although the Waiver is already complex, there may be added complexity due to additional requirements imposed by nations. This section provides an overview of some possible requirements based on what countries have implemented so far.

---

[23] Countries that have amended their laws include Switzerland, EU Member States, Canada, India, Korea and Norway. *See, e.g.,* WTO, Annual Review of the Decision on the Implementation of Paragraph 6 of the Doha Declaration on the TRIPS Agreement and Public Health, WTO Doc IP/C/42, para.5 (2006) (European Communities); WTO, Annual Review of the Decision on the Implementation of Paragraph 6 of the Doha Declaration on the TRIPS Agreement and Public Health, WTO Doc IP/C/37, paras. 5-7 (2005) (noting national implementation by Canada, India, and Korea); WTO, Annual Review of the Decision on the Implementation of Paragraph 6 of the Doha Declaration the TRIPS Agreement and Public Health, WTO Doc IP/C/33, para.4 (2004) (Norway); Swiss Patent Law 40(d); *see also* AIPPI, Report Q202, at 3 https://www.aippi.org/download/comitees/202/GR202switzerland.pdf.

1. Mandatory Prior Negotiation with Patent Owner

Some countries may require prior negotiation with the patent owner in all cases, including a compulsory license for export. This is not required under TRIPS. To the contrary, TRIPS explicitly permits compulsory licenses to be issued without prior negotiation in cases of national emergency, extreme urgency, or public noncommercial use. A national emergency in an importing country would seem to be a plausible basis for waiving the usual TRIPS requirement.[24] Nonetheless, TRIPS also does not mandate that countries always waive the prior negotiation requirement in these circumstances. Accordingly, domestic legislation that requires a proposed generic manufacturer to always first seek a voluntary license before being granted a compulsory license for export is consistent with TRIPS—yet imposes an unnecessary hurdle and delay to the supplying of generic drugs.[25] An additional issue may exist if domestic legislation requires the patent owner to be informed of not just the drug desired, but also the country to which the drug will be sold. As discussed in Chapter 5, companies sometimes retaliate against countries that issue compulsory licenses. No retaliation should exist against a country that is exercising its rights under the Waiver, but retaliation has nonetheless occurred with basic compulsory licenses issued consistent with TRIPS.

2. Limited Types of Drugs

Another way that countries can impose more limits on the existing Waiver provision is by significantly limiting the specific drugs that may be licensed. As noted earlier, the Waiver does not limit the scope of potential drugs; however, some countries have imposed such conditions. For example, Canada's initial legislation excluded most antiretrovirals. Although Canada's legislation has been amended, it is not the only country to use such a restrictive approach. For example, China currently limits drugs made under the Waiver to those for *infectious* diseases.[26] Although this plainly covers HIV, not every condition that plagues poor countries is infectious. Indeed, health policy advocates have noted that noninfectious, yet chronic conditions such as cancer and heart disease are major issues for such countries. However, treatment for such chronic conditions may be excluded by

---

[24] Carlos M. Correa, Trade Related Aspects of Intellectual Property Rights: A Commentary on the TRIPS Agreement 333 (2007).

[25] This is the case in Canada. The EU, on the other hand, does permit compulsory licenses for export to be issued without prior negotiation. Council Regulation 816/2006, Compulsory Licensing of Patents Related to the Manufacture of Pharmaceutical Products for Export to Countries with Public Health Problems 2006 O.J. (L 157) 1.

[26] Ellen't Hoen & Suerie Moon, China: State Intellectual Property Office Order #37, http://cptech.org/ip/wto/p6/china-order37.html (unofficial translation of the Chinese Patent Act with commentary); *see also* Tove Iren S. Gerhardsen, *TRIPS Public Health Amendment Questioned; China Implements Decision*, Intell. Prop. Watch (Dec. 14, 2005), *available at* http://www.ip-watch.org/weblog/index.php?p=178 (noting that China implemented the WTO Waiver effective January 2006).

nations that want to exclude so-called life-style medications for conditions such as hair loss or erectile dysfunction. While such conditions may seem to bear little relation to a serious medical illness usch as cancer, patent owners may consider all the conditions to be similar non-infectious diseases. Alternatively, even if they see a distinction, they may still fear that permitting exceptions for cancer would inevitably lead to exceptions for all drugs, including the life-style ones. To add to the complexity, there are some chronic, but not necessarily life-threatening conditions such as seasonal allergies, skin conditions or asthma that may also make it more challenging to draw lines regarding what drugs should be included – regardless of the fact that the Waiver does not place any limitations on conditions. The lack of conditions in the Waiver simply moves the discussion about eligible drugs (and conditions) to the domestic forum.

3. Eligible Countries—and Additional Requirements

Some countries may also limit the use of the Waiver to certain countries. For example, some exporting countries limit exports to only WTO member states.[27] Although the Waiver procedure is clearly applicable to WTO member states, there is also nothing in the Waiver itself that requires nations to so limit its application.

National laws may impose additional procedural hurdles. For example, Canada requires importing countries to provide notification of intent to use the Waiver to Canada in addition to the TRIPS Council.[28]

In addition, for countries that agreed to use the Waiver as importing countries only in the case of an emergency or extreme urgency, Canada requires such countries to provide a certified copy of the notice to the TRIPS Council that states the member is faced with a national emergency or situation of extreme urgency, such that it lacks adequate manufacturing capacity.[29]

4. Eligible Importer

Although article 31bis refers to importing countries, there is actually nothing in the provision that requires the country itself to place the order for drugs. However, some countries do require that exported drugs be requested by the government rather than by an aid organization, such as Doctors without Borders (MSF), which frequently purchases drugs to provide to countries in need.[30] This can create unneeded complexity—especially for

[27] For example, although Canada will consider exporting to countries outside the WTO, it will only do so if the country is eligible for official development assistance, declares a national urgency, or has other circumstances of extreme urgency.

[28] Canada Patent Act, art. 21.03.

[29] Canada Patent Act, art.21.04(3)(iv).

[30] Canada permits nongovernmental organizations to purchase drugs, but only if they have permission from importing countries. Canada Patent Act, art. 21.04(2)(f) (requiring the compulsory license applicant to

importing countries that may have national laws that require competitive bidding to purchase drugs.

5. Limited Scope of License

a. Duration

Some countries have imposed limitations on the time period for which a compulsory license may be issued and on the quantity of drugs to be licensed. For example, Canada currently limits the duration of a compulsory license to two years, after which a country that desires to continue to import additional quantities of the same drug must repeat the entire process.[31] However, neither the basic TRIPS provision on compulsory licenses or the Waiver impose any time limit that would require repetition of all the steps of using the Waiver.

b. Single Country

A license can also be limited with respect to being provided for a single country. Although it is true article 31bis envisages that a notification must be made for each country, that does not mean a compulsory license must also be so limited. In fact, it is likely difficult for a company to create economies of scale if it must seek a compulsory license for each request from an individual country.

6. Additional Regulatory Approval

Another potential procedural hurdle that may exist is if exported drugs must be approved by the national authority. For example, Canada will not authorize a compulsory license for a drug unless it has been deemed to satisfy the regulatory approval process for domestic use drugs. However, this has been criticized as creating an extra layer of procedure—at least with respect to drugs that are already prequalified by the WHO. Many developing countries may lack the internal infrastructure to review drugs for safety and efficacy such that they rely on the WHO prequalification as verification of safety and quality. Accordingly, so long as WHO has already prequalified a drug, there would be no need from a safety perspective for the exporting country to duplicate this examination.

7. Remuneration

Another issue is the amount of remuneration. Nothing in TRIPS provides numerical guidelines for remuneration. TRIPS article 31 speaks of "adequate" remuneration, and article 31bis does nothing to change that standard. However, lack of specificity under

---

provide the name of the governmental entity or the "entity permitted by the government of the importing country" to which the product is to be sold).

[31] Canada Patent Act, art. 21.09, 21.12

national laws could pose a problem if potential generic companies do not know what the potential revenue might be. On the other hand, if national laws set a royalty rate that is considered too low, companies would still not have an incentive to engage in a compulsory license for export.

B. CASE STUDY: CANADA/RWANDA

This section evaluates the single use that has been made of the article 31bis/Waiver process to help assess its utility. Canada issued a compulsory license to Apotex to make an HIV drug for export to Rwanda. Interestingly, the sequence of events turned out to be a bit backwards from that contemplated under article 31bis. Whereas the Waiver seems to contemplate the importing country will first signal its intent to use the process (through formal notification), that actually happened fairly late in the sequence of events and only through the actions of a persistent intermediary—the humanitarian group Doctors Without Borders, more frequently referred to as MSF, the acronym of their French name.[32]

The story begins in May 2004, a year after the Waiver is first adopted by TRIPS Council. At that time, Canada had amended its laws to permit licenses for export (although they would not become effective until May 2005), and MSF had publicly committed to use Canada for its projects.[33]

About six months after the amendments were enacted, the generic manufacturer Apotex enters the picture. Aptotex agrees to produce a fixed dose combination of three HIV medications (to be sold as Apo-Triavir) for export pursuant to the Waiver. This was not something Apotex had previously made, so several months were required to create a prototype. By the time Apotex had a prototype in place (April 2005), the amended laws were essentially in effect. However, Canadian laws impose restrictions on what patented compounds can be licensed for export; in particular, only drugs listed on a schedule are permitted (essentially limited to products on the WHO List of Essential Medicines).[34] Although each of the individual drug components of Apo-Triavir was listed, Apo-Triavir was not. Accordingly, an amendment was required, which happened within a few months.

---

[32] Although the Rwanda case is an example of a completed use of the Waiver process, the sequence of events may be typical of what is required for the Waiver to be completed. In India, a generic company filed three requests for a compulsory license to export pursuant to the Waiver, but ultimately withdrew those requests when no importing country could be found. *See* Department of Industrial Policy and Promotion, Compulsory Licensing, Discussion Paper, at 16 (Aug. 24, 201), *available at* www.dipp.nic.in/CL-DraftDiscussion.doc (citing interview with Natco Pharma representative).

[33] *E.g.*, MÉDECINS SANS FRONTIÈRES, NEITHER EXPEDITIOUS, NOR A SOLUTION: THE WTO AUGUST 30TH DECISION IS UNWORKABLE (2006), *available at* http://www.doctorswithoutborders.org/news/hiv-aids/WTO_chretien.pdf.

[34] Canada Patent Act 21.03, Schedule 1.

Although Apotex had agreed to make the drugs under license and the laws were amended to permit a license for export, *no country* had requested export of the drugs at the time. Apotex filed an application to the regulatory agency for approval of its drug (in December 2005), as required under national laws, while MSF tried to find a country willing to test-drive the new Waiver. Regulatory approval was granted in six months pursuant to a fast-track procedure. However, it took a full year from the date of regulatory approval for MSF and Apotex to find a country willing to use the Waiver.

In July 2007, Rwanda was identified as such a country, and it made a formal notification to the WTO of its intent to import Apo-Triavir.[35] Its notification stated its intent to import 260,000 packs of a fixed dose combination of Zidovudine, Lamivudine, and Nevirapine to treat HIV/AIDS over the course of two years.[36] However, its notification also explicitly stated that "because it is not possible to predict with certainty the extent of the country's public health needs, we reserve the right to modify the foregoing estimate as necessary or appropriate."[37]

As an LDC, Rwanda does not yet need to provide product patents on drugs under TRIPS. In addition, unlike some other LDCs, Rwanda had not yet entered into any bilateral or multilateral agreement requiring such protection. Accordingly, Rwanda did not need to issue a compulsory license to import the drugs from Apotex as there were no relevant patents in Rwanda at the time.

In July 2007, Apotex sought a royalty-free license from four separate patent owners for use of nine patents. Although voluntary negotiations are not strictly required under TRIPS, they are mandatory under Canada's implementation of the Waiver procedure. Apotex informed each of the patent owners of the amount of drugs it sought to make solely for export and the price per drug US $0.40 per table; it also indicated it was doing so at no profit, something that was important to these companies. Two of the patent owners were amenable to their drug being licensed for export under the Canadian export law, although they did not give voluntary authorization. Patent owner Boehringer Ingelheim even suggested a list of terms that it considered more favorable than was required by Canadian law; for example, it was willing to grant the license for the entire patent term, rather than the two-year limit under Canadian law.

Apotex filed for a compulsory license on August 28, 2007. Its application listed the desired drug to be exported and all of the patent owners and relevant patent numbers.

---

[35] Although Rwanda agreed to use the Waiver, it was not the country that MSF originally planned to approach. *E.g.*, Holger P. Hestermeyer, *Canadian-Made Drugs for Rwanda: The First Application of the WTO Waiver on Drugs and Medicines*, ASIL Insight, Dec. 10, 2007.

[36] Council for Trade-Related Aspects of Intellectual Property Rights, Notification under Paragraph 2(c) of the Decision of August 30, 2003 on the Implementation of Paragraph 6 of the Doha Declaration on the TRIPS Agreement and Public Health—Rwanda, IP/N/9/RWA/1 (July 19, 2007).

[37] *Id.*

The company also certified it had sought a voluntary license by consulting the patent owners at least thirty days prior to the application, but without success, and included all of the correspondence with the patent owners. Apotex was granted a compulsory license just two weeks after its application was filed.

However, Apotex could not immediately begin making drugs for shipment to Rwanda as the company was apparently still in the process of developing the active ingredients. An even bigger hurdle was that Apotex did not yet have a purchase request from Rwanda. Thus, Apotex's compulsory license merely permitted it to be considered in Rwanda's open bidding process in October 2007 (a public tender is apparently required under Rwandan law).[38] There was no guarantee Apotex would receive the bid as competitors from India where the drug was not patented were also being contemplated. However, Rwanda apparently was willing to consider a supplier beyond India to ensure a "quality generic manufacturer."[39]

While Apotex awaited the results of the Rwandan bidding process, Canada made its official notification to the WTO (on October 8, 2007) that it had issued the compulsory license pursuant to the Waiver. Canada's notification specified the licensed drug, the licensee (Apotex), the quantities, and the receiving country (Rwanda); it also noted the Apotex website where information on the shipment would be posted.[40]

Apotex was successful at winning the bid in May 2008; it began production and delivered the first shipment within a few months. Notably, its final bid was significantly lower than the price it initially planned on making at no cost; it is reported to ultimately have supplied the drugs for a mere $0.195 per tablet, which is about half of its original "no profit" price.[41] Some have suggested that Apotex clearly accepted the loss in money so as to be competitive with other generic manufacturers.[42]

The second and final shipment occurred one year later. Although Rwanda was interested in additional shipments, no others were possible because Canadian laws specified a maximum amount of drugs to be exported, such that the process would need to be restarted for another shipment. At that point, Apotex was not willing to undergo the process again.

---

[38] *Canadian WTO Notification Clears Path for Rwanda to Import Generic HIV/AIDS Drug*, Bridges Wkly Trade News Dig., Oct. 10, 2007, http://ictsd.org/i/news/bridgesweekly/6568/.

[39] *Id.*

[40] Council for Trade-Related Aspects of Intellectual Property Rights, Notification under Paragraph 2(c) of the Decision of August 30, 2003 on the Implementation of Paragraph 6 of the Doha Declaration on the TRIPS Agreement and Public Health—Canada, IP/N/10/CAN/1 (Oct. 5, 2007).

[41] Canadian HIV/AIDS Legal Network, Canada's Law on Compulsory Licensing for Export, Update, Dec. 3, 2009, http://www.aidslaw.ca/publications/interfaces/downloadFile.php?ref=1585.

[42] Amir Attaran, *Why Canada's Access to Medicines Regime Can Never Succeed*, 60 U. New Brunswick L. J. 150 (2010).

Although drugs were delivered to Rwanda pursuant to the Waiver, the process proved to be cumbersome. Delivery of the necessary drugs required the intervention and assistance of both MSF and Apotex. Indeed, no additional shipments were made to Rwanda under this procedure. Instead, a generic Indian company was able to provide the same drugs at the same price charged by Apotex—but without triggering the complexities of the Waiver.

Of particular note, Apotex was willing to make the drug *at no profit*—a situation that cannot necessarily be expected for all shipments of goods pursuant to article 31bis. In addition, although Apotex was willing to provide the drugs without profit, there were definite downsides to shipping drugs from Apotex's home country of Canada as Canadian laws impose additional restrictions beyond those under the Waiver rules.

IV. Evaluation

A. CURRENT CRITICISMS

A critical question is whether the Waiver is effective in providing low-cost drugs to poor countries. Although this is a key issue, there is no consensus.

Developing countries and those in support of greater access to medicine have repeatedly suggested the Waiver is ineffective. They repeatedly note that the Waiver has only been used once in over six years—and that this only occurred through strong prompting by MSF that required not only persuading a country to rely on the system to import drugs, but also helping to modify Canada's laws to permit export of the desired HIV drug. Although Canada's official position is that the use that the Waiver as well as its national implementing laws are effective, others are more critical. For example, the Canadian company that exported that drug has publicly stated it would *not* be willing to do so again because the procedure was so cumbersome; however, some suggest that the company may be unwilling because it lost money when it had to bid against Indian companies making generic versions.[43] There is no financial incentive and if anything, a financial disincentive for companies to pursue a compulsory license for export as they will be only able to make a limited quantity of drugs and thus not have the usual large-scale production advantage that generally exists with supplying generic drugs. In addition, some countries have apparently been interested in using the Waiver as importing countries, but

[43] Apotex, Press Release, *CAMR Federal Law Needs to Be Fixed if Life-Saving Drugs for Children are to be Developed*, May 14, 2009, http://www.apotex.com/global/about/press/20090514.asp (stating that "in its current form it's not workable for us," but noting an interest in developing generic HIV treatments for children if the Canadian law were simplified); *see also* Canada, Report on the Statutory Review of Sections 21.01 to 21.19 of the Patent Act (2007), http://www.camr-rcam.gc.ca/review-reviser/camr_rcam_report_rapport-eng.pdf.

dissuaded because of the complexity of the requirements.[44] In addition to the rare use of the Waiver, some are also concerned that the still-incomplete ratification of article 31bis, as well as minimal modification of national laws to permit export, indicate this "solution" is imperfect, such that the WTO should revisit the system.

Developing countries continue to press for greater discussion of article 31bis and possible consideration of alternatives, but there is some resistance from developed countries who contend there is no problem. For example, some developed countries (United States, Switzerland, and Canada) suggest countries simply do not need the Waiver because they are able to obtain necessary drugs through the normal market or humanitarian aid. In addition, such countries note that the fact the system has been used indicates it works.[45] Although developing countries recently succeeded in discussing this issue at the WTO, the discussion itself was not without opposition.[46] For example, the United States was initially opposed to any discussion, and even when it ceded to the consensus, was only willing to do so based on the representation that there would be a single meeting with no written record.[47]

Despite opposition, developing countries have continued to raise concerns with the TRIPS Council. They suggest that a system to support developing countries should be low cost and easy to use and that the current Waiver does not have either of these attributes. In fact, Venezuela reportedly suggested that it was ironic that a system designed to help developing countries is disliked by those countries and simultaneously defended fiercely by developed countries.[48] Developing countries are nonetheless hoping for further discussion

---

[44] *E.g.* Kaitlin Mara, *TRIPS Council Discusses Efficacy of ACTA, Public Health Amendment*, INTELL. PROP. WATCH (Oct. 29, 2010) (noting that Ghana wanted to import drugs under the Waiver); WTO, 2010 News Items, Little-Used Par.6 system will have its day, WHO tells Intellectual Property and Health Review, Oct. 27-27, 2010, http://www.wto.org/english/news_e/news10_e/trip_26oct10_e.htm (noting that a least developed country wanted to import medicines from India, but was dissuaded by the complexity of the procedures).

[45] The United States has alternatively suggested that the number of uses is the wrong metric to consider in effectiveness and that the relevant issue is the extent to which meicines reach affected populations – although the United States has not provided any data on the purported relevant issue. Statement of the United States to WTO TRIPS Council on the Paragraph 6 System, Agenda Item F, Annual Review of the Paragraph 6 System, Oct. 27, 2010, *available at* http://keionline.org/node/1007.

[46] *E.g.*, Kaitlin Mara, *Efficacy of TRIPS Public Health Amendment in Question at WTO*, INTELL. PROP. WATCH (Mar. 1, 2010), http://www.ip-watch.org/weblog/2010/03/01/efficacy-of-trips-public-health-amendment-in-question-at-wto/; Kaitlin Mara, *Medicines Access Again Captures Attention at WTO as Progress Urged in Round*, INTELL. PROP. WATCH, (Oct. 30, 2009), http://www.ip-watch.org/weblog/2009/10/30/medicines-access-again-captures-attention-at-wto-as-progress-in-round-urged/; Kanaga Raja, *Members Discuss Implementation of TRIPS "Para 6" Solution*, Third World Network, Feb. 16, 2010, http://www.twnside.org.sg/title2/wto.info/2010/twninfo100212.htm.

[47] Kaitlin Mara, *WTO Members to Consider Review of TRIPS Public Health Amendment*, INTELL. PROP. WATCH, (Feb. 11, 2010), http://www.ip-watch.org/weblog/2010/02/11/wto-members-to-consider-review-of-trips-public-health-amendment/.

[48] Kaitlin Mara, "*TRIPS Council Discusses Efficacy of ACTA, supra* note 44.

and a possible workshop open to a number of stakeholders, as well as non-govenrmental organizations within the WTO context. Whether such a workshop will transpire is unclear since developed countries, such as the United States and EU are opposed.

A complete assessment of the utility of the Waiver provision may be premature while generic versions of many drugs are currently still available from India. For example, at the time Apotex was willing to make drugs under license for export to Rwanda, the same drugs could have been bought on the open market from an Indian company for the same price—but without all the logistical hassles. Indeed, MSF, who had helped to get both Apotex and Rwanda involved in utilizing the Waiver, has stated that it is not inclined to repeat the process. Although India was required to begin providing patent protection for drug compounds in 2005, whether some of the newer drugs will remain unpatented—-and thus still available cheaply to poor countries—is an open question (as discussed in Chapter 4).

B.NEXT STEPS?

Although some might be inclined to suggest that article 31bis be scrapped in lieu of a more streamlined and efficient process, that seems politically unlikely. More simplistic options were previously rejected as not properly protecting the interests and rights of the pharmaceutical industry against the threat of parallel imports of goods intended for low-income countries. One popular prior proposal by developing countries was a clarification that a compulsory license for export to a needy country would necessarily be a "limited exception" under article 30.[49] However, that was not satisfactory to all members. An important issue for countries with a strong pharmaceutical industry remains the fear that low-cost drugs might be diverted to wealthy markets. This is seen as a problem because these manufacturers generally make their profit by selling the same drug at the highest possible price within each market. If low-cost drugs enter into wealthy markets, that can undercut the manufacturer price and possibly place pressure on the manufacturer to lower it.

One important issue could potentially be the subject of WTO focus and have a major impact. In particular, the current article 31bis scheme does not provide for economies of scale, such that companies lack any commercial incentive to make drugs under a compulsory license for only a minimal profit. Generic companies generally make money by pricing low, but selling large quantities. Article 31bis requires low prices, but does not permit selling of large quantities as licenses must be provided on a country-by-country basis. Although poor countries in a regional trade agreement are permitted to export their

[49] *E.g.*, Developing Country Draft Ministerial Declaration, IP/C/W/312, ¶ 9; Communication from the Permanent Mission of Brazil et al., *Paragraph 6 of the Doha Declaration of the TRIPS Agreement and Public Health*, IP/C/W/355 (June 24, 2002) (on behalf of Bolivia, Brazil, Cuba, China, Dominican Republic, Ecuador, India, Indonesia, Pakistan, Peru, Sri Lanka, Thailand, and Venezuela); *see also* Haochen Sun, *A Wider Access to Patented Drugs Under the TRIPS Agreement*, 21 B.U. INT'L L.J. 101, 127–36 (2003) (analyzing article 30 option).

drugs to one another, that still does not help provide a commercial incentive to the generic companies who are needed to create low-cost drugs for export.

The utility of article 31bis may also be improved if more nations enact simplified procedures implementing its laws as to exporting countries pursuant to article 31bis. As noted earlier, a major issue is the extent to which national laws currently impose *additional* restrictions that limit the ease of use. Canada has considered proposals to amend its law to promote greater use of the Waiver.[50] In particular, the proposal includes a streamlined "single license" solution that would have no limit on its duration, the maximal quantity of medicines, or the countries to which the drug would be exported. In addition, the proposal suggests a compulsory license could be sought before there is a specific request from an importing country. This initial request could create more certainty for generic manufacturers and enable them to benefit from economies in scale by potentially providing for multiple countries. This process would also be more consistent with the realities of purchasing drugs for developing countries—as noted in the Apotex situation, countries often bid competitively for drugs, but a company needs a compulsory license in hand to know what price it can bid. In addition, by starting the process with the compulsory license in an exporting country, it may be easier to identify multiple countries interested in an ongoing use of a compulsory license system. However, although these suggestions do indeed address many of the prior hurdles in the single use of the Waiver to date, they have faced resistence in Canada and have not even been contemplated in any other countries.[51]

[50] Bill S-232, Fortieth Parliament, 2d Sess. (2009); Bill C-393, Fortieth Parliament, 3d. Sess. (2009); *see also* Richard Elliott, *Making CAMR Work: Streamlining Canada's Access to Medicines Regime,* Brief to the Senate Banking, Trade and Commerce Committee regarding Bill S-232 (Oct. 21, 2009), http://www.aidslaw.ca/publications/interfaces/downloadFile.php?ref=1563.

[51] *E.g.*, Gloria Galloway, *Access to Medicines Bill Gets a Grim Prognosis*, GLOBE & MAIL, Oct. 26, 2010, http://www.theglobeandmail.com/news/politics/access-to-medicine-bill-gets-grim-prognosis/article1773866/.

III The Evolving Framework

8 An Overview of "TRIPS-Plus" Standards

This chapter provides an overview of international developments since TRIPS that further impact access to medicine. In particular, some countries are now required to adopt higher standards than those required by TRIPS. These "TRIPS-Plus" standards are predominantly implemented as a result of bilateral or regional trade agreements instigated by wealthy countries since the conclusion of TRIPS. Although they are dubbed "TRIPS-Plus," they have no formal relation to TRIPS. Rather, the term is used to indicate that these requirements go beyond the minimum obligations imposed under TRIPS.

This chapter explains common themes among TRIPS-Plus requirements that impact the cost of drugs. This thematic approach aims to highlight how drug costs may be impacted, as well as how and why countries adopt such requirements; however, it is not intended to provide a comprehensive review of every enacted or pending TRIPS-Plus requirement.

By the conclusion of this chapter, a reader should be able to answer the following questions:

- If TRIPS already requires nations to provide patents, what does an agreement imposing TRIPS-Plus standards require?

- Can a bilateral free trade agreement imposing TRIPS-Plus standards impact countries that are not signatories to the agreement?
- How do TRIPS-Plus requirements impact current access to medicine?

I. Background 224
A. Significance of TRIPS-Plus 225
B. Adoption of TRIPS-Plus Standards 226
II. TRIPS-Plus Requirements 228
A. Enhancing Patent Protection 228
1. PCT 228
2. Pipeline Protection 229
3. The Big Picture of PCT & Pipeline Protection 231
B. Specific Patentability Requirements 232
1. Patentability Requirements 232
2. Limited Reinforcement of Patentability Requirements 234
C. Patent Rights and Remedies 235
1. Patent Term Extensions 235
2. Compulsory License Limits 237
3. Limited Revocation 239
4. Limited Exceptions to Patent Rights 239
5. Limitations on Parallel Imports 240
D. Regulatory Protection of Drugs 241
1. Types of Regulatory Protection 242
2. Impact on Access to Medicine 243
E. Extending Protection of Drugs 246
1. Minimizing Price Controls 246
2. Encouraging Use of Patented Drugs 248
III. Controversy Concerning TRIPS-Plus 249
IV. Conclusion 251

I. Background

This chapter first explains how TRIPS-Plus requirements generally impact access to medicine. Then, the chapter discusses requirements regarding patent and regulatory laws that impact the availability of lower-priced generic drugs, followed by requirements that directly limit the ability of countries to control the costs of patented drugs. Each section begins with a brief recap of the relevant TRIPS requirement (if any), followed by the corollary TRIPS-Plus requirement.

A. SIGNIFICANCE OF TRIPS-PLUS

The phrase "TRIPS-Plus" refers to laws that restrict flexibility under TRIPS or provide more protection than the TRIPS requires.[1] TRIPS-Plus requirements are best understood in comparison to the original TRIPS requirements. For example, whereas TRIPS is silent on whether third parties may oppose patent applications (thus permitting member states to do so), a TRIPS-Plus requirement would specifically preclude such oppositions. Alternatively, a definition of "new" that includes new uses of known compositions is also a TRIPS-Plus requirement because TRIPS lets member states decide what they want to consider "new."

Both of these TRIPS-Plus requirements could potentially impact access to medicine. For example, if third parties are prohibited from opposing patent applications, erroneous patents are more likely to issue. Although improper drug patents can be contested later, evidentiary thresholds may be higher. In addition, the public could be improperly denied access to medicine while an erroneous patent is challenged. Similarly, if "new" were defined to include new *uses* of old medicines, this could expand the scope of inventions that will be priced out of reach for consumers in the short term.

Even though the number of countries that have adopted TRIPS-Plus standards is modest compared to WTO membership, each country that does so affects other nations. For example, Jordan's TRIPS-Plus standards impact everyone seeking a patent in Jordan. TRIPS requires that members provide equal treatment to applicants from all member states such that national patent laws will provide the same treatment to all applicants. Accordingly, Jordan cannot provide better patent treatment to US citizens than citizens of a different WTO country. This means that although Jordan's TRIPS-Plus patent requirements were adopted to comply with a free trade agreement (FTA) with the United States, those patent rules would affect nationals of other countries as well. For example, a Japanese citizen who applied for a Jordanian patent would benefit from the increased patent protection negotiated by the United States, even though Japan was not a party to the U.S.-Jordan FTA.

[1] Sometimes the term TRIPS-Plus is used to refer to standards adopted by a country in advance of when it needs to comply with TRIPS. For example, least developed countries that are members of the WTO do not have to provide patent protection for drugs until at least 2015 (as discussed in Chapter 3). However, LDCs that already provide such patent protection are considered to have adopted TRIPS-Plus rules.

B. ADOPTION OF TRIPS-PLUS STANDARDS

Although any nation could theoretically decide to adopt TRIPS-plus standards because it believes increased protection is to its advantage, these standards are more typically adopted as a result of external pressure. In particular, developing countries often accept TRIPS-plus standards as a result of political or economic pressure from other nations. This pressure may exist in the form of threatened trade sanctions or withdrawal of trade preferences. The US, for example, has a number of domestic laws that permit it to potentially impose economic sanctions for intellectual property laws that are deemed inadequate—even if in full compliance of current international agreements, such as TRIPS.[2] The US can withdraw trade benefits or impose duties on goods from a country that fails to provide "adequate and effective" protection for intellectual property rights—as determined unilaterally by the US. Under US law, the US Office of Trade Representatives is required to provide an annual report listing countries with inadequate intellectual property protection, which is commonly referred to as the "Special 301" Report.[3] A country that is listed in this report is often then subject to increased pressure to modify its laws. Even for countries not listed, there may be pressure to enter into a specific bilateral or multilateral FTA[4] that promotes increased trade while also mandating increased protection of patent and related rights.[5] The United States has initiated and concluded many FTAs, and the EU has recently been negotiating them.

Any country that wants to improve its export market may be vulnerable to pressure to modify its patent rules. Developing countries are often eager to increase the marketability of their exports and thus especially vulnerable to pressure from a country with a desired export market, such as the United States or the EU. However, FTAs are not concluded exclusively with developing countries. For example, Australia entered into an FTA with the United States, and Canada is contemplating an FTA with the EU.[6]

---

[2] *E.g.*, 19 U.S.C. § 2242(d)(4)(2000) (noting that a country can be deemed to have inadequate protection of intellectual property rights "notwithstanding the fact that the foreign country may be in compliance" with TRIPS).

[3] 19 U.S.C. § 2242(a)(2000).

[4] Although TRIPS-Plus standards often exist in free trade agreements, they are not the only type of agreement that may incorporate such standards. An alternate vehicle for TRIPS-Plus standards are bilateral investment agreements.

[5] TRIPS-Plus rules may also be created through multilateral agreements focusing solely on intellectual property rights. Thus far, however, no such agreements have been concluded involving patents. Negotiations on a Substantive Patent Law Treaty (SPLT) under the auspices of the World Intellectual Property Organization (WIPO) that would establish uniform patent standards for both developed and developing countries (that are members of WIPO) have been stalled for several years. The proposed Anti-Counterfeiting Trade Agreement (ACTA) addresses enforcement of patent and other intellectual property.

[6] U.S.-Austl. Free Trade Agreement, art. 17.9(1)–(2), U.S.-Austl., May 18, 2004, T.I.A.S. No. 6422, [hereinafter U.S.-Austl. FTA], *available at* http://www.ustr.gov/trade-agreements/free-trade-agreements/australian-fta/final-text; EUROPEAN COMMISSION, INTERNATIONAL AFFAIRS: FREE TRADE AGREEMENTS, http://ec.

The negotiation and implementation of TRIPS-Plus standards bears a similarity to the genesis of TRIPS. In both cases, patent owners such as multinational pharmaceutical companies play an instrumental role in seeking stronger global patent rights. In addition, both involve a combination of "forum shifting" and the promise of increased trade in goods.[7] As noted in Chapter 3, the United States "shifted" the forum for seeking higher intellectual property standards to a trade framework after its initial efforts to amend an existing intellectual property agreement (the Paris Convention) failed. Developing countries had no incentive to agree to higher intellectual property standards that they viewed as detrimental to their interests. However, requiring intellectual property standards as part of a broader agreement that promised greater access to wealthy markets has been a successful strategy both with the original TRIPS agreement and subsequent FTAs with TRIPS-Plus standards.

In addition to shifting forums to their advantage and using the inducement of increased trade, wealthy countries have encouraged countries to agree using unilateral political and economic pressure. For example, concurrent with the negotiation of TRIPS, the United States unilaterally pressured some countries to agree to higher patent standards as a step toward minimizing objections to these standards with the broader group of countries involved in the WTO.[8] In addition, some developing countries agreed to TRIPS in part because of a belief that its conclusion would end unilateral trade pressure to increase heightened protection intellectual property rights. However, pressure to adopt heightened standards seems to have actually *increased* rather than decreased.

The process of establishing TRIPS-Plus standards differs in one major respect from the negotiation of the TRIPS. Notably, the WTO/TRIPS was negotiated by many developing countries. In that context, a few developing countries could often represent the interests of the broader group; for example, India played a major role. However, FTAs are negotiated with a smaller subset of countries. Sometimes when negotiations break down among a group of countries, the powerful country that initiated the FTA may elect to negotiate with individual countries. By pursuing countries individually, the country with lucrative markets has greater bargaining power and is more likely to obtain its desired results. Although there are always disparities in bargaining power, the power disparity

---

europa.eu/enterprise/policies/international/facilitating-trade/free-trade/index_en.htm (Apr. 31, 2010) (listing FTAs under negotiation).

[7] Forum shifting is a recognized strategy used by countries and organizations to seek the optimal reception for their goals. *See, e.g.*, John Braithwaite & Peter Drahos, Global Business Regulation ch. 24, (2000); Laurence Helfer, *Regime Shifting: The TRIPs Agreement and New Dynamics of International Intellectual Property Lawmaking*, 29 Yale J. Int'l. L. 1 (2004)

[8] *See, e.g.*, Peter Drahos with John Braithwaite, Information Feudalism: Who Owns the Knowledge Economy? 99–107 (2002). Although some countries immediately changed their laws in response to trade pressure, others faced sanctions. For example, tariff penalties were imposed on Brazilian imports in 1987 when the United States comprised twenty-five percent of Brazil's export market. By 1990, Brazil's laws were in compliance with the United States desires.

with FTAs is often much more distorted than in the WTO negotiation context because of the limited number of countries involved. The problem is particularly severe given that a number of scholars think that developing countries as a whole were at an unfair disadvantage in negotiating the WTO and TRIPS. In addition, because FTAs are often negotiated in secret, with the text usually made public only after agreements have been signed, there may be limited opportunities to consider implications.[9]

II. TRIPS-Plus Requirements

This section provides an overview of TRIPS-plus requirements. It begins with overall procedures for obtaining patents that enhance patent protection. The section then turns to requirements that have parallels in the TRIPS agreement and addresses them in roughly the same order as Chapter 3 (Introduction to TRIPS). For example, what must be patented will be discussed before patent rights and remedies. Then, non-patent issues that complement patent protection for drug companies (data exclusivity and price control limits) will be discussed.

A. ENHANCING PATENT PROTECTION

This section addresses two separate issues that expand patent protection for drug companies by requiring procedures that promote patent protection. First, many FTAs require developing countries to become signatories of the Patent Cooperation Treaty (PCT), which has important implications for the number of patent applications that will likely be filed. Second, some developing countries were pressured to adopt so-called "pipeline protection" that provides patents in cases not required by TRIPS. Each of these will be discussed in more detail below.

1. PCT

Many FTA require developing countries to recognize the PCT. This is an international agreement to which a large number of countries are signatories. The PCT provides patent

[9] For example, although the EU and India are reportedly expected to conclude a free trade agreement by Spring 2011, as of late 2010 the negotiations were being criticized for lack of transparency. *See, e.g.*, Monika Ermert, *Lack of Transparency in EU-India FTA Talks Spurs Requests for Halt*, IP WATCH MICROBLOG (Sept. 3, 2006, 6:59 PM) http://www.ip-watch.org/weblog/2010/09/03/lack-of-transparency-in-eu-india-fta-talks-spurs-requests-for-halt/?utm_source=daily&utm_medium=email&utm_campaign=alerts; *Does EU-India FTA Serve Mutual Interests? Policymakers, NGOs Disagree*, INTELL. PROP. WATCH (Nov. 12, 2010), available at http://www.ip-watch.org/weblog/2010/12/10/policymakers-ngos-disagree-does-eu-india-fta-serve-mutual-interests/. Although negotiations were launched in 2007, the first leaked text in 2009 noted that the EU would provide a "revised text" for the controversial provision on data exclusivity. Draft EU-India Free Trade Agreement art. 10 (Feb. 24, 2009), http://file.wikileaks.org/file/eu-india-fta-feb-2009.pdf.

owners, such as large pharmaceutical companies, with an easy and cost-effective mechanism to globally file patent applications. While individual nations still examine whether an application meets national criteria of patentability, a PCT application streamlines the process with an initial single application; national examination occurs later.[10] While this may seem a small procedural detail, it may have significant implications. Countries that are not members of the PCT are likely to have few patents filed, even if drugs are permitted to be patented. This was indeed the case in Jordan which signed an early FTA with the US that did not require adherence to the PCT; as discussed later, most companies did not immediately seek patent protection on drugs in Jordan because it was considered expensive.[11]

2. PIPELINE PROTECTION

Some developing countries that complied with the TRIPS requirement to provide full patent protection of drugs also adopted what is referred to as "pipeline protection" as a result of political pressure from developed countries.[12] Pipeline protection essentially provides patents to inventions (in the pharmaceutical area) that were considered "in the pipeline" of development when new patent laws providing for patentability of drug compositions were adopted. Importantly, pipeline protection grants patents for compounds that would otherwise not have been patentable–either before or after the new patent laws. Before the new laws, the drugs would not be patentable because the laws did not provide protection for pharmaceutical products. However, even after national laws recognized pharmaceuticals as patentable, these drugs would not have been deemed patentable because they would have failed the novelty requirement; this is because the drugs would have been known in other countries. For example, if Brazil began granting patents on pharmaceuticals in 1997, a drug that was invented in 1993 would fail to meet the novelty requirement in 1997. However, because Brazil provides pipeline protection, the 1993

---

[10] If an application is filed pursuant to the PCT, it provides the applicant with several benefits. First, the applicant can initiate a request for a patent in all countries that are members of the PCT; however, the high costs of many parallel national applications are deferred for a period of months and sometimes years. The lag time also enables an applicant to delay a decision concerning which countries are desirable for patent protection. Second, the applicant is entitled to a preliminary examination of its patent application through the PCT, which, if negative, may enable the applicant to elect not to pursue some or all national applications.

[11] Oxfam International, *All Costs, No Benefits: How TRIPS Plus IP Rules in the US Jordan FTA Affect Access to Medicine* 7-8 (Oxfam Briefing Paper No.102, 2007), *available at* http://www.oxfam.org.uk/resources/issues/health/downloads/bp102_trips.pdf. In addition, as noted by Professor Drahos, Vietnam has seen a rapid increase in the number of patent applications since it became of a member of the PCT in 1993 whereas Thailand received less than fifty applications in 2007. PETER DRAHOS, THE GLOBAL GOVERNANCE OF KNOWLEDGE; PATENT OFFICES AND THEIR CLIENTS 269-79 (2010).

[12] *E.g.*, CHRISTOPHER SCOTT HARRISON, THE POLITICS OF THE INTERNATIONAL PRICING OF PRESCRIPTION DRUGS 103-10 (2004) (describing US pressure on Argentina and Brazil to adopt pipeline protection).

drug could be considered in the "pipeline" of development and thus be eligible for pipeline patent protection.

As discussed in Chapter 3, developing countries that did not previously provide patent protection on drugs were required under TRIPS to provide patent-like protection if they took advantage of a five year transition period. In particular, during the transitional time, the developing country was required to accept "mailbox applications" to be later examined when patent law recognized drugs as patentable, and also to provide exclusive marketing rights (EMR) to some of these mailbox applications. However, no such requirement exists for countries that elect (albeit often under political pressure) to prematurely grant product patents. Moreover, pipeline patent protection actually provides even more protection than an EMR in that it provides full patent rights, rather than the more limited rights of an EMR (as discussed more fully in Chapter Three).

Although a number of countries have provided pipeline patent protection, the exact parameters of such protection vary.[13] In Mexico, pipeline patents were granted for drugs that were already patented elsewhere so long as the drug had not yet been marketed in Mexico.[14] In contrast, pipeline patents are available on a more limited basis in Brazil. In particular, pipeline patents are only available for drugs that are subject to a foreign patent application, but not yet granted a patent at the time a pipeline patent is requested; in addition, the protection is only available if the drug is not yet marketed anywhere.[15] Although Brazil's pipeline protection was narrower, it was still controversial–it was subject to a constitutional challenge for removing information that currently exists in the public domain.[16] While that challenge did not succeed, more limited challenges to the term of specific pipeline patents have been effective. In 2010 alone, three blockbuster drugs were found to have shorter terms than desired by the patent owner. Pfizer's patent on Viagra was held to terminate in June 2010 (rather than in 2011), while its patent for the cholesterol drug sold as Lipitor was held to terminate in August 2010, rather than December 2010. Most recently, a court found that Novartis's patent on the cancer drug it sells as Gleevec will terminate in 2012 (rather than 2013).[17] In each of these cases, a court

---

[13] Countries that provided pipeline protection include Brazil, Thailand, Argentina and Uruguay.

[14] Technically, the Mexican law is not a traditional TRIPS-Plus protection since the pipeline protection was required pursuant to the North American Free Trade Act. NAFTA 1709(4). However, NAFTA was negotiated concurrently with TRIPS and the pipeline protection is similar.

[15] Lei No. 9.279, de 15 Maio de1997, Regula Direitos e Obrigacoes Relativos a Propiedade Industrial, t. 230, Maio 1997 (Braz.), *available at* http://www.araripe.com.br/law9279eng.htm#titulo8.

[16] *E.g.*, Claudia Jurberg, *Challenge Raised to Constitutionality of Brazilian Pipeline Patents*, INTELL. PROP. WATCH (Jan. 22, 2008), http://www.ip-watch.org/weblog/2008/01/22/challenge-raised-to-constitutionality-of-brazilian-pipeline-patents/. This challenge existed despite the fact that the Brazilian law bars pipeline protection if a third party has made "serious and effective preparations" to make the drug.

[17] *See, e.g.*, Claudia Jurberg, *Brazilian Generic Drug Registration Sets Standard for 'Pipeline' Patents*, INTELL. PROP. WATCH (May 13, 2010), http://www.ip-watch.org/weblog/2010/05/13/brazilian-generic-drug-

found that the twenty-year patent term should not begin with the date the pipeline patent was granted, but, rather, the earlier international filing date.[18]

3. THE BIG PICTURE OF PCT & PIPELINE PROTECTION

As discussed above, procedures for obtaining patents promote patent protection. After all, even if domestic laws offer the type of patent protection desired, that protection is elusive if there are too many logistical hurdles to obtaining patent protection. The ability to use a PCT application removes such hurdles and thus makes even the basic patent protection under TRIPS more useful. Pipeline protection also helps enhance protection, but in a different way. In particular, pipeline protection enables companies to obtain patent protection on some drugs that would have otherwise been barred from consideration. Pipeline protection provides patent protection for a time when desired patent laws have not yet taken effect. Without pipeline protection, companies would be denied patents in countries that had only newly adopted patent laws. Before moving on to address specific requirements of patentability, the following questions should help to reinforce the issues discussed above.

Q: Does a PCT application mean that a single application will result in a global patent?

A: No. A PCT application is essentially a common application that will—if successful—result in a number of different patents granted under different national laws.

Q: Does a PCT application mean that there are uniform global patentability requirements applied to an invention?

A: No. The PCT only governs the process of obtaining a patent. The laws of each country granting a patent apply to all patents – whether filed pursuant to the PCT, or as a single application.

Q: Does a country provide pipeline patent protection for all patents?

A: No. Pipeline patent protection is usually only provided for inventions within a small window of time. Usually, the timeline for such protection terminates when new drug inventions are eligible for patentability (under newly effective patent laws).

Q: Does pipeline patent protection provide duplicative protection?

---

registration-sets-standard-for-pipeline-patents/; Ed Taylor, *Novartis Loses Court Battle over Lifespan of Brazilian Patent on Gleevec*, BNA PAT. TRADEMARK & COPYRIGHT L. DAILY, Oct. 29, 2010.

[18] In some cases, there was a dispute about what constitutes the earlier international filing date. For example, in one case, the first international application was abandoned and the patent owner wanted to use the later filing date as the relevant date for calculating the term from which the twenty years should run.

A: No. Inventions that qualify for pipeline patent protection would not otherwise be eligible for patent protection.

Q: Does a country that grants a pipeline patent evaluate whether the invention satisfies patentability standards?

A: No. Pipeline patent protection is granted by a country that does not examine the invention; rather, protection is based on the fact that the invention is patented or subject to a patent application in another country.

B. SPECIFIC PATENTABILITY REQUIREMENTS

TRIPS-plus requirements set specific patentability requirements that enable patents to be granted to a broader scope of inventions than required under TRIPS. As discussed in Chapter 3, nations must generally provide patents on all inventions (including drugs) that meet TRIPS patentability requirements, but TRIPS does not define the key terms of patentability. In particular, while TRIPS requires that patents be generally granted on inventions that are "useful," "new," and have an "inventive step," TRIPS does not define these terms.[19] Thus, under TRIPS, nations have discretion to define these terms and thus decide what to patent.

TRIPS-Plus patentability requirements tend to impact access to medicine by expanding what is patentable in two ways.[20] First, many FTAs often limit a country's discretion to decide what is "new" and thus patentable. In addition, many FTAs limit discretion regarding procedures to reinforce patentability requirements.

1. Patentability Requirements

a. Limited Prior Art; Modified Definition of What is "New"

Under TRIPS, member states can freely define what they consider "new." Although it may seem this term can have only one meaning, national patent laws have historically provided different standards.[21] In some countries, an invention is only new if it has never been disclosed to the public in any fashion before the patent application is filed.[22] However, other countries would consider an invention new and deserving of a patent if

[19] The application must also adequately explain the invention such that others with a similar scientific background could make and use it. TRIPS art. 29(1).

[20] There are some requirements that do not directly impact access to medicine, such as those that plants and animals be patentable rather than excluded from patentability. TRIPS art. 27(3).

[21] Whether or not the invention in a patent application is new is generally compared with *prior art*—a term that usually references publicly available information before the patent application. However, there are presently differing standards on when an invention is new in different countries.

[22] This is considered an "absolute novelty" regime. This system encourages inventors to promptly file their inventions.

the inventor had disclosed the invention a limited time before the application.[23] For example, a scientist who published an article describing an invention would be barred from a patent in a country that considered an invention new only if it was not previously disclosed, but the same scientist could obtain a patent in a country that considered inventions publicly disclosed to still be new. Thus, the same invention could be patented in one country but not another. Interestingly, both of these standards are consistent with TRIPS but have dramatically different results.

Some FTAs impose a specific definition of when an invention is new. In particular, a number suggest that an invention must be considered "new" if the invention was publicly disclosed by the inventor no longer than twelve months prior to the date of the patent application.[24] This definition will permit more inventions to be considered patentable. Although this may be helpful to scientists who need not worry about whether the timing of a scientific publication will preclude a patent, it also has implications for access to medicine. In particular, a rule that is more permissive toward patents means that more patents are likely to result—which in turn can increase the number of patented drugs to which patients have only limited access.

b. Required Protection of New Uses

A related TRIPS-Plus requirement often seen in an FTA is that new *uses* of previously known compounds should be considered new and thus patentable.

To understand why a company would want to patent a new use, consider this example: the active ingredient in AZT, zidovudine, was initially studied as a potential cancer treatment. It was later discovered as useful for treating HIV, and a US patent was granted because the United States considers new uses of known compounds to be patentable.[25] However, if the United States did not consider such new uses to be patentable, AZT would have originally been available at a greatly reduced cost, as the original patent on the active ingredient had expired.

New uses are often found for old compounds, including those that have previously been sold as blockbuster drugs. For example, the active ingredient in the antidepressant

[23] This is considered a "relative novelty" regime that is more accommodating of the realities of traditional practice in which scientists primarily share knowledge through publications rather than patents. Under this system, the scientists would not lose all patent rights because they preferred to first publish rather than patent the invention.

[24] U.S.–Pan. Trade Promotion Agreement, U.S.-Pan, art. 15.9(7), June 28, 2007 [hereinafter U.S.-Panama TPA], *available at* http://www.ustr.gov/trade-agreements/free-trade-agreements/panama-tpa/final-text; U.S.-Korea Free Trade Agreement, U.S.-Korea, art. 18.8(7), June 30, 2007 [hereinafter U.S.-Korea FTA], *available at* http://www.ustr.gov/trade-agreements/free-trade-agreements/korus-fta/final-text; United States-Morocco Free Trade Agreement, U.S.-Morocco, art 15.9(8), June 15, 2004 [hereinafter US-Morocco FTA], *available at* http://www.ustr.gov/trade-agreements/free-trade-agreements/morocco-fta/final-text.

[25] U.S. Patent No. 4,724,232 (Feb. 9, 1988). In addition, the use of AZT to increase the T-lymphocyte count of HIV patients was subject to a separate patent. U.S. Patent No. 4,818,750 (April 4, 1989).

sold as Prozac was later found to be useful in treating severe premenstrual symptoms. Both the active ingredient in Prozac, as well as the later-discovered use, were patented and provided market exclusivity to its patent owner.

New uses that are patented can sometimes prevent generic companies from making old drugs that have newly discovered uses. This is particularly true in nations that link regulatory approval of generics, as further explained later in this chapter.

2. Limited Reinforcement of Patentability Requirements

Before discussing how TRIPS-Plus agreements limit the ability to reinforce patentability requirements, it is necessary to review how patents are issued and sometimes revoked.

As noted earlier, a patent will only issue after a patent application is filed with a national patent office that determines a patent is appropriate. However, it is widely known that issued patents do not always meet the standards of patentability—many patents are later found invalid during litigation when the patent owner attempts to enforce the patent against others.

Some countries provide opposition proceedings to enable third parties to assist in ensuring that patentability requirements are met. Third parties are often competitors and peers in the same field of technology who are likely to have superior knowledge of the relevant prior art, including prior art that may not be easily accessible to patent examiners. Patent examiners have ready access to patents and published patent applications. However, they are less likely to have access and knowledge concerning published articles, as well as other relevant prior art (such as commercial sales).

Countries have provided two types of opposition proceedings: oppositions to a patent application, and oppositions to a granted patent (these are usually referred to as "pre-grant" or "post-grant" oppositions). Both types serve the same goal of promoting patent validity and the public interest. To the extent they even provide any opportunity to challenge a patent, developed countries tend to allow only post-grant oppositions. The existence of a pre-grant opposition can be important because issued patents are often given a presumption of validity, making them more difficult to challenge.

a. Pre-Grant Oppositions

TRIPS does not prevent countries from permitting challenges to patent applications and patents. Rather, TRIPS requires that *if* such challenges exist, they must comply with the same general requirements for enforcement of all patent rights. In particular, a challenge must be fair, not unduly complicated or costly, and not involve unreasonable time limits or unwarranted delays.[26]

[26] TRIPS art. 62(4); TRIPS art. 41.

Many FTAs preclude countries from using a pre-grant opposition to enhance the patent office's ability to issue valid patents. In particular, FTAs commonly state that if third parties are allowed to participate, it can only be *after* a patent issues.[27]

Although eliminating pre-grant oppositions does not change the standards of patentability, it still plays an important role in discussions concerning access to medicine. In particular, if issued patents have a stronger presumption of validity than pending patent applications, it may be more difficult to challenge a patent. This does not mean the invention is any more deserving of a patent—just that the procedural presumptions may favor the status quo. However, procedures that do so may result in improper protection of drugs that should be in fact more easily accessed without a patent premium.

C. PATENT RIGHTS AND REMEDIES

TRIPS requires that patents provide certain rights as well as remedies. With respect to patent rights, member states must generally give patent owners the right to exclude all others during the patent term, subject to exceptions to patent rights under TRIPS articles 30 and 31.

Most FTAs increase patent rights while limiting patent exceptions previously available under TRIPS. In particular, most FTAs mandate increased patent terms while restricting the use of compulsory licenses and parallel imports. All of these changes can have a marked impact on the availability of lower-cost generic drugs.

1. Patent Term Extensions

A key issue for access to cheaper medicines is the term of patents. After all, while the patent is in effect, generic drugs cannot be made and sold, as that would violate the patent right.

TRIPS requires that the term of protection cannot end before twenty years from the filing date of the patent application. Most countries provide a patent term based on the filing date of the patent application (twenty years minus the time of patent examination), such that the longer the patent examination process takes, the shorter term during which

[27] *E.g.*, U.S.-Morocco FTA, art. 15.9.(5) ("Where a Party provides proceedings that permit a third party to oppose the grant of a patent, a Party shall not make such proceedings available before the grant of the patent"); *see also* U.S.-Korea FTA, art. 18.8(4); U.S.–Singapore Free Trade Agreement, U.S.-Sing., art. 16.7.4, May 6, 2003 [hereinafter U.S.-Singapore FTA], *available at* http://www.ustr.gov/trade-agreements/free-trade-agreements/singapore-fta/final-text; U.S.-Bahrain Free Trade Agreement, U.S.-Bahr., art. 14.8.4, Jan. 11, 2006 [hereinafter U.S.-Bahrain FTA], *available at* http://www.ustr.gov/trade-agreements/free-trade-agreements/bahrain-fta/final-text; U.S.-Oman Free Trade Agreement, U.S.-Oman, art. 15.8.4, Jan. 19, 2006 [hereinafter U.S.-Oman FTA], *available at* http://www.ustr.gov/trade-agreements/free-trade-agreements/oman-fta/final-text.

the patent owner can actually enforce the patent. In addition, the *effective* patent term for a drug patent can be much shorter because the drug patent is of no commercial utility unless and until the drug is approved for sale—something that typically happens several years after patent issuance. This results in drug patents having an average effective term of eleven years in contrast to seventeen for all other patents.[28] A WTO panel has previously noted that owners of drug patents are not entitled to the same patent term as owners of all other patents. However, this panel conclusion is moot for countries that are required to adopt patent extensions under FTAs.

Many FTAs require that patent terms be *extended* beyond the TRIPS requirement (that the term end no earlier than twenty years from the patent application filing date). The FTAs generally suggest or require the term be extended to "compensate" for two different possible types of delays, either with respect to regulatory approval or review of the patent application itself.[29] Typically, countries are required to extend the patent term if there has been an "unreasonable" delay in either the review of a patent application or an application for regulatory approval.[30]

An important issue is assessing what constitutes an "unreasonable" delay—as well as how much compensation is required. Some FTAs contain specifics on these issues. For example, one FTA negotiated by the United States defines an unreasonable delay in the patent examination as taking more than five years for the patent to issue from the date of filing (or three years from request for examination), whichever is later.[31] However, it does not state what would be an "unreasonable curtailment of the effective patent term" as a result of the marketing approval process.[32]

---

[28] WTO Panel Report, *Canada—Patent Protection of Pharmaceutical Products*, WT/DS114/R (Mar. 17, 2000).

[29] *E.g.*, U.S.-Jordan Free Trade Agreement, U.S.-Jordan, art. 4(23), Oct. 24, 2000, 41 I.L.M. 63 [hereinafter U.S.-Jordan FTA], *available at* http://www.ustr.gov/trade-agreements/free-trade-agreements/jordan-fta/final-text (extension for regulatory delay only); U.S.-Singapore FTA, art. 16.7(7), 18.8(4)(a) (extension for patent or regulatory delay); U.S.-Chile Free Trade Agreement, art. U.S.-Chile, arts. 17.9(6), 17.10(20(a)), June 6, 2003, [hereinafter U.S.-Chile FTA], *available at* http://www.ustr.gov/trade-agreements/free-trade-agreements/chile-fta/final-text (extension for patent or regulatory delay); U.S.-Austl. FTA, arts. 17.9(8), 17.10(4) (extension for patent or regulatory delay); U.S.-Morocco FTA, art. 15.9(7); 15.10(3) (extension for patent or regulatory delay); Free Trade Agreement, Eur.-Korea, art. 10.35(2), Oct. 15, 2009 [hereinafter EU-Korea FTA], *available at* http://trade.ec.europa.eu/doclib/press/index.cfm?id=443&serie=273&langId=en (extension for regulatory delay).

[30] In addition, for countries that approve drugs based on prior approval in another country, patent term extensions must be provided if there was a delay in the prior country's approval process. *E.g.*, U.S.-Bahrain FTA, art. 15.6(b)(ii).

[31] Central America-Dominican Republic-United States Free Trade Agreement, art. 15.9, Aug. 5, 2004 [hereinafter CAFTA], *available at* http://www.ustr.gov/trade-agreements/free-trade-agreements/cafta-dr-dominican-republic-central-america-fta/final-text.

[32] *Id.* art. 15.10

The patent term extensions mandated in FTAs may actually result in longer patent terms in developing countries than in wealthy markets (such as the United States). In the United States, where there is a strong generic drug lobby to balance the interests of PhRMA, the patent term extension in cases of regulatory delay is subject to a complex formula with an absolute maximum effective term of fourteen years from the time of marketing approval.[33] The FTA requirements, however, often require extensions and do not impose a maximum time period.[34] One exception is the EU-Korea agreement, that specifies the extension of patent protection may not exceed five years.[35]

Although Congress made patent term extensions merely permissive rather than mandatory for Peru and Panama as one of a number of concessions to promote public health, that goal may be elusive.[36] Peru and Panama are expected to make their "best efforts" to process patents and marketing applications expeditiously to avoid "unreasonable delay," as well as agree to cooperate with the United States in achieving these objectives. Notably, although the word *may* is usually not obligatory, the requirement of continued cooperation might make this only a cosmetic change—especially as countries that have signed FTAs would not be immune from trade pressure for having inadequate intellectual property laws.

2. Compulsory License Limits

As discussed extensively in other chapters, although compulsory licenses are controversial, they are clearly permissible under TRIPS, which also permits member states to decide what subject matter may be subject to them. TRIPS primarily governs the *procedure* of how such licenses are issued, not the substance of what may be licensed.

a. Limited Subject Matter

Many FTAs restrict compulsory licenses to a narrow number of situations. Notably, these limited situations are similar to what some developed countries had previously proposed during TRIPS negotiations. In addition, these grounds have sometimes been (erroneously) suggested by the pharmaceutical industry as the only permissible grounds for compulsory licenses under TRIPS. Thus, some FTAs are making the pharmaceutical industry's desires the new reality.

---

[33] 35 U.S.C. § 156(c)(3).

[34] Moreover, they are unlikely to do so where there is no substantial domestic generic drug industry to lobby for such changes.

[35] EU-Korea FTA, art. 10.35(2).

[36] Whereas prior FTAs stated that countries "shall" provide a patent term extension, Peru and Panama's FTAs were to be modified to state that they "may" provide such an extension for drug patents.

Most FTAs limit compulsory licenses to the following grounds: national emergency, situation of extreme urgency, or public noncommercial use. All these grounds are permissible under TRIPS, but they are clearly not the *only* permissible grounds; rather, countries have flexibility under TRIPS to consider others. For example, India permits compulsory licenses in other situations, including when the patented invention is not made available to the public at a reasonably affordable price. If India entered into a FTA that set forth only the most typical grounds, it would need to strike this ground of compulsory licensing from its patent laws.

In addition, although some FTAs permit compulsory licenses for lack of so-called "local working," they may explicitly define local working in a way that negates this traditional ground. As discussed in Chapter 5, many nations historically imposed compulsory licenses if the patent owner failed to "work" its invention in the patent-granting country by manufacturing it locally. Such failure to work was considered an abuse of the patent right that justified imposition of a compulsory license. As noted in Chapter 5, there is some controversy concerning whether a compulsory license can be issued on this ground. However, this is a moot issue for countries that enter into FTAs that explicitly define importation as adequate working.

b. Remuneration

Another TRIPS-Plus requirement with respect to compulsory licenses is a modification to the amount of remuneration provided to the patent owner. Whereas TRIPS only requires "adequate" compensation, an FTA may instead demand "reasonable and entire" compensation.[37] This is not an issue of semantics. During negotiation of the compulsory licensing provision under TRIPS, there was a proposal that remuneration should be "adequate to compensate the right holder *fully*" for the license.[38]

Even if a compulsory license can issue under the more restrictive TRIPS-Plus requirements, the license may be of no actual value in countries that also impose TRIPS-Plus regulatory requirements that bar approval of generic drugs during the patent term, as further explained in the section on regulatory protection of drugs.

---

[37] U.S.-Singapore FTA, art. 16.7(6)(b)(ii); *see also* U.S.-Austl. FTA, art. 17.9(7) (requiring patent owner be provided "reasonable compensation"). Although this is a TRIPS-Plus standard that favors the interests of patent owners, it is not yet a uniform standard in FTAs. Several FTAs negotiated after the "reasonable and entire" compensation standard was included contained no specific mention of compulsory licenses, which suggests the TRIPS standard of "adequate" compensation would apply. *See generally* U.S.-Chile FTA; CAFTA; U.S.-Morocco FTA; U.S.-Panama TPA; Trade Promotion Agreement, U.S.-Peru, Apr. 12, 2006 [hereinaft er U.S.-Peru TPA], available at http://www.ustr.gov/trade-agreements/free-trade-agreements/peru-tpa/final-text; U.S.-Colom. Trade Promotion Agreement, Nov. 22, 2006, available at http://www.ustr.gov/trade-agreements/free-trade-agreements/colombia-fta/final-text; U.S.-Korea FTA.

[38] *See* Chairman's Report, *Status of Work in the Negotiating Group* ¶ 5A.3.9, MTN.GNG/NG11/W/76 (Jul. 23, 1990) (noting this as one of several suggestions for the compensation standard).

3. Limited Revocation

TRIPS does not dictate when a patent may be revoked. TRIPS contemplates patent revocation may occur, but only requires that there be an opportunity for judicial review[39] of any decision to do so.[40] Therefore, under TRIPS, nations have discretion to decide the grounds for revoking patents.

A typical TRIPS-Plus provision limits revocation primarily to patentability requirements.[41] Such a restriction prevents nations from revoking a patent on traditional grounds, such as when a compulsory license issued for lack of local working still fails to make an invention publicly available. In addition, restricting revocation to patentability requirements prevents countries from developing or using new public interest reasons. For example, if India entered into an FTA that limited revocation grounds solely to patentability requirements, India would no longer be able to revoke a patent if the patented drug were not made available to consumers at a reasonable cost. This would obviously remove a potentially significant Indian lever in the patentability system to ensure consumer access to reasonably priced drugs.

4. Limited Exceptions to Patent Rights

TRIPS permits member states to provide "limited exceptions" to patent rights pursuant to a number of ambiguously phrased and undefined conditions.[42] A WTO panel has previously held that a company may make a limited quantity of a patented drug as part of seeking regulatory approval to sell a generic version of the drug after the patent expires (as

---

[39] There is some dispute concerning whether this should be interpreted literally to mean review by a court, or whether any review by a higher authority, such as a higher administrative authority, is permissible. *See* Carlos M. Correa, Trade Related Aspects of Intellectual Property Rights: A Commentary on the TRIPS Agreement 342 (2006). In either case, however, it is clear the TRIPS requirement is a modest one of imposing only a need for review rather than setting any substantive standards.

[40] TRIPS art. 32 ("An opportunity for judicial review of any decision to revoke or forfeit a patent shall be available."). There is also one additional requirement that exists for patents revoked when a compulsory license for failure to work is deemed inadequate. In such cases, the revocation cannot occur until at least two years after the compulsory license is issued. Paris Convention for the Protection of Industrial Property art. 5(a)(3), Mar. 20, 1883, *revised* July 14, 1967, 21 U.S.T. 1583, 828 U.N.T.S. 305 [hereinafter Paris Convention].

[41] Many FTAs limit grounds for revocation to patentability requirements and possibly fraud of some type—without the option to revoke a patent where a compulsory license has been inadequate. *See, e.g.*, U.S.-Korea, art. 18.8(4) ("Each party shall provide that a patent may be revoked only on grounds that would have justified a refusal to grant the patent. A Party may also provide that fraud, misrepresentation, or inequitable conduct may be the basis for revoking a patent or holding a patent unenforceable."). Similar language appears in a number of other FTAs. *E.g.*, U.S.-Morocco FTA, art. 15.9(5); U.S.-Austl., art. 17.9(5); U.S.-Singapore FTA, art. 16.7(6)(b)(ii); U.S.-Chile FTA, art. 17.9(5); U.S.-Peru TPA, art. 16.9(4).

[42] TRIPS art. 30 ("Members may provide limited exceptions to the exclusive rights conferred by a patent, provided that such exceptions do not unreasonably conflict with a normal exploitation of the patent and do not unreasonably prejudice the legitimate interests of the patent owner, taking account of the legitimate interests of third parties.").

a limited exception to the usual patent right to exclude others from making a patented invention). However, the WTO panel ruling left open the possibility that limited exceptions could also include a situation where a generic company makes a limited quantity of drugs toward the end of the patent term to ensure no delay in generic competition the very day the patent expires. Although the panel held that Canada's particular law was inadequate because it failed to provide any limit on the amount of drugs made, a country that permitted only limited production might be able to come within the exception.

A TRIPS-Plus requirement would limit the scope of possible exceptions to patent rights, for example, by listing certain categories as the *only* possible exceptions to patent rights.[43] Accordingly, an FTA that permitted an exception only for regulatory approval would be more limited.[44] While an exception for regulatory approval is clearly permissible under TRIPS, the danger of an FTA limiting exceptions to patent rights is that they not only preclude exceptions that are generally recognized as useful, but also bar countries from crafting exceptions to situations that may be difficult to anticipate.[45]

5. Limitations on Parallel Imports

As discussed in Chapter 2, a nation can reduce drug costs by permitting *parallel imports* of a patented product (as an exception from patent liability) when the product was originally sold by the patent owner in another country. However, the ability to consider parallel imports to be noninfringing is limited by some FTAs.

Some FTAs explicitly state the patent owner has the right to bar imports of a patented product—including when the product was previously sold in another country.[46] Often the agreements specify that members must bar imports *at least* in cases where the patent owner has placed contractual restrictions on distribution.[47] This means multinational companies that sell drugs can effectively prevent unauthorized distribution via parallel imports through contractual restrictions. For example, if Glaxo sells drugs below cost in

---

[43] Thus far, restricting permissible exceptions to patent rights beyond the compulsory realm is the exception, rather than the norm, but is included for discussion here as FTAs are ever-evolving.

[44] U.S.-Jordan FTA, art. 19.

[45] An FTA that restricts exceptions from patent rights to regulatory approval would bar a common exception for "experimental use" that enables scientists to use patented inventions in the process of creating new innovations. In the absence of such an exception, there is the danger that scientists will be delayed in pursuing research or elect not to pursue certain avenues of research.

[46] U.S.-Morocco FTA, art. 15.9. This FTA permits parties to limit the restriction to cases where the patent owner by contract has placed restrictions on importation.

[47] *E.g.*, U.S.-Austl. FTA, art. 17.9(4) ("Each party shall provide that the exclusive right of the patent owner to prevent importation of a patented product... without the consent of the patent owner shall not be limited by the sale or distribution of that product outside its territory, at least where the patentee has placed restrictions on importation by contract or other means."); U.S.-Morocco FTA, art. 15.9. This FTA permits parties to limit the restriction to cases where the patent owner has placed restrictions on importation by contract.

Zimbabwe, it can do so pursuant to a contract that specifies that those drugs cannot be subsequently resold in another country. The contract would preclude an enterprising business from buying low-cost drugs in Zimbabwe and then selling them at a higher price in another market. This could be either a violation of the patent right to bar imports or a breach of contract—FTAs have utilized both possibilities. For example, one agreement specifies that the patent owner will have a cause of action for breach of a contract regarding distribution of a patented product regardless of whether the alleged breach occurred and even if the breach occurred by someone without actual knowledge that there was a contractual bar on distribution, as long as that person *should have known* further distribution was unauthorized.[48]

The focus on using contracts to limit parallel imports is one way to prevent undesired imports without addressing a nation's view on international exhaustion of patent rights. After all, as discussed in Chapter 2, most nations do not consider a patent owner's rights "exhausted" with respect to a good that was sold subject to a contractual restriction.

D. REGULATORY PROTECTION OF DRUGS

TRIPS also provides protection beyond patents. In particular, under TRIPS article 39(3), clinical data that is submitted to national authorities for the purpose of obtaining approval to sell a new drug must be protected from "unfair commercial use." As discussed in Chapter 3, there is great controversy concerning the meaning of this phrase. However, this controversy is moot for countries that have adopted TRIPS-Plus requirements that explicitly bar generic companies from relying on clinical data of another (commonly known as *data exclusivity*). An additional TRIPS-Plus requirement restricts regulatory approval for generic drugs if the generic drug might arguably infringe a patent (referred to as *patent linkage*). Each of these will be briefly discussed; however, Chapter 9 provides a more complete discussion of the genesis of these protections, issues of concern, and how to minimize harms of such protection. The focus here is to simply highlight these issues to provide a complete picture of TRIPS-Plus requirements.

---

[48] U.S.-Singapore FTA, art. 16.7(2) ("Each party shall provide a cause of action to prevent or redress the procurement of a patented pharmaceutical product, without the authorization of the patent owner, by a party who knows or has reason to know that such product is or has been distributed in breach of a contract between the right holder and a licensee, regardless of whether such breach occurs in or outside its territory.")

1. Types of Regulatory Protection

a. Data Exclusivity

Most FTAs limit a nation's ability to approve generic drugs based on reliance on the clinical data of another—what is called data exclusivity.[49] This essentially means that the first company to submit clinical data has exclusive rights over it for a certain period of time; during the term of exclusivity, no company is allowed to rely on the data in seeking approval to sell a generic drug.

Data exclusivity protects clinical data associated with a drug, regardless of whether the drug is protected by a patent. In fact, data exclusivity may provide a drug manufacturer with commercial exclusivity in the marketplace even for an unpatentable (or patent expired) drug. While data exclusivity does not technically forbid another competitor from creating its own clinical data (unlike a patent), most companies do not do so. Accordingly, through data exclusivity a company may practically obtain commercial exclusivity for a new drug even without patent protection. Data exclusivity is thus a powerful tool to provide exclusivity in the marketplace without a patent and without the necessity of satisfying any of the patentability requirements. In addition, no additional application is generally necessary, as data exclusivity is usually automatically granted with approval of a drug for marketing/sale.

b. Patent Linkage

Many FTAs impose a second requirement on regulatory laws beyond data exclusivity. In particular, they require countries to adopt what is called *patent linkage*. Although discussed in greater detail in Chapter 9, patent linkage in its strongest form means that a regulatory authority is precluded from approving a generic drug if that drug would infringe a patent. The application to sell the generic does not infringe but will nonetheless be denied if the manufacture or sale of the generic *would* infringe.[50] Often, the FTAs not only preclude approval of a generic, but also require the regulatory authorities to notify the patent owner of the company that sought to sell a generic.[51]

---

[49] *E.g.*, U.S.-Singapore FTA, art. 16.8; U.S.-Austl. FTA, art. 17.10; U.S.-Morocco FTA, art. 15.10; U.S.-Korea FTA, art. 18.9; *see also* Karin Timmermans, *Monopolizing Clinical Trial Data: Implications and Trends*, 4 PLoS MED. 206, 208 (2007) (summary table comparing data exclusivity provisions in FTAs); Dolores Cullen, *Data Protection: The New IP Frontier—An Overview of Existing Laws and Regulation*, 5 J. GENERIC MED. 9 (2007).

[50] *See, e.g.* U.S.-Singapore FTA, art. 16.8(4)(c); U.S.-Chile FTA, art. 17.10(2)(b); CAFTA, art. 15.10(3); U.S.-Austl. FTA, art. 17.10(5); U.S.-Morocco FTA, art. 15.10(4). Pursuant to the 2007 Congressional agreement, the Panama and Peru FTAs do not demand that the drug regulatory agency withhold approval of a generic during the patent term, but still require procedures to permit expeditious resolution of patent infringement or validity disputes. In addition, the patent owner must still be promptly notified if an application is filed to market a generic version of the patented drug.

[51] *E.g.*, U.S.-Chile FTA, art. 17.10(2)(c); CAFTA, art. 15.10(3); U.S.-Austl. FTA, art. 17.10(5); U.S.-Morocco FTA, art. 15.10(4).

Notably, although patent linkage is premised on possible patent infringement, the regulatory authority that enforces patent linkage is not an expert on patents. To the contrary, the regulatory authority is generally knowledgeable about safety and efficacy of drugs and lacks expertise on when a patent might be infringed. Given this lack of expertise, regulatory authorities tend to defer to the perceived experts–the patent owners—to assess when a generic drug might infringe. Some assert the patent linkage requirement thus improperly turns the government into an enforcement arm of the patent owner. In addition, some critics suggest patent linkage encourages sequential filing of frivolous patents that are allegedly associated with a drug to delay the entry of generic drugs, as each new patent would have a later expiration date.

Patent linkage is definitely a TRIPS-Plus requirement. TRIPS itself provides no requirement that even suggests such linkage. As noted above, the only TRIPS obligation with respect to regulatory authorities is that they protect data from unfair competition.

2. Impact on Access to Medicine

a. Overall Impact

TRIPS-Plus requirements that mandate regulatory protection may, in combination with other factors, delay access to generics, especially for developing countries. Data exclusivity provisions generally calculate the term of data exclusivity for a given drug from the first date the drug is approved in a domestic market. The data exclusivity term—which directly impacts when generic versions may enter the market—is also impacted by a non-legal issue. In particular, many companies will first introduce a new drug in a wealthy country (such as the United States), but delay introduction in less lucrative markets.[52] The decision to first sell a drug in the most profitable market is a logical business decision. However, if data exclusivity is not based on the first global approval, the delayed approval of a drug in poorer countries may delay entry of generics in such countries. For example, hypothetical company Mirck may begin marketing a new HIV drug in the United States in 2006, such that when a five year period of exclusivity expires in 2011, a generic version might be available in the United States. However, if Mirck does not seek approval to market that same drug in Brazil until 2012, a generic company could be precluded from obtaining approval to market a generic version until after the data exclusivity expires in Brazil. Under the usual five-year term of exclusivity, the generic would not be available in Brazil until 2017 - long after it was available in the United States.

The delayed availability of lower cost generics for countries that enter into FTAs with data exclusivity provisions is not hypothetical. One recent study noted that generic versions of a number of drugs will be possible in the United States before Guatemala because

[52] There may be no financial incentive to do so if the market is not expected to be profitable. Moreover, from the corporate perspective, there are clear dangers to promptly seeking approval because that could provide the basis for compulsory licenses of drugs as well as parallel imports.

of earlier expiration of patents or data exclusivity.[53] It also noted that drugs protected by data exclusivity were substantially more expensive than those in the same therapeutic class; for example, the protected insulin drug sold as Lantus cost 846 percent more than the unprotected drug isophane insuline, and the protected antibiotic sold as Invanz cost 362 percent more than the unprotected meropenem.[54] The study cautions that its results are limited to those drugs sold in the public sector—the only sector for which public data were readily available. Nonetheless, the public sector accounts for 70 percent of drugs sold in Guatemala and covers poor citizens who are more likely to face economic barriers to obtaining access to medicine.[55]

The 2007 Congressional compromise directly addressed this problem, at least for the then-pending FTA with Peru and Panama. In particular, although still requiring data exclusivity, Congress suggested that if marketing approval is granted in reliance on prior approval by the United States, and that if Peru or Panama promptly grants regulatory approval for the same drug, the exclusivity period will be concurrent with the US period.[56] In other words, even if approval was sought in Peru after it was sought in the United States, the data exclusivity period would not extend beyond the date of exclusivity in the United States. This modification would thus result in lower-cost generics being available more quickly than under the terms of the former FTA.

b. Patent versus Regulatory Protection

Even if FTAs require both patent and regulatory protection, they will not necessarily all have the same immediate impact on access to medicine. In particular, as will be explained below, data exclusivity may initially play a more significant role in limiting access to medicine in developing countries that had not previously provided patent protection on drugs. Eventually, however, the interplay among data exclusivity, increased patent protection, and patent linkage will all operate together to provide a stronger barrier to short-term access to medicine than is necessary under TRIPS.

In a country that is new to providing patent protection on drugs, data exclusivity may be the only realistic method of protecting drugs immediately after the FTA is concluded. Although patent protection would technically be available, drugs that have already been patented in other jurisdictions would likely be unpatentable because they would fail to meet the requirement that they be "new." In particular, a prior patent by another country would be considered prior art that would establish the drug was no longer new.

---

[53] Ellen R. Shaffer & Joseph E. Brenner, *A Trade Agreement's Impact on Access to Generic Drugs*, 28 Health Affairs 957, 961, 964–65 (Aug. 25, 2009), http://content.healthaffairs.org/cgi/content/abstract/28/5/w957.

[54] *Id.* at 962.

[55] *Id.* at 965.

[56] The partner country is required to grant approval within six months of the application.

Data exclusivity, however, can provide market exclusivity even for a drug that cannot be patented. There are no requirements to obtain market exclusivity other than submitting an application to sell a drug in the country, together with clinical tests establishing the drug is safe and effective. If the applicant is granted permission to sell a drug, data exclusivity is automatically granted. This means that during the term of data exclusivity, no generic equivalent will be approved based on the same data. As generic companies generally lack resources to undergo their own clinical tests, data exclusivity can provide a period of market exclusivity even where there is no patent protection.

Data exclusivity (rather than patent protection) is likely to be the most significant barrier to access to medicine in developing countries that are new to providing patent protection. For example, in Jordan, a developing country that historically did not provide protection on drug compounds, only five out of over one hundred newly marketed medicines were patented subsequent to a 2001 FTA.[57] Multinational drug companies were nonetheless able to dominate the market; as of 2007, almost 80 percent of the new drugs had no generic competition. These drugs were protected solely by data exclusivity.

Yet, patents may subsequently be just as important as data exclusivity, and perhaps even more so with respect to the duration of market protection. Several years after patent protection on drugs is provided, patents are more likely to be granted on drugs. At this time, companies would be seeking patent protection in the developing country (such as Jordan)[58] at the same time it was pursuing global protection. [59] Importantly, the roughly simultaneous filing of patent applications will ensure that an invention in Jordan is not precluded for receiving a patent because of a prior application in another country.

Although both patents and data exclusivity can provide market protection, patent protection may provide longer protection—especially under TRIPS-Plus rules that require extensions of patent terms. For example, many data exclusivity laws provide a term of five years whereas the effective term of patents at generally about eleven years is nearly double that.

In addition, once patents are more commonly granted on drugs in developing countries, there may be more delay in the introduction of generic drugs if there are multiple patents filed on different aspects of a marketed drug. In other words, even though data exclusivity may have expired for the active ingredient of a drug, if other aspects of the drug are patented, that patent may still prevent entry of a generic drug. This is especially true for

---

[57] *See* Oxfam International, *All Costs, No Benefits: How TRIPS Plus IP Rules in the US Jordan FTA Affect Access to Medicine* 7-8 (Oxfam Briefing Paper No.102, 2007), *available at* http://www.oxfam.org.uk/resources/issues/health/downloads/bp102_trips.pdf.

[58] *Id.*

[59] There is actually an international agreement that helps applicants avoid barring themselves from patentability. In particular, pursuant to the Paris Convention, applicants are given one year from the first globally filed application to file other similar applications while retaining the same filing date. Paris Convention, art. 4.

countries that also require patent linkage because the generic manufacturer may be denied approval so long as there is an unexpired patent on any aspect of the original drug.

The intersection of patent and regulatory laws can be complex; thus, they are described in further detail in Chapter 9.

E. EXTENDING PROTECTION OF DRUGS

The latest aspect of TRIPS-Plus requirements increases the value of patents primarily through a two-pronged strategy of minimizing the use of price controls on patented drugs and promoting new (and typically more expensive) drugs to consumers. Both of these measures complement patent protection by increasing the value of patents without directly changing patent laws. Both of these strategies seem aimed at imposing the US marketing and sales model (that results in the highest profit margins) on other countries. Although this may have benefits for patent-owning companies, it may also increase costs and distort the path of innovation.

1. Minimizing Price Controls

An important recent trend in FTAs is to minimize a nation's ability to impose price controls on drugs. This can enhance the ability of patent or regulatory protection to result in increased profits. For example, in the United States, where there are no national price controls, the cost of patented drugs is higher than in other countries. To pharmaceutical companies, minimizing (or even eliminating) price controls in other nations is desirable for increasing profitability. Although the profit incentive is clear, companies as well as FTAs typically frame the issue in terms of recognizing the value of innovative pharmaceuticals.

The U.S.-Australia FTA was the first to introduce provisions to impact price controls. Before the FTA, patented drugs were estimated to cost up to twice as much in the United States as in Australia,[60] with Australia having one of the cheapest drug costs globally.[61] The cost differential is likely attributed to Australia's Pharmaceutical Benefit Scheme (PBS) providing reimbursement for drugs based on their relative cost-effectiveness. Drugs do not need to be listed on the PBS to be sold; however, as most consumers buy drugs through the PBS (because the PBS substantially subsidizes costs), companies tend to seek listing on the PBS even though that limits the maximum sale price. The PBS will not permit a drug to be listed for reimbursement unless it is more effective, safer, or both. In addition, even if a drug is included for listing, the reimbursement for the drug will be

[60] *E.g.*, AUSTL. PRODUCTIVITY COMM'N, NO. 0107004, INTERNATIONAL PHARMACEUTICAL PRICE DIFFERENCES 79 fig. 3.1 (2001).

[61] K. Lokuge & Richard Denniss, *Trading in Our Health System? The Impact of the Australia-US Free Trade Agreement on the Pharmaceutical Benefits Scheme* 16 (The Austl. Inst., Discussion Paper No. 55, 2003), *available at* https://www.tai.org.au/file.php?file=discussion_papers/DP55.pdf.

capped at the price of existing comparable drugs if it does not establish improved effectiveness, safety, or both. The PBS system is thus a type of reference pricing that aims to provide greater reimbursement for drugs that are more socially valuable.

According to drug companies, a system such as the PBS that imposes a cap on prices (rather than permitting the company to sell drugs at whatever price the market will bear), fails to allow proper recoupment of the company's research and development costs. In particular, it has been suggested that Australian citizens were improperly free riding on patients in other countries who paid higher prices for drugs. Indeed, one of the negotiating objectives by the United States was to eliminate the PBS prices—and to seek to dismantle similar reference pricing in other developed countries.[62]

The U.S.-Australia FTA does not require that Australia eliminate its PBS but does introduce a number of measures that may have a substantial impact on how the PBS operates. For example, the FTA requires increased transparency of the PBS reimbursement process, as well as an opportunity for patent owners to independently challenge PBS listing decisions.[63] In addition, the FTA creates a new "Medicines Working Group" to address the "importance of pharmaceutical research and development to continued improvement of healthcare outcomes."[64] The creation of this group leaves open the possibility for continued pressure to modify the PBS process in a way that would better promote the interests of patent owners. In addition to these measures for influencing the PBS process, the PBS could be directly challenged pursuant to the dispute resolution procedures in the FTA. Unlike TRIPS, the US-Australia FTA permits a party to initiate a dispute when there is no violation of any explicit provision, but there is an impairment of expected benefits; this is called a "nonviolation" dispute.[65] Accordingly, the US could raise a dispute based on ambiguous statements in the FTA concerning the mutual agreement to "recognize the value of innovative pharmaceuticals."[66]

The subsequent U.S.-South Korea FTA goes even farther in specifying considerations in setting reimbursement.[67] The agreement requires that reimbursement amounts for

---

[62] *See* INT'L TRADE ADMIN., U.S. DEP'T OF COMMERCE, PHARMACEUTICAL PRICE CONTROLS IN OECD COUNTRIES: IMPLICATIONS FOR U.S. CONSUMERS, PRICING, RESEARCH AND DEVELOPMENT, AND INNOVATION 52 (2004), *available at* www.ita.doc.gov/drugpricingstudy. However, a trend towards emulating the general lack of price controls may have undesirable consequences beyond higher drug costs. In particular, without price controls, there may be an incentive to promote inefficient research into similar versions of profitable products rather than innovative and socially valuable new products.

[63] U.S.-Austl. FTA, Annex 2-C, 2(f).

[64] *Id.*, Annex 2-C, 3.

[65] US-Austl. FTA, art. 21.2(c).

[66] *E.g.*, Buddima Lokuge & Thomas Faunce, *Trade Disputes and the Pharmaceutical Benefits Scheme: Constructive Ambiguities, Non-Violation Nullification Disputes and the Australia US Free Trade Agreement* 18 (Centre for Governance of Knowledge & Dev., Working Paper, 2004), *available at* http://cgkd.anu.edu.au/menus/PDFs/TradedisputesandthePharmaceuticalBenefitsScheme.pdf.

[67] The agreement also goes beyond the FTA with Australia in extending the same conditions to medical *devices*.

drugs be based on "competitive market-derived prices," which suggests the price would be what the pharmaceutical industry wishes to charge in the absence of any limits. South Korea enables the reimbursement amount to be alternatively determined, but the "value" of patented products is expressly noted as a factor that must be recognized in the reimbursement amount. In addition, pharmaceutical companies are allowed to have additional input into determination of the reimbursement amount. First, such companies may seek a higher reimbursement than comparative products based on evidence of safety or efficacy.[68] In addition, after an initial reimbursement decision is made, a company is given a second opportunity to seek a higher reimbursement.

The limitation on national ability to control costs through limiting reimbursement amounts, as in the Australia and South Korea FTAs, appears to be a new trend. Although only a limited number of FTAs include such provisions thus far, the pharmaceutical industry seems supportive of this additional approach beyond increased patent and regulatory protection so as to maximize profitability. Although there is general recognition access to medicine is an important issue, one "practical" means to achieve balance is to insist that developed countries pay more for drugs. For example, the global profitability of drugs could be increased if countries besides the United States did not use price controls. Indeed, it has been suggested that a "middle ground" for the issue of global access to medicine would be for developed nations to commit to ensuring that government purchasing authorities pay an "adequate price to encourage research" and cover research costs for products that are of primary value to developing countries.[69]

2. Encouraging Use of Patented Drugs

A second strategy that has received less attention, but may also be important, is a requirement in FTAs that permits companies to disseminate information about their patented drugs to consumers. In particular, companies may provide information directly to consumers through an Internet site about the drug. Although the information must include a balance of risks and benefits, it nonetheless seems to essentially be advertising on the web and introduces the same issues that exist with direct advertising to consumers. In the United States, advertisements have played a pivotal role in inducing consumers to switch to newer-patented—and thus more expensive—drugs. In particular, as noted in Chapter 1, pharmaceutical companies sometimes will introduce a newly patented drug that is only a slight variation of an older compound just as the older compound is about to lose patent protection. If consumers are persuaded to switch to the newer-patented drug, this has a positive impact on profits.

---

[68] U.S.-Korea FTA, art. 5.2.

[69] Letter from Pfizer CEO and Professor John Barton to Honorable Max Baucus (Aug. 13, 2008), *available at* http://media.pfizer.com/files/news/kindler_baucus_letter_081308.pdf.

Increased information to consumers may seem like an innocuous requirement that can only promote better consumer decision making. However, consumers are never in an ideal position to make effective decisions on their own, as they do not have the expertise necessary to evaluate whether a new drug is actually an improvement. This stands in contrast to the experts involved in reference-pricing systems such as Australia's PBS who can appropriately decide whether a new drug actually provides a worthwhile improvement over existing medicines—including medicines that are available at low cost because they are no longer protected by patents. The increased information companies may provide to consumers is also a novel way to influence them even in countries that do not officially permit direct-to-consumer advertising of prescription medicine.[70] The web-based information could have the same effect of inducing consumers to switch to a new drug that is more expensive than a generic.

III. Controversy Concerning TRIPS-Plus

Whether TRIPS-Plus standards are appropriate for developing countries is questionable.[71] Proponents of these standards often proclaim that stronger patent rights promote innovation as well as direct investment. These same claims were made in support of the original TRIPS agreement—but the actual benefits have yet to be realized.[72] Similarly, countries that have adopted TRIPS-Plus agreements have not reported increased benefits, despite what proponents of such agreements have suggested. For example, although the USTR has stated that there has been a "blossoming of Jordan's pharmaceutical industry" since increased patent rights were introduced,[73] the evidence is ambiguous. There have been no

[70] U.S.-Austl. FTA, Annex 2-C(5) (dissemination of information).

[71] In fact, some have even suggested that poor countries should be exempted from intellectual property requirements altogether—although this is obviously not realistic in the wake of TRIPS. *E.g.*, Alan V. Deardorff, *Should Patent Protection be Extended to All Developing Countries?*, 13 World Econ. 497 (1990). An alternative, but similarly unrealistic proposal would be to have patent-owning companies choose to protect their inventions in either industrialized or developing countries, but not both. Jean Olson Lanjouw, *Beyond TRIPS: A New Global Patent Regime,* Center for Global Development, CGD Brief, Aug. 2002, *available at* http://www.cgdev.org/content/publications/detail/2860/.

[72] Many have noted the argument between stronger rights and increased foreign direct investment (FDI) is weak. *E.g.*, F.M. Scherer & S. Weisburst, *Economic Effects of Strengthening Pharmaceutical Patent Protection in Italy,* 26 Int'l Rev. Indus. Prop & Copy. L. 1009 (1995); E. Mansfield, Intellectual Property Protection, Direct Investment, and Technology Transfer, Int'l Finance Corporation, Discussion Paper, No. 27 (1995); *see also* A.S. Kirim, *Reconsidering Patents and Economic Development: A Case Study of the Turkish Pharmaceutical Industry,* 13 World Dev. 219 (1985); Frederick M. Abbott, *The International Intellectual Property Rider Enters the 21st Century,* 29 Vand. J. of Trans. L. 471, 474 (1996) (noting countries with the weakest levels of intellectual property rights (IPR) protection—the People's Republic of China, Taiwan, Brazil, Argentina, Thailand, etc.—over the past decade have routinely been recipients of the largest net FDI inflows).

[73] *E.g.*, Marilyn Chase & Sarah Lueck, *In New Trade Pacts, U.S. Seeks to Limit Reach of Generic Drugs,* Wall St. J., July, 6, 2004, at A1.

new production facilities established by major multinational firms in Jordan since the FTA.[74] In fact, although there is a local pharmaceutical industry, it existed before the FTA. In addition, Jordanian firms note that R&D levels remain unchanged since the FTA. Jordan's experience is not unexpected: prior independent studies suggested that increasing patent standards for developing countries may not promote local innovation as many poor countries do not have an adequate infrastructure to develop new medicines.[75]

Although the trade-off of increased prices for possible increased innovation may make sense in wealthy countries, it may be less appropriate in developing ones. Not only is there little prospect of either increased innovation or foreign direct investment, but stronger protection may have an adverse effect on access to medicine in countries that have only limited resources. In addition, higher prices as a result of patent protection may be more easily sustained in wealthy countries where most people are covered by health insurance such that they pay only a small fraction of the retail cost of drugs. In contrast, many citizens in developing countries lack insurance, and thus their out-of-pocket costs for medicine may actually be higher than citizens in wealthy countries.

Although some countries may feel pressure to adopt TRIPS-Plus standards, the possible impact on access to medicine is not wholly unrecognized—at least by the country forced to adopt the agreement. Sometimes a nation will enter into an agreement demanding TRIPS-Plus standards *despite* opposition from consumer groups or local industry. Occasionally public opinion will be strong enough to thwart a proposed agreement; however, this is the exception rather than the norm.[76] More typical are side letters to FTAs that purport to assert the agreement has no impact on the ability of a country to take "necessary measures to protect public health by promoting access to medicine for all."[77] Although these may sound reassuring, they likely have little to no legal weight. Moreover, the side letters do not permit countries to ignore the actual requirements of the agreement that may in fact make it difficult to promote access to medicine by mandating that more medicines be patented.

TRIPS-Plus requirements also deserve close attention as there has generally been a constant "ratcheting up" of standards, with the potential to substantially and significantly

---

[74] Hamed El-Said & Mohammed El-Said, *TRIPS-Plus Implications for Access to Medicines in Developing Countries: Lessons from Jordan-United States Free Trade Agreement*, 10 J. W. INTELL. PROP. 438, 453–55 (2007).

[75] *E.g.*, UK COMMISSION ON INTELLECTUAL PROPERTY RIGHTS, INTEGRATING INTELLECTUAL PROPERTY RIGHTS AND DEVELOPMENT POLICY, ch. 2 (2002).

[76] For example, Thailand has thus far resisted efforts to enter into an FTA, although it continues to be subject to political pressure and has been consistently listed on the US Priority Watch list.

[77] *See, e.g.*, Letters between Minister Taïb Fassi Fihri, Del. for Foreign Aff. and Cooperation, Kingdom of Morocco, and Robert B. Zoellick, Trade Rep., United States (June 15, 2004), *available at* http://www.ustr.gov/trade-agreements/free-trade-agreements/morocco-fta/final-text (courtesy translation of the understanding of chapter 15 of the U.S.-Morocco FTA, *supra* note 14).

impact public health.[78] In other words, each subsequent agreement containing TRIPS-Plus standards generally imposes *higher* standards than the last one. Increasing standards of protection generally also increases barriers to current access to medicine. Many developing countries have limited resources as well as serious public health challenges. Accordingly, to the extent a developing country adopts a TRIPS-Plus standard that requires more protection for patents, more drugs are likely to be patented—and priced out of reach. This ratcheting up phenomenon has been a trend, but it need not continue. For example, in 2007, the US Congress bowed to public pressure and announced it would modify a then-pending FTA with Peru and Panama to promote public health.[79]

IV. Conclusion

Although developing countries retained some flexibility to tailor patent laws to their social priorities (as discussed in Chapter 4), TRIPS-Plus requirements may be rapidly eroding that protection. This is particularly true not only because FTAs that impose TRIPS-Plus requirements have multiplied, but because there has been a general trend toward increasing the required protection. Even if subsequent FTAs do not further increase protection, the current ones already impose much higher standards than TRIPS requires.

Subsequent chapters further explore some of the TRIPS-Plus requirements introduced here. For example, Chapter 9 provides a more in-depth discussion of TRIPS-plus regulatory requirements and also discusses concrete measures that countries can try to take to minimize the health impact of such requirements. Chapter 10 discusses a TRIPS-Plus requirement in EU law that threatens international trade in low-cost generic drugs and has ramifications beyond the EU, as there is a pending international agreement under negotiation that would have a similar impact for member countries.

---

[78] *E.g.*, Peter Drahos, *Bits & Bips—Bilateralism in Intellectual Property*, 4 J. World Intell. Prop. 791 (2001); Susan K. Sell, *The Global IP Upward Ratchet, Anti-Counterfeiting and Piracy Enforcement Efforts: The State of Play* (IQsensato, Occasional Papers No. 1, 2008), http://www.iqsensato.org/wp-content/uploads/Sell_IP_Enforcement_State_of_Play-OPs_1_June_2008.pdf.

[79] *E.g.*, Peru & Panama FTA Changes 8 (May 10, 2007), *available at* http://waysandmeans.house.gov/Media/pdf/110/05%2014%2007/05%2014%2007.pdf. However, developing countries that negotiated FTAs prior to 2007 have thus far not been successful in securing modified agreements.

9 Beyond Patents: Protecting Drugs through Regulatory Law

Although patents are the prototypical type of protection that most people consider applicable to protecting drugs, patents are just the most-established and well-known method available to protect drugs from competition. However, there are other types of mechanisms in regulatory laws that provide patent-like protection or further strengthen existing patent protection.

This chapter provides an in-depth explanation of two types of regulatory laws that provide important protection to drug companies: data exclusivity and patent linkage. As noted in earlier chapters, not all countries currently provide such protection, but these mechanisms are highly desired by drug companies and thus are frequently included in FTAs that impose TRIPS-Plus requirements. Because these protections may be additional hurdles to generic competition, a thorough understanding of these requirements (as well as how they intersect with patent protection) is important to anyone interested in promoting access to medicine. By the conclusion of this chapter, a reader should be able to answer the following questions:

- How can an unpatentable drug be protected from competition?
- Why might data exclusivity pose a problem for access to generic drugs?
- Is the rationale for data exclusivity applicable to developing countries?

I. Introduction 255
II. Data Exclusivity 256
A. Overview 256
1. Basic Term of Data Exclusivity 257
2. Additional Terms of Exclusivity 257
3. Does Data Exclusivity Pose an Additional Bar for Patented Drugs? 258
B. Genesis of Data Exclusivity 260
C. Evaluating the Case for Global Data Exclusivity 262
1. Challenging Assumptions in Support of Data Exclusivity 263
2. Additional Concerns about Data Exclusivity for Developing Countries 268
3. Minimizing the Effects of Data Exclusivity 269
III. Patent Linkage 273
A. Overview 273
1. Middle Ground? 275
2. Implications of Patent Leakage 275
B. Genesis of Patent Linkage 277
C. Patent Linkage in the Global Arena 278
1. Mediating the Harm of Patent Linkage 280
2. Minimal Patent Linkage 280
3. Mandatory Patent Linkage, but Limited Scope of Patents 280
4. Mandatory Patent Linkage, but with Procedural Safeguards to Protect the Public Interest 281
IV. Patents Plus—The Complete Picture 281
A. Patents Only 282
B. Patents Plus Data Exclusivity Only 283
C. Patents Plus Data Exclusivity and Patent Linkage—with Challenges Permissible 283
D. Patents Plus Data Exclusivity and Patent Linkage—No Exceptions 283
V. Concluding Thoughts 284

I. Introduction

This chapter discusses data exclusivity[1] and patent linkage, both of which are considered important types of protection to the pharmaceutical industry. Although each type of protection will be discussed in more detail, a brief overview of each may be helpful in understanding how these mechanisms complement traditional patent protection. In brief, data exclusivity prevents a potential generic company from relying on the clinical data submitted by an originator company when the generic company wants to establish bioequivalence during the period of exclusivity. Although there are many rationales cited, a common one is that data exclusivity is necessary to prevent "free riding" on an originator's investment. The second type of protection, patent linkage, bars a national regulatory agency from approving a generic drug if it is "linked" to one or more existing and unexpired patents. Patent linkage is argued to be a logical method to ensure that regulatory approvals do not unduly promote patent infringement. Data exclusivity and patent linkage are independent types of protection that may be available in a given nation. Data exclusivity is more widely adopted among developed countries, but both types of protection are strongly encouraged (if not required) among many developing countries pursuant to FTAs, as discussed in Chapter 8.

Whether these types of regulatory protection exist in a given country may have tremendous impact on the availability of generic drugs. In particular, if neither protection exists, a generic company can seek to obtain regulatory approval for a generic drug during the patent term. Alternatively, in the worst-case scenario, data protection or patent linkage may provide an absolute bar to generic drugs even where a patent is subject to a compulsory license. Moreover, patent linkage could bar generic entry even when a generic company has generated its own data and is not relying on that of the original patent owner. Accordingly, it is important to understand each type of protection as well as how to mitigate its possible harm. To best understand the impact of such protections, this chapter provides a separate overview of each protection, its genesis, and concerns about expanding global protection. The chapter concludes with a summary of how the regulatory protections can work in combination with patents to impede market entry of generic drugs.

---

[1] Although this chapter uses the phrase *data exclusivity*, the same protection is often referred to as *test data exclusivity* or *market exclusivity*. Some have confusingly used the term *data protection* to refer to both data exclusivity and what is required under TRIPS, although TRIPS does not in fact require exclusivity (as discussed in Chapter 3).

II. Data Exclusivity

A. OVERVIEW

A company that develops a new drug[2]—typically called a *pioneer* or *originator* company—may obtain a competitive advantage through regulatory laws that provide data exclusivity. As noted in Chapter 1, a company seeking to sell a new drug must submit extensive clinical data to establish that the drug is safe and effective. Subsequent companies that want to sell an equivalent to the previously approved drug can—subject to data exclusivity— rely[3] on the data submitted by the originator company and only submit more limited data to establish its proposed drug is bioequivalent to what has already been deemed safe and effective. A bioequivalent drug is deemed equivalent in that the active ingredient and dosage is the same as a previously approved drug, such that the regulatory agency infers the drug is equally safe and effective as the originally approved drug. Bioequivalent drugs are thus considered equivalent and interchangeable for treating the same condition as the original drug.

In essence, during the period of data exclusivity, generic companies do not apply for approval to sell a generic version of an existing drug by relying on the original data submitted by the originating company. This is true even if the drug is unpatented, or the patent has expired. The company seeking to sell a generic version can theoretically create and submit its own data. However, generic companies typically wait for data exclusivity to expire. Not only is clinical data expensive and time-consuming to generate, but conducting duplicate clinical tests of drugs known to be safe and effective may be unethical.

Where applicable, data exclusivity provides powerful market exclusivity akin to patent protection—but without the many requirements attendant to patent protection. No innovation is necessary to obtain data exclusivity. Indeed, an unpatentable drug can be granted data exclusivity. For example, a new use of a known drug could be granted data exclusivity, but patent protection may be denied if a country only considers new compounds, but not new *uses* to be patentable. In addition, obtaining data exclusivity is simpler than patent protection because no additional application beyond the basic application to obtain market approval is necessary. In a country that recognizes data exclusivity, this exclusivity is automatically granted when marketing approval for the drug is granted. Moreover, whereas a patent may be challenged as invalid for failing to meet the requirements

[2] The term *new* here is used to refer to drugs that need clinical tests before the company making them can seek their approval, as explained in Chapter 1.

[3] Although they can rely upon the conclusions of the previous data in establishing bioequivalence to the previously established safe and effective drug, the subsequent companies do not have actual access to this data.

of patentability, there are no requirements for data exclusivity that can be challenged as being inadequately met.

1. Basic Term of Data Exclusivity

Because data exclusivity delays the sale of generic drugs, its length is an important issue. Countries that presently provide data exclusivity set terms that generally last for five to ten years from the date of regulatory approval of the drug. During the term of data exclusivity, applications for generic versions that seek to rely on another's data will not be approved. Some countries will accept applications at some point during this period, but others refuse to do so.[4] If generic applications are not accepted until after the end of exclusivity, the originator company receives an additional de facto period of exclusivity for the time it takes the regulatory agency to review the application.[5]

2. Additional Terms of Exclusivity

Some countries provide additional periods of exclusivity based on submission of additional clinical data. A common additional period of exclusivity is for a new use, indication, or dosage for a previously approved compound that is accompanied by additional clinical studies.[6] For example, an injection of a compound marketed as Botox to treat

---

[4] For drugs approved since 2006 in the EU, the originator is granted up to ten years of exclusivity in the marketplace. The EU will not accept generic applications until the last two years and will not approve any such generic applications until the full ten years has expired. In the United States, generic applications are not generally accepted during the basic five-year term, but they can be submitted in the last year (after four years of exclusivity) if the applicant certifies the patent(s) for the reference drug is either not infringed or invalid.

[5] It has been suggested that it may take up to eighteen months to approve a generic application, such that a five-year data exclusivity actually creates commercial exclusivity for over six years after the reference drug is approved. *E.g.*, JUDIT RIUS SANJUAN, CONSUMER PROJECT ON TECH. DISCUSSION PAPER NO. 1, US AND EU PROTECTION OF PHARMACEUTICAL TEST DATA 6 (2006), *available at* http://www.cptech.org/publications/CPTechDPNo1TestData.pdf.

[6] Less common, but also possible periods of exclusivity exist when companies engage in studies for pediatric use, or when companies develop and seek approval of "orphan" diseases. *E.g.* 21 U.S.C. § 355(a) (2006) (six-month period of exclusivity for completion of pediatric studies requested by FDA); 21 U.S.C. § 360(cc) (2006) (seven-year term of exclusivity from the date of marketing approval for "orphan" diseases); Regulation 141/2000 of the European Parliament and of the Council of 16 Dec. 1999 on orphan medicinal products, 2000 O.J. (L 18) 1. However, both exclusivities have been criticized for resulting in high prices and enabling companies to manipulate the laws to protect drugs that are relevant to a large population. *See, e.g.*, Robert Steinbrook, *Testing Medications in Children*, 347 NEW ENG. J. MED. No. 18, at 1462, 1466-67 (Oct. 31, 2002); *Anticompetitive Abuse of the Orphan Drug Act: Invitation to High Prices: Hearings before the Subcomm. on Antitrust, Monopolies & Business Rights of the Senate Commn. on the Judiciary*, 102d Cong., 2d Sess. 198 (1992); Marlene E. Haffner, *Orphan Products—Ten Years Later and Then Some*, 49 FOOD & DRUG L.J. 593, 596–98 (1994).

wrinkles is a new use of a drug previously known to treat seizures.[7] Countries differ with respect to whether to grant such periods, as well as the length of the additional term and when during the term a generic application (related to the new data) can be submitted.[8] However, as discussed in Chapter 8, many free trade agreements (FTAs) require exclusivity for new uses. Even where such additional exclusivity periods exist, applications for generic versions of the originally approved drug may be accepted and approved before the additional term expires. However, in cases where the new version (with the additional period of exclusivity) provides a distinct advantage to a patient, such as a preferable dosing scheme, the additional term of exclusivity may confer a strong commercial advantage that enables its manufacturer to compete even against the lower-priced generics. In such a case, the expiration of the original period of exclusivity may have no real impact on consumers if the newer version is perceived as a substantial improvement.

3. Does Data Exclusivity Pose an Additional Bar for Patented Drugs?

Although data exclusivity is clearly of great importance for unpatentable (and patent-expired) drugs, an important question is whether such exclusivity poses any substantial barrier to generic competition when a drug is in fact patented. At first glance, there would not seem to be any problem as patent terms are generally close to seventeen years whereas data exclusivity terms are less than ten. However, as noted in Chapter 1, the *effective* patent term—the time when the patent term is in force and the patented drug has been approved for sale—averages eleven years.[9] The effective patent term is thus much closer to the actual period of data exclusivity. Although some patented drugs may not gain additional commercial protection from data exclusivity, those that do obviously limit access to lower-cost generic alternatives.[10] Moreover, for drugs that are unpatented or whose patent

---

[7] Alternatively, if a previously approved antiretroviral is reformulated so that it no longer needs refrigeration, it could qualify for a new period of exclusivity.

[8] The term of the additional exclusivity based on new clinical data varies. In the United States, after the initial five-year period expires, an application with new clinical data for a previously approved chemical may result in an additional three years of exclusivity. Three additional years are possible for *each* new use or dosage based on new clinical data. The United States will receive and evaluate generic applications during this period, but no generic approvals are effective until after the full three-year period has elapsed. The EU provides one year of market exclusivity for new indications that establish "significant clinical benefit" over existing indications. Council Directive, 2004/27/EC, art. 10.1, ¶ 2, 2004 O.J. (L 136) 34-57. Unlike the United States, no additional periods of exclusivity are provided.

[9] *E.g.* Henry G. Grabowski & John M. Vernon, *Effective Patent Life in Pharmaceuticals*, 19 INT'L J. TECH. MGMT. 98 (2000) (studying drugs approved between 1990 and 1995).

[10] Some suggest that only a very few high-selling drugs gain additional commercial advantage from data exclusivity beyond patent protection. *E.g.*, MEIR PEREZ PUGATCH, INTELLECTUAL PROPERTY AND PHARMACEUTICAL DATA EXCLUSIVITY IN THE CONTEXT OF INNOVATION AND MARKET ACCESS 13 (2004) *quoting* IMS Health, Data Exclusivity: The Generics Market's Third Hurdle (2001); ICTSD-UNCTAD Dialogue on Ensuring Policy Options for Affordable Access to Essential Medicines (Oct. 2004),

expires at or shortly after market approval, data exclusivity could be the only legal bar to entry of a generic version.

For patented drugs, data exclusivity is most likely to last beyond the patent term where the patent application, drug development, or regulatory approval process was lengthy. In each of these cases, the effective patent term is diminished as most countries calculate the patent term from the date of filing. Although drug development and regulatory approval processes do not directly impact the patent term, they do impact the *effective* patent term, which is the only term of relevance to the manufacturer of a patented drug as the drug cannot be commercialized without regulatory approval.

A brief review of the impact of data exclusivity will be provided here in the form of some FAQ before discussing the broader international scene, as well as more detailed arguments for and against data exclusivity.

Q: Is data exclusivity the same as patent exclusivity?

A: No. Data exclusivity is provided by a different set of laws than patent laws. A drug may be protected by data exclusivity even if it is not patentable.

Q: What does a company need to establish to be allowed data exclusivity?

A: Unlike patent protection, data exclusivity is not something that must be applied for. Rather, for countries that provide it, data exclusivity is automatically conferred on successful applicants for marketing approval of a new drug. Applicants need not establish anything other than the usual requirements for new drug approval (safety and effectiveness).

Q: What is the effect of data exclusivity? What happens during this time?

A: A firm that has data exclusivity is effectively free from competition by identical products during the term of exclusivity. Although data exclusivity does not preclude a second company from seeking approval, the inability to rely on prior tests to establish bioequivalence presents a substantial hurdle to entry of generic drugs. Most potential generic companies wait until the exclusivity period expires to avoid the time and expense of creating a duplicate set of clinical tests. The drug may have competition from other drugs within the same class (i.e. beta-blocker, antihistamine, or SSRI antidepressant). Still, no direct competition is a significant competitive advantage.

Q: What happens in countries without data exclusivity?

A: If a country does not grant data exclusivity, the only legal issue for competitors is patentability. If a country provides no data exclusivity, a generic drug manufacturer can immediately apply to sell its generic version. However, even if the generic is approved, it may be subject to litigation for patent infringement.

---

http://www.iprsonline.org/unctadictsd/bellagio/docs/Pugatch_Bellagio3.pdf. However, if those drugs are important to patients, it is nonetheless a problem.

Nonetheless, this is a risk a generic manufacturer may be willing to take if the patent is believed to be invalid.

B. GENESIS OF DATA EXCLUSIVITY

Data exclusivity is a relatively recent concept in the international arena. The first country to provide modern data exclusivity for drugs was the United States. Earlier US practice provided originators with greater protection of their data by considering it a trade secret of infinite duration, although there was no official law requiring this practice. The current limited-time data exclusivity for drugs was introduced in 1984 as part of the Hatch-Waxman Act, which made a number of changes to patent and regulatory laws.[11] Data exclusivity was considered to be part of a balance that was intended to promote innovation by pioneer companies while simultaneously encouraging greater entry of generic drugs to the market.

Data exclusivity is best understand as one of the measures intended to protect the interest of pioneer companies in light of other changes that aided generic companies under the Hatch-Waxman Act. This Act expedited the entrance of generics in two ways. First, it created an exception to patent infringement if a generic company was only making a limited amount of the patented drug to obtain regulatory approval.[12] This exception is often referred to as the "Bolar exception," and has since been broadly adopted in other countries. Second, it established the abbreviated new drug application procedure (ANDA) that permits a generic company to rely on the data of the pioneer applicant, such that the generic applicant only needed to submit limited data regarding its drug's bioequivalence to the pioneer.[13] Although this is now common in most countries, it was not at the time. Prior to 1984, generic companies generally needed to create their own clinical studies to enter the market because the pioneer data was considered a trade secret by the government and thus subject to indefinite protection. Not surprisingly, few companies incurred this expense—even after patents expired.[14] To address pioneer companies' concerns about the ability of generic companies to free ride on the time and expense

---

[11] Drug Price Competition and Patent Term Restoration Act of 1984, Pub. L. No. 98-417, 98 Stat. 1585, 1585-1605 (codified as amended at 21 U.S.C. § 355 (2006)).

[12] This exception is often referred to as regulatory review exception the "Bolar exception," and has since been broadly adopted in other countries.

[13] The same legislation allowed generic companies to apply for marketing approval *during* the patent term to help expedite the entry of generic medicines in the market, yet also provided research-based pharmaceutical companies with extensions of their patent term for delays in marketing approval.

[14] For example, between 1962 (when slightly different rules existed) and 1984, there were 150 drugs whose patents had expired, but for which there were no generics. *See* Gerard J. Mossinghoff, *Overview of the Hatch-Waxman Act and Its Impact on the Drug Development Process*, 54 FOOD & DRUG L.J. 187, 187 (1999) (citing H.R. REP. NO. 98-857, pt. 1 (1984).

of their data, data exclusivity was created. To further safeguard the interest of originator companies, patent terms were extended in cases where the regulatory approval process decreased the effective term of patent protection. An additional safeguard for originator companies was "patent linkage," as will be explained in more detail later in this chapter.

Although data exclusivity limits immediate generic competition, the ability to eventually rely on the data has resulted in more generic drugs. Before Hatch-Waxman, only two of the top thirteen drugs had generic competitors within one year of patent expiration, whereas for the same number of drugs with patents expiring between 1990 and 1993, eleven had generic competition within a mere two months of patent expiration.[15]

Although the United States was the first to provide data exclusivity, the EU soon followed suit.[16] In the EU, data exclusivity was granted for different reasons. In part, it was seen as important because not all EU Member States provided the scope of patent protection desired by the pharmaceutical industry; in particular, Spain and Portugal did not provide product patents to pharmaceuticals at that time. In addition, industry representatives suggested that not only should the EU adopt exclusivity, but that a longer period of exclusivity should be adopted to help its industries obtain a competitive edge over the United States.[17] However, the argument about supporting EU industries does not apply in the global market for pharmaceuticals as the same exclusivity applies to all companies.

Before considering whether other countries should adopt data exclusivity, the following questions recap the original rationale for data exclusivity. These points may be useful to consider in contrast to current arguments for requiring developing countries to adopt data exclusivity, as discussed in the next section.

---

[15] David Reiffin & Michael R. Ward, *Generic Drug Industry Dynamics*, 6 (F.T.C. Working Paper, 2002) (citing Anna Cook, Congressional Budget Office, How Increased Competition from Generic Drugs Has Affected Prices and Returns in the Pharmaceutical Industry (1998) and R. Caves, M. Whinston & M. Hurwicz, *Patent Expiration, Entry, and Competition in the U.S. Pharmaceutical Industry*, *in* Brookings Papers on Economic Activity: Microeconomics 1 (1991), *available at* http://www.ftc.gov/be/workpapers/industrydynamicsreiffenwp.pdf).

[16] Council Directive 87/21/EEC of Dec. 22, 1986, art. 2, 1986 O.J. (L 015) (amending Council Directive 65/65/EEC of 26 Jan. on the Approximation of Provisions Laid Down by Law, Regulation or Administrative Action Relating to Proprietary Medicinal Products, O.J. (022)). The EU data exclusivity legislation has since been revised. *See* Directive 2004/27/EC of the European Parliament and of the Council of March 31, 2004, O.J. (L 136) 34-57. Somewhat confusingly, two different periods of exclusivity apply to drugs. Generic applications submitted after 2005 are subject to harmonized data protection, but applications submitted before then are subject to the 2001 rule.

[17] For a detailed discussion of the politics behind enacting data exclusivity in the EU, see Sandra Adamini et al., *Policy Making on Data Exclusivity in the European Union: From Industrial Interests to Legal Realties*, 34 J. Health Pol. Pol'y & L. 979 (2009).

Q: Was data exclusivity introduced in the United States because drug companies needed more protection beyond patents?

A: No. Data exclusivity was introduced as part of a package of changes to US law. Data exclusivity in conjunction with other changes in the law actually enabled more generic competition because originators previously received infinite protection of clinical data, such that generic companies could not rely on earlier data.

Q: Did the EU introduce data exclusivity because companies needed better protection than provided under then-existing patent laws?

A: In part. At the time the EU introduced data exclusivity, some EU members did not provide patent protection on drugs, such that data exclusivity was the only method of protection for investment in originator drugs. However, that is no longer the case.

C. EVALUATING THE CASE FOR GLOBAL DATA EXCLUSIVITY

An important current issue is the appropriate scope of data exclusivity worldwide. As discussed in Chapter 3, TRIPS permits, but does not require data exclusivity. Many developed countries already provide some form of it, but the current issue is whether developing countries should be required to adopt it, as some have already had to do pursuant to FTAs (as discussed in Chapter 8).

Although patents are more commonly known and criticized for their impact on access to medicines, data exclusivity is the next frontier in debates about this topic.[18] As with patent protection, arguments for or against data exclusivity are mostly made by those who would benefit from such protection (i.e., research-based drug companies) or who would suffer from it (generic drug companies and those who believe there should be broader access to medicine). For a snapshot of the debates, consider the two following quotes:

> *[P]atent and trademark protection are not the entire story. . . . [T]he protection of crucial information generated by painstaking analysis in the drug and vaccine development process [is essential].*
>
> —Harvey Bale, Director-General, International Federation of Pharmaceutical Manufacturers Association[19]

---

[18] Professor Brook Baker is strongly critical of data exclusivity for developing countries and starkly suggests "[d]eath by patent is being reinforced by death by registration." Brook K. Baker, *Ending Drug Registration Apartheid: Taming Data Exclusivity and Patent/Registration Linkage*, 34 AM. J. L. & MED. 303, 343 (2008).

[19] INT'L FED'N PHARM. MFRS. & ASS'NS, ENCOURAGEMENT OF NEW CLINICAL DRUG DEVELOPMENT: THE ROLE OF DATA EXCLUSIVITY, Forward (2000) [hereinafter IFPMA, ENCOURAGEMENT].

For originator pharmaceutical companies, the expansion of data exclusivity provisions has become one of the main ways of extending market protection and blocking generic competition. Data exclusivity is seen now as the principal means of extending market protection for new indications, pharmaceutical forms and other variations, especially where these are not innovative enough to gain patent protection.

—Greg Perry, Director General of the European Generic Medicines Association[20]

This section first challenges common assumptions regarding the need for data exclusivity in general, as well as its extension to developing countries. Then, this section moves on to discuss specific issues in existing or pending laws, as well as possible ways to mediate unnecessary harm to access to medicine in cases where some type of data exclusivity seems inevitable.

1. Challenging Assumptions in Support of Data Exclusivity

a. Does it Provide a Crucial Incentive for Drug Development?

Research companies claim that without data exclusivity, no new drugs, let alone generic versions, would ever be brought to market. They assert there would simply be inadequate incentive to develop new drugs because other companies could free ride on their investment. These companies repeatedly emphasize the lengthy and costly drug development process: that it takes up to fifteen years, and costs around $1 billion to develop a single drug.[21] Although these data cannot be independently verified and thus the numbers are likely somewhat inflated, there is no serious question that drug development involves significant financial investment.[22]

---

[20] Valerie Junod, *Drug Marketing Exclusivity under United States and European Union Law*, 59 FOOD & DRUG L.J., 479, 480 (2004) (citing Greg Perry, *Data Exclusivity—A Major Threat to Access to Affordable Medicines*, *in* BUSINESS BRIEFING: PHARMAGENERICS 16 (2002).

[21] *E.g.*, IFPMA, ENCOURAGEMENT, *supra* note 19, at 1 ($500 million); INT'L FED'N PHARM. MFRS. & ASS'NS, A REVIEW OF EXISTING DATA EXCLUSIVITY LEGISLATION IN SELECTED COUNTRIES, intro. at 3 (5th rev. ed. 2005), *available at* http://www.ifpma.org/documents/NR2799/DataExclusivity_2005.pdf ($800 million and ten to fifteen years to complete); Adams & Branter, *Estimating the Cost of New Drug Development: Is it Really $802 Million?*, 25 HEALTH AFFAIRS 427 (2006) ($868 million, varying between $500 million and $2 billion by drug type and company). Moreover, they note that only about three of ten drugs have sales revenue that equal or exceed average development costs. *E.g.*, Henry G. Grabowski & John M. Vernon, *Returns on R&D on New Drug Introductions in the 1980s*, 13 J. HEALTH ECON. 383 (1999).

[22] Others cite much lower numbers. *E.g.*, Donald W. Light, *Misleading Congress About Drug Development*, 32 J. HEALTH POLITICS, POL'Y & L. 897 (2007) (suggesting a median of $300 million); Christopher P. Adams & Van B. Branter, *Estimating the Cost of New Drug Development: Is It Really $802 Million?*, 25 HEALTH AFFAIRS 420 (2006); COMM'N ON INTELLECTUAL PROP. RIGHTS, INNOVATION, & PUB. HEALTH, WORLD HEALTH ORG., PUBLIC HEALTH, INNOVATION, AND INTELLECTUAL PROPERTY RIGHTS 75–76 (2006) (estimating cost at $115–240 million).

However, although drug development may be costly, the proper question is whether data exclusivity is needed *in addition* to patent protection. Notably, drug companies cite the same costs in asserting the need for patent protection and data exclusivity. Although the costs of generating clinical data is allegedly 60–70 percent of the total cost of drug development,[23] that cost is already accounted for in the arguments for patent protection. For patentable drugs, data exclusivity does not provide any additional incentive to create better drugs or better clinical data. After all, a patented drug is of no value unless and until approved for sale. Moreover, as drug companies develop drugs for global sale, the existence of maximum protection in the most lucrative markets should adequately protect their sales.

Sometimes the pharmaceutical industry suggests that data exclusivity is necessary as a safety net for drugs that are unpatentable, or subject to more modest protection to ensure that there remains an incentive to develop new drugs. In other words, the industry assumes that patent protection necessarily promotes innovation, such that for drugs that are not patentable, alternatives should exist. Sometimes the argument for data exclusivity is couched in terms of a need for companies to be secure in their research investments. At the extreme, some might suggest that even if patent protection exists, data exclusivity is needed *in addition* to ensure that companies can be secure in investing in risky and expensive development processes. To some extent, data exclusivity may provide greater security to companies as unlike patent protection, it cannot be challenged, albeit the protection it provides is for a shorter term.

A related argument that is sometimes made is that data exclusivity provides an incentive to promote incremental research that benefits patients. Although clinical studies cannot meet the standards of patentability, some have argued that such trials nonetheless constitute a type of research that ultimately is beneficial to patients—but might not be carried out without the security provided by data exclusivity.[24] However, this incentive argument is misplaced because clinical trials are required to market a drug. Data exclusivity does not result in more or improved trials of a drug. This argument might have some relevance for data exclusivity regarding additional uses of existing drugs, but there are other problems with providing data exclusivity for new uses.

The incentive argument is especially questionable with respect to additional terms of data exclusivity for new *uses* of previously approved drugs. Although drug companies

---

[23] *E.g.*, Henry Grabowski, Patents and New Product Development in the Pharmaceutical and Biotechnology Industries 5, fig.1 (unpublished manuscript, Duke Univ.) (July 2002), http://en.scientificcommons.org/1043605.

[24] IFPMA, Encouragement, *supra* note 19, at 1; Manuel Campolini, Experiences of the Pharmaceutical Industry in Europe on Approval Procedures and Data Protection, EU-ASEAN Workshop on Data Protection, 5–6 (Apr. 2004), http://www.ecap-project.org/fileadmin/ecapII/pdf/en/activities/regional/data_protection_2004/campolini_data_protection_word.pdf (representing views of INTERPAT, a Swiss-registered association of major pharmaceutical companies).

tend to cite the same data for all discussions of exclusivity, the data necessary to obtain approval for a new use or formulation is likely to be substantially less costly because less data is necessary (as discussed in Chapter 1).

Even if there were an argument that pioneer companies deserve the incentive provided by data exclusivity, that argument has less weight in the global arena. Some question whether data exclusivity in all countries is necessary when it is already common in most developed countries, where pharmaceutical companies make the vast majority of their sales and profits. This issue parallels the argument against global patent protection; namely, pharmaceutical companies are arguably already provided adequate incentives to develop drugs from patent and data exclusivity protection in developed countries. In fact, some scholars have argued that requiring similar legal protection from developing countries would provide financial windfalls without any additional incentives—while at the same time unnecessarily limiting access to medicine for some global citizens with the greatest need.[25]

b. Is Data Exclusivity Important to Prevent Improper Free Riding?

Drug companies assert that they need and deserve a period of exclusivity before a competitor can free ride on their data—especially as the generic companies can price their drugs at a lower cost while still making a profit due to their minimal initial investment. The drug companies note that the cost of more limited testing by generic companies to show bioequivalence to a previously approved drug is only about $1–2 million—in comparison to the oft-cited $1 billion figure for a new drug.[26]

However, the free-riding rationale may be a red herring. First, it is a false claim that the only alternative to data exclusivity is complete free riding. As discussed later, a second entrant could be required to pay for use of the data. In addition, free riding off the work of the originator arguably occurs in other industries and is considered legitimate competition.[27] The drug industry may have more expensive investments. However, it is a false presumption that the only fair solution is complete exclusivity.

In addition, the lower financial investment of generic companies is also a red herring. The originator has the opportunity to recoup costs as the only manufacturer of a patented drug. In contrast, although generic companies have lower initial investments, their

[25] *E.g.*, Jerome H. Reichman with Catherine Hasenzahl, UNCTAD-ICTSD Project on IPRs and Sustainable Development, Non-Voluntary Licensing of Patented Inventions 6 (UNCTAD-ICTSD 2003).

[26] IFPMA, Encouragement, *supra* note 19, at 7 (estimating $1 million); Henry G. Grabowski et al., *Entry and Competition in Generic Biologics*, 28 Managerial & Decision Econ. 439, 443 (2007) (estimating in the low end of a $2 million to $200 million scale).

[27] Carlos Maria Correa, *Protection of Data Submitted for the Registration of Pharmaceuticals: Implementing the Standards of the TRIPS Agreement* 27–29 (2002), South Centre, http://apps.who.int/medicinedocs/pdf/h3009ae/h3009ae.pdf.

profits will also be lower due to increased competition. Generic companies must compete not only against the originator (who may continue to have some market dominance) but also with other generic companies.

Moreover, the argument that pioneer companies will be at an unfair advantage if generic companies can free ride on their data is especially unpersuasive in the global context. As noted earlier, global companies make most of their profits in countries that already provide data exclusivity. Therefore, the issue is whether it is unfair for companies in developing countries to free ride on their efforts when the originator company already has exclusivity in its dominant and most profitable markets. Moreover, the lack of data exclusivity in developing countries should have only minimal financial impact for originator companies as their profits in these countries are small. Although originator companies have little to gain financially in developing countries, the implications of data exclusivity in these countries is exacerbated because citizens often have only minimal resources. This issue is analogous to the argument that permitting poor countries to free ride on innovations of other countries by not providing patent protection is an overall positive for global welfare considering the minimal income lost in these countries versus the possible social utility.[28] Although this argument in the context of patents is now a moot point in light of TRIPS, there is no reason to follow the same path with respect to data exclusivity.

c. Is the Data Exclusivity Incentive Especially Important Now?

IFPMA, the international organization promoting the interests of multinational companies, has suggested that data exclusivity is increasingly important due to a variety of factors, including providing certainty in the risky business of drug development as an increasing number of compounds are denied patent protection, with only data exclusivity remaining to support that development.[29] In addition, PhRMA has suggested that drug developers are incurring greater financial costs because clinical trials are more lengthy and costly.[30]

---

[28] *E.g.*, Frederic M. Scherer, *A Note on Global Welfare in Pharmaceutical Patenting*, 27 World Economy 1127, 1127 (2004); Sean Flynn, Aidan Hollis & Mike Palmedo, *An Economic Justification for Open Access to Essential Medicine Patents in Developing Countries*, 37 J.L. Med. & Ethics 184, 185 (2009).

[29] *E.g.*, IFPMA, Encouragement, *supra* note 19, at 2 ("more and more compounds which are not patent protected (for whatever reason) are being developed and thus data exclusivity in some instances is the only available intellectual property right").

[30] *E.g.*, Organisation of Pharmaceutical Producers of India, OPPI Position Paper, Regulatory Data Protection—A Building Block for Pharmaceutical R&D 1 (2008), http://www.indiaoppi.com/OPPIpositionDataProtection.pdf (asserting the average number of clinical trials has doubled since the 1980s); Henry Grabowski, *Data Exclusivity for New Biological Entities*, Working Paper, 33 (2007), http://econ.duke.edu/Papers/PDF/DataExclusivityWorkingPaper.pdf (table showing that mean length of trials has increased from thirty-two to ninety-eight months over the last twenty years).

However, neither of these assertions withstands serious scrutiny. With respect to the purported decline in patent protection, that could be attributable to less innovation in the industry. Indeed, some cynics suggest that pharmaceutical companies use data exclusivity to extend protection for drugs at a time when their innovation has been decreasing and they have few potential blockbusters in the pipeline.[31] Moreover, even if patent protection is less available, that does not necessarily mean the social cost of providing exclusivity in the marketplace is worthwhile. IFPMA and PhRMA frequently note that data exclusivity was important to the development of Bristol Myer Squibb's anticancer drug marketed as Taxol, which does not have any patents on its active ingredients, yet incurred costs in excess of $500 million in testing.[32] However, there is contrary information suggesting that the federal government—not Bristol Myers—sponsored the clinical tests and footed the earlier development costs.[33] Even if Bristol Myers incurred costs, they are likely dwarfed by actual profits; sales for the *single year* of 2000 are reported to exceed $1.5 *billion*. In fact, Bristol Myers was summoned before Congress to justify its price on Taxol, and there is an entire book dedicated to the issue of whether Bristol Myers unduly profited from this single drug.[34]

In addition, the statement that costs are increasing due to more expensive clinical trial tests is simply not confirmed. The frequently cited Grabowski report is based on proprietary and unverifiable data. Verifiable data has not documented any trend in increase in clinical trial time.[35]

d. Does Data Exclusivity Only Provide Modest Protection?

Another argument for data exclusivity is that it actually offers only a modest amount of protection. In PhRMA's view, in a world without data exclusivity, data would be protected indefinitely and generic companies would *never* be permitted to rely on the original data.[36] In other words, PhRMA assumes the only alternative to data exclusivity is the situation that previously existed in the United States where not only was clinical data considered a trade secret, but regulatory laws did not permit generic companies to rely on the original data in seeking approval of generic versions, such that generic companies had to duplicate existing test data.

---

[31] *E.g.*, Dolores Cullen, *Data Protection: The New IP Frontier—An Overview of Existing Laws and Regulations*, 5 J. Generic Med. 9, 9 (2007); Pugatch, *supra* note 10, at 10–12.

[32] *E.g.*, IFPMA, Encouragement, *supra* note 19, at 7.

[33] *See* Consumer Project on Technology, *Disputes Involving Paclitaxel, A Cancer Drug Sold under Different Brand Names, Including Taxol*, http://www.cptech.org/ip/health/taxol/ (last visited Sept. 4, 2010) (collecting sources).

[34] Jordan Goodman & Vivien Walsh, The Story of Taxol: Nature and Politics in the Pursuit of an Anti-Cancer Drug (2001).

[35] Salomeh Keyhani et al., *Are Developing Times for Pharmaceuticals Increasing or Decreasing?*, 25 Health Affairs 461, 461-67 (2006) (analyzing clinical trial periods for drugs approved from 1992 to 2002).

[36] *E.g.*, IFPMA, Encouragement, *supra* note 19, at 2.

Although it is true that PhRMA agreed to data exclusivity in exchange for permitting generic companies to rely on their data in the United States, this balance is not required. This is evident by other countries approaching the issue differently. India, for example, does not currently provide data exclusivity, and thus permits companies to immediately seek approval of generic versions of previously approved drugs. Of course, an approved generic manufacturer may nonetheless be subject to patent infringement, but some companies may be willing to take the risk if they believe the patent is invalid.

e. Is Data Exclusivity Consistent with Competition and Promotion of Generics?
PhRMA does not view data exclusivity as barring competition. In particular, PhRMA suggests that competitors can engage in their own independent clinical tests if they do not want to wait until the end of data exclusivity.[37]

Those in favor of generic drugs note that the claim the generic industry could do its own testing is unrealistic and has negative social consequences. As a subsequent supplier of the identical drug, the generic company would never be able to price its drug high enough to recoup its costs. Moreover, repeating clinical tests is not only an expensive waste of resources, but is unethical if patients are denied access to a drug known to be superior.

f. Is Data Exclusivity Actually Beneficial to Developing Countries?
PhRMA suggests that global expansion of data exclusivity would benefit all. In particular, PhRMA has indicated that countries who provide data exclusivity may reap benefits, such as attracting foreign direct investment and increased research and development collaboration. However, there is no evidence thus far that this is the case[38] and available evidence suggests the opposite is true. For example, after introducing patent protection and data exclusivity pursuant to an FTA, Jordan reported no such benefits.[39] In addition, it may be worthwhile to note that PhRMA made the same claim earlier in support of TRIPS with regard to raising patent standards globally, but there is not yet convincing evidence that most developing countries have benefited from increased patent protection.

2. Additional Concerns about Data Exclusivity for Developing Countries

There is some irony to the fact that a compound may be denied patent protection because it is not adequately innovative, yet be entitled to stronger protection than patents through

[37] *E.g.*, *id.* at 3.

[38] Robert Weissman, *Public Health-Friendly Options for Protecting Pharmaceutical Registration Data*, INT'L. J. INTEL. PROP. MGMT. 113, 117 (2006).

[39] *See, e.g.*, Hamed El-Said & Mohammed El-Said, *TRIPS-Plus Implications for Access to Medicines in Developing Countries: Lessons from Jordan-United States Free Trade Agreement*, 10 J. WORLD INTELL. PROP. 438, 452-57 (2007); OXFAM INT'L, ALL COSTS, NO BENEFITS: HOW TRIPS-PLUS INTELLECTUAL PROPERTY RULES IN THE US-JORDAN FTA AFFECT ACCESS TO MEDICINES 7 (Oxfam Briefing Paper, 2007), *available at* http://www.oxfam.org.uk/resources/issues/health/downloads/bp102_trips.pdf.

data exclusivity. After all, unlike patent protection, data exclusivity is ironclad: there is no way to challenge data exclusivity. Generic companies and consumers must simply wait until the term expires. Data exclusivity also provides additional protection to patented drugs by essentially eviscerating existing exceptions to patent rights. For example, although countries who comply with the procedural requirements under TRIPS (see Chapter 5) are permitted to override the patent owner's desires by issuing compulsory licenses on patented drugs, the ability to make and sell the patented drug is completely undermined if that drug is still protected by data exclusivity.

Data exclusivity may have pernicious effects in developing countries that do not exist in the most profitable countries. Although it is true that data exclusivity is for only a limited time, there is generally no requirement that the originator seek permission to sell its drug in all countries simultaneously. Accordingly, a company could first seek marketing approval in wealthy countries, but delay seeking similar approval in countries with a more modest ability to pay. This would result in a delay in when new drugs are available to the poorer countries. Moreover, if these poorer countries also grant data exclusivity, their citizens will not have access to low-cost generics until long after consumers in wealthy countries have such drugs. In fact, this is often the situation.[40]

Moreover, all arguments of pioneer companies should be considered against the potential costs for citizens of the developing world. Citizens in these countries have limited resources, such that data exclusivity may exacerbate existing health crises for the many who are unable to purchase monopoly-priced drugs. Whereas drug costs for citizens in wealthier countries are often substantially subsidized by public or private insurance, citizens in poor countries often must pay the entire cost of the medicine. In addition, as pharmaceutical companies profit from a global market and already have data exclusivity in the wealthiest markets, there does not seem to be a strong case to charge higher prices to the poorest citizens through data exclusivity.

3. Minimizing the Effects of Data Exclusivity

Even though there are serious concerns about data exclusivity, national laws that provide it are rapidly being adopted with little public awareness of the laws themselves, let alone their impact. As noted in Chapter 8, many countries have already succumbed to pressure to enter FTAs that contain provisions mandating data exclusivity. As a complete rejection of data exclusivity may not be realistic, this section instead explores possible ways to

---

[40] *E.g.*, Baker, *supra* note 18, at 310-11; Ellen R. Shaffer & Joseph E. Brenner, *A Trade Agreement's Impact on Access to Generic Drugs*, 28 HEALTH AFFAIRS, 957, 961, 964–65 (Aug. 25, 2009), http://content.healthaffairs.org/cgi/content/abstract/28/5/w957; *see also* MINORITY STAFF OF H.R. COMM. ON GOVT. REFORM, 109TH CONG., TRADE AGREEMENTS AND ACCESS TO MEDICATIONS UNDER THE BUSH ADMINISTRATION, PREPARED FOR REP. HENRY A. WAXMAN 7 (2005).

minimize the harm of data exclusivity if trade pressures force developing countries to adopt some form of it.

a. Cost Sharing Instead of Exclusivity

The ideal, yet most ambitious approach to promoting access to medicine would be to institute a cost-sharing approach, rather than complete exclusivity. For example, drawing upon the model used for agricultural approvals, exclusivity could be replaced with a cost-sharing model. To address the issue of free riding, the generic company would agree to share in some of the costs of producing the original data.[41] This approach is popular among public health advocates but generally opposed by the pharmaceutical industry and yet to be adopted by any country.[42] Resistance is unsurprising in that this approach would necessarily result in lower profits. However, the resistance may also be a function of the fact that compensation under this approach would require documentation of actual costs; pharmaceutical companies have been notoriously resistant to providing a clear picture of costs of research and development of individual drugs.[43] They repeatedly claim it is impossible to calculate costs for specific drugs as they engage in broad-based research that may be relevant to multiple drugs. They also assert that only looking at costs for commercialized drugs is unfair because most of their expense is spent on the many compounds that are ultimately not marketable. Despite this resistance, the recent FTA between the EU and Korea seems to adopt a cost-sharing approach.[44] Although this is impressive, it may have no impact on Korean citizens as Korea's FTA with the United States requires data exclusivity without the opportunity for a cost-sharing approach.[45]

---

[41] A complete cost sharing model with no period of exclusivity is actually more generous to generic companies than the current scenario under US law that guarantees the agricultural originator complete exclusivity for the first ten years and only requires cost-sharing in years 11–15. U.S. Federal Insecticide, Fungicide, and Rodenticide Act (FIFRA), 7 U.S.C. § 136a(c)(1)(F).

[42] *E.g.*, Jerome H. Reichman, *The International Legal Status of Undisclosed Clinical Trial Data: From Private to Public Good*, *in* NEGOTIATING HEALTH: INTELLECTUAL PROPERTY AND ACCESS TO MEDICINES 135, 145-46 (Pedro Roffe et al. eds., 2006); Weissman, *supra* note 38, at 117; JUDIT RIUS SANJUAN ET AL., CONSUMER PROJECT ON TECH., POLICY BRIEF NO. 1, A COST SHARING MODEL TO PROTECT INVESTMENTS IN PHARMACEUTICAL TEST DATA 3-8 (2006); *see also* TAHIR AMIN ET AL, INITIATIVES FOR MEDICINES, ACCESS & KNOWLEDGE (I-MAK), THE IMPACT OF ARTICLE 39.3 IN INDIA: A PRACTICAL PERSPECTIVE 13–17 (July 2006) (discussing different variations of cost sharing with a view toward considering the best option for India).

[43] This issue has been noted as a major stumbling block to a compensatory model by Professor Jerome H. Reichman, Lecture, *Rethinking the Role of Clinical Trial Data in International Intellectual Property Law: The Case or a Public Goods Approach*, 13 MARQ. INTELL. PROP. L. REV. 1, 34-35 n.151 (2009).

[44] Free Trade Agreement Between the EFTA States and the Republic of Korea, Annex XIII, art. 3, Ice.-Liech.-Nor.-Switz., Dec. 15, 2005 (permitting a party to permit "applicants to rely on such [clinical] data if the first applicant is adequately compensated").

[45] Free Trade Agreement Between the United States of America and the Republic of Korea, art. 18.9, U.S.-Kor., Jun. 30, 2007.

However, even if data exclusivity (rather than cost-sharing) is mandatory, steps can be taken to minimize its potential impact on public health, as detailed below.[46]

b. Avoid Data Exclusivity for Unpatented Drugs

One step that can be taken to minimize the harm of data exclusivity is to bar the use of exclusivity for drugs that are not patented. Although this may seem an odd approach given that data exclusivity is most valuable *precisely* for compounds that are unpatentable, this is a critical global issue for countries where there is currently no requirement to provide patents on drugs. Even under TRIPS, LDCs currently need not provide patent protection over products, so they can legally make generic versions of drugs patented elsewhere—unless they have agreed to provide data exclusivity pursuant to an FTA. In addition, even in countries that already provide patents on products, limiting data exclusivity for patented drugs can minimize costs. Accordingly, countries interested in the broadest possible access to unpatented drugs should exclude *unpatented drugs* from data exclusivity.

c. No Exclusivity without Reliance

Data exclusivity could also be limited to countries that actually require submission of clinical data with applications to sell a new drug. Although most developed countries require those seeking to sell a new drug to submit clinical data of safety and efficacy, some developing countries do not because they lack the necessary infrastructure to evaluate such information. However, some of these developing countries are nonetheless required to grant data exclusivity based on clinical data submitted *in a different country*. For example, the company with first marketing approval in Chile could be granted a period of market exclusivity based on its submitting clinical data not in Chile, but in the United States. Permitting the first entrant in Chile to preclude generic competitors based on data filed in the United States seems to be a tenuous approach. After all, why should a generic applicant be denied the ability to submit an application that relies on data submitted in a *different* country—especially as data exclusivity is likely already granted in the other country.

d. No Protection for Public Data

The argument for data exclusivity is also especially weak with respect to data that is publicly available, in which there should not be any improper reliance or claims of free riding. This may seem intuitive, but some FTAs have attempted to provide protection for *any data* submitted rather than data that is undisclosed except to the regulatory agency.[47]

---

[46] For a detailed discussion of model laws, see Carlos M. Correa, *A Model Law for the Protection of Undisclosed Data*, *in* INTELLECTUAL PROPERTY AND SUSTAINABLE DEVELOPMENT: DEVELOPMENT AGENDAS IN A CHANGING WORLD 370 (Ricardo Melendez-Ortiz & Pedro Roffe, eds., 2009).

[47] *E.g.*, *E.g.*, United States-Singapore Free Trade Agreement, art. 16.8(1), U.S.-Sing., May 6, 2003, *available at* http://www.ustr.gov/trade-agreements/free-trade-agreements/singapore-fta/final-text; Free Trade Agreement,

e. Limit Duration of Data Exclusivity

As noted earlier, the duration of data exclusivity is of essential importance with respect to its impact on access to medicine. To facilitate access to medicine, the duration should be as short as possible. Although most data exclusivity rules provide a five- to ten-year term, there is no evidence these terms correlate with the expense of clinical trials, or how long it would take to recoup costs.[48] In fact, even in countries that already provide data exclusivity, some have suggested a five-year period may already provide too much protection.[49] Accordingly, to the extent possible, countries desirous of promoting access to lower-cost generics should seek a minimum term.

The duration of exclusivity can also be limited by linking it to other important events. For example, if data exclusivity were required to begin from the first worldwide approval of the drug, that would be helpful in addressing a current problem where companies can delay seeking marketing approval and thus delay access to generic drugs in some markets. This delay, when combined with data exclusivity, serves to prolong the period during which generic drugs are unavailable. The term of data exclusivity could also be set to expire at the same time as the corresponding patent term such that data exclusivity does not extend commercial exclusivity beyond the patent term.

f. Limit Scope of Data Protection

There are two important issues with regard to limiting the scope of protection for data exclusivity. First, the least onerous amount of data exclusivity would protect only data related to "new" entities and deny protection for new *uses or dosages* of already known entities. In addition, the scope can be further limited by narrowly defining what is considered to be a "new" entity. For example, a country could consider a compound to be new only if it was never granted approval in *any* country worldwide. Alternatively, a compound could be deemed not new if it was publicly known, regardless of whether it was previously subject to regulatory approval.

g. Additional Safeguards

Beyond the above suggestions to limit the scope of data exclusivity, additional steps can be taken to mediate its usual blunt and exclusive protection. As noted earlier, data exclusivity generally cannot be challenged; it also may nullify the utility of patent exceptions, such as compulsory licenses. However, that need not be the case. For example, data exclusivity provisions could include exceptions that mirror existing patent exceptions, such as

---

art. 15.10(1)-(2), U.S.-Morocco, Jun. 15, 2004, *available at* http://www.ustr.gov/trade-agreements/free-trade-agreements/morocco-fta/final-text.

[48] Weissman, *supra* note 38, at 128 (suggesting the term is arbitrary).

[49] *E.g., Statement of James Love on Health Registration Data Exclusivity, Biomedical Research, and Restrictions on the Introduction of Generic Drugs, before S. Subcomm. on Labor, Health and Human Services of the S. Comm. on Appropriations*, 105th Cong. (Oct. 21, 1997), *available at* http://www.cptech.org/pharm/senhregd.htm.

establishment of procedures to ensure that if a patent is subject to a compulsory license, data exclusivity may be similarly waived. In addition, data exclusivity can exist together with new exceptions to waive the exclusivity in certain cases where public health is at risk. Data exclusivity could also be terminated after a drug reaches a certain sales target or profit. This would likely be strongly opposed by the pharmaceutical industry, but it is a possibility nonetheless. In addition, especially for developing countries that are pressured to adopt data exclusivity, the most pernicious impacts can be alleviated by requiring companies to promptly seek approval for new drugs shortly after US approval.[50]

III. Patent Linkage

A. OVERVIEW

The term *patent linkage* refers to a practice by some national regulatory authorities of denying approval of generic drugs that are "linked" to an existing patent. In other words, if the generic drug would allegedly infringe a patent if sold in the market, it will be denied approval. Importantly, denial is not based on the usual regulatory standards of drug safety and efficacy but solely on a presumed *potential* patent problem.

In countries that practice patent linkage, regulatory authorities typically deny approval of generic versions of drugs when those drugs are on a list of drugs linked to current patents. This list reflects assertions by self-interested patent owners regarding what patent(s) are violated by making or selling certain drugs.[51] There is no independent assessment of whether the listed patent(s) are relevant, let alone valid. This is not surprising as regulatory authorities have no competency in the area of patents.

According to patent owners, patent linkage is an efficient means for preventing patent infringement.[52] In their view, patent linkage is a logical method to prevent unnecessary, expensive, and uncertain litigation against generic manufacturers. Patent owners assert that patent linkage is preferred over traditional legal patent enforcement (for patent infringement) because once a generic drug is on the market, it will inevitably and improperly erode the market share of the patent owner while the parties await a trial on the merits. Although patent owners can generally seek preliminary injunctions before trial,

[50] Chile has taken this approach. Ley No. 19.996, Titulo VIII, Articulo 91(e), Marzo 11, 2005, DIARIO OFICIAL [D.O.] (Chile).

[51] For example, in the United States, this information is in what is referred to as the "Orange Book." Although the US regulatory agency compiles the list, the data solely reflects information submitted by applicants without any independent review. In addition, the FDA has no legal duty to analyze whether the listed patents are actually relevant. *See* Alphapharm v. Thompson, 330 F. Supp. 2d 1, 9 (D.C. Cir. 2004).

[52] *E.g.*, PHARMACEUTICAL RESEARCH AND MANUFACTURERS OF AMERICA [PHRMA], SPECIAL 301 SUBMISSION 2010, at 14 (2010), *available at* http://keionline.org/sites/default/files/USTR-2010-0003-0245.1.pdf.

they contend that this is an inadequate solution because the granting of such injunctions is uncertain and depends on a judicial weighing of equities.[53] Moreover, patent owners consider patent linkage to be a necessary procedural safeguard to ensure that drug regulatory agencies do not contribute to infringement of patent rights granted by another government entity.[54]

However, generic manufacturers—and their sympathizers—paint a different picture. In their view, patent linkage results in an improper extension and enforcement of patent rights.[55] An application to sell a generic drug is not a traditional form of patent infringement. Nonetheless, the generic applicant is barred from approval where patent linkage exists. In effect, patent linkage is more powerful than traditional patent infringement, where normally a patent owner must initiate a lawsuit and is not guaranteed an injunction before trial. With patent linkage, the patent owner need take no legal action to preclude the generic drug from entering the market. Accordingly, some suggest that the usual evidentiary presumption of validity granted to patents in traditional litigation is replaced by an absolute guarantee of patent validity and a de facto injunction. Patent linkage may thus shield patents from a proper judicial determination of validity. This is highly problematic as issued patents are not necessarily valid. Indeed, more than half of litigated patents are found invalid.[56] Moreover, the fact that patent linkage is imposed by an agency without the competence to assess patent issues means there is no protection of the public interest from invalid patents and the resulting higher prices that may occur.[57]

---

[53] Patent owners typically do not mention that if they prevail, they will be entitled to damages to compensate them, including lost profits during the period of infringement; in addition, they do not mention that if they prevail and obtain a permanent injunction, their market share will be entirely restored.

[54] PhARMA, *supra note* 52, at 14.

[55] *E.g.*, Carlos M. Correa, *Expanding Patent Rights in Pharmaceuticals: The Linkage between Patents and Drug Registration*, *in* The Development Agenda: Global Intellectual Property and Developing Countries 247 (Neil Weinstock Netanel ed., 2009); Judit Rius Sanjuan, Consumer Project on Tech. Discussion Paper No. 2, Patent-Registration Linkage 6 (2006), http://www.cptech.org/publications/CPTechDPNo2Linkage.pdf (criticizing patent linkage).

[56] This is true not only for patents in general, but for pharmaceutical patents in particular. *E.g.*, Fed. Trade Comm., Generic Drug Entry Prior to Patent Expiration: An FTC Study 13 (2002) (finding that generic applicants successfully challenged pharmaceutical patents as invalid or not infringed in 73% of the cases). In addition, considering that some have suggested that a presumption of validity is too strong even where patents are litigated, patent linkage is especially problematic if there is no opportunity to legally challenge the patent. Fed. Trade Comm. To Promote Innovation: The Proper Balance of Competition and Patent Law and Policy ch. 6, at 26–28 (2003) (suggesting lower evidentiary threshold for all litigated patents).

[57] *See, e.g.*, World Health Organization, Briefing Note - Access to Medicines: Data Exclusivity and Other "TRIPS-Plus" Measures 3 (Mar. 2006).

Patent linkage may also exacerbate the existing practice of "evergreening." As noted in Chapter 1, pharmaceutical companies often seek to make incremental changes to drugs that may have only modest social utility, but may help extend commercial profits. Patent linkage may increase this evergreening phenomena by providing an incentive for companies to sequentially patent different aspects of a commercial drug to extend the time before generic manufacturers can enter the market.

1. Middle Ground?

Although the above paragraphs paint the most extreme views of patent linkage, there may be some middle ground. In particular, a nation could apply the general principle of patent linkage while simultaneously permitting generic manufacturers to challenge whether a patent is either invalid or not infringed in litigation. The United States and Canada both provide an opportunity for generic manufacturers to challenge the patent validity at some point during the term of data exclusivity.[58] However, the United States is unusual in offering an incentive to the first generic applicant in the form of a six-month period of exclusivity protecting it from competition from other generic companies.

In addition, although the type of patent linkage typically suggested in FTAs and by the United States requires an absolute bar on approval of generic applicants, patent linkage need not be absolute. For example, Australia was required to adopt a form of patent linkage pursuant to an FTA with the United States, but its patent linkage only requires *notification* to the patent owner and not a mandatory stay of approval.[59] If patent linkage only requires notification, it will not result in a de facto injunction. The generic manufacturer may still be subject to a patent infringement suit. However, in a traditional patent infringement suit, a generic company at least has the opportunity to challenge a patent on its merits. In addition, Australia introduced an additional safeguard to public interest by requiring a patent owner initiating an infringement suit to certify a good faith basis in a reasonable prospect of success, subject to fines of up to $10 million if that certification is later found to be false or misleading.[60]

2. Implications of Patent Linkage

Any form of patent linkage may impact traditional exceptions to patent rights. In particular, patent linkage may limit the ability to use the regulatory review exception as well as compulsory licenses, both of which are exceptions to patent rights consistent with

---

[58] Although some other countries allow a generic applicant to challenge patent validity, the United States is the only country that provides a period of exclusivity to the first generic applicant who is successful in such a challenge.

[59] US-Australia Free Trade Agreement, art. 17.10.4(b), U.S.-Austl. May 18, 2004.

[60] *Therapeutic Goods Act 1989* (Cth) § 26C(3)–(5) (Austl).

TRIPS. As noted in other chapters, many nations have an exception to standard patent rights that permits a generic company to make limited quantities of what would be an infringing drug during the term of a patent. This permits the generic company to do the testing necessary to obtain regulatory approval to sell a generic *after* the patent expires. The regulatory review exception is considered necessary because the patent owner would otherwise have a de facto extension of exclusivity for the several years typically needed for the generic company to do the test, then apply for and receive regulatory approval. However, patent linkage may limit the ability of a generic company to engage in such tests unless there is an exception to patent linkage to permit this activity.[61] Similarly, the ability of nations to issue compulsory licenses could be hampered without an express exception to patent linkage in the case of compulsory license. After all, even if a compulsory license is granted on a patented drug, that license is of no utility unless and until regulatory approval is granted to sell the licensed drug.

Before discussing the genesis of patent linkage, as well as its introduction into the international realm, a few points are worth emphasis:

Q: What is the effect of patent linkage?

A: Patent linkage prevents approval of generic drugs that would arguably violate patent rights if actually sold.

Q: What is the harm of patent linkage if it simply prevents approval of products likely to infringe a patent?

A: Patent linkage may improperly prevent approval of generic drugs that would not be found to infringe if actually adjudicated; in addition, if there is no mechanism to challenge the relevance, let alone the validity of the associated patent, the patent will be given more protection than it deserves.

Q: What happens in countries without patent linkage?

A: If there is no patent linkage, a national regulatory authority can approve a generic version of a patented drug at any time—subject to any other restrictions, such as data exclusivity. Notably, regulatory approval of a generic will not provide immunity against a patent infringement suit. However, a generic manufacturer generally cannot challenge the validity of a patent if there is patent linkage.

[61] The European Generic Medicines Association has suggested that the negative implication for patent linkage on use of the regulatory review (Bolar) exception is one reason to avoid adopting a linkage requirement, although some European countries (Hungary and Slovakia) have adopted some linkage schemes. *See, e.g.*, EUROPEAN GENERICS ASS'N, PATENT RELATED BARRIERS TO MARKET ENTRY FOR GENERIC MEDICINES IN THE EUROPEAN UNION 4.1, at 23 (Kristof Roox, ed., 2008), *available at* http://www.marcasepatentes.pt/files/collections/pt_PT/1/178/EGA%20Report%20IP%20Barriers%20Generic%20Medicines.pdf.

B. GENESIS OF PATENT LINKAGE

As with data exclusivity, patent linkage originated in the United States in the same legislation that created data exclusivity: the 1984 Hatch-Waxman Act. As stated previously, this law provided generic manufacturers with two things. First, after a period of data exclusivity, generic companies were permitted to rely on clinical data of pioneer companies so that they no longer had to create and submit duplicative tests. Second, generic companies were provided an exception from patent infringement for making the patented invention to establish the bioequivalence necessary for obtaining marketing approval. This exception enabled generic drugs to get on the market faster as it would otherwise take several years to obtain regulatory approval of a generic after the patent had already expired. In exchange for these benefits, originator companies received not only a limited term of data exclusivity, but an extension of the patent term and patent linkage.

Patent linkage provides a balance to both data exclusivity and the exception from patent infringement for generic drugs. Previously, generic companies could not begin to take steps to obtain approval until after the patent expired, and they had to do their own tests. Now, generic companies could, after a period of data exclusivity, piggyback on the test results of originator companies and apply for approval during the patent term. But patent linkage prevents approval of a generic during the patent term.

The potential harm of linking to patents, however, is mediated under US law. In particular, the United States provides a nuanced system that permits generic applicants to challenge a linked patent as being either invalid or not infringed during the last year of data exclusivity.[62] The mechanism by which this happens is that the generic applicant certifies as part of its application that the linked patent is either invalid or not infringed, and the patent owner is informed of this information.[63] The patent owner can then sue the generic manufacturer for patent infringement (as by statute an application with such a certification is considered patent infringement).[64] So long as the patent owner promptly sues, the regulatory authority may no longer take action on the generic application;[65] it must cease action for thirty months until the patent is declared invalid or not infringed, or until the patent expires—whichever happens first.

A generic company may be inclined to challenge patents not only because the companies win in over 70 percent of the cases,[66] but because US law provides an incentive

[62] This is referred to as a "Paragraph IV certification" in reference to the relevant statutory provision. 21 U.S.C. § 355(j)(2)(A)(vii)(IV) (2006). Every applicant for a generic drug must certify one of four things in its application: that there is no patent, that the relevant patent has expired, that the patent will expire, or that the patent is invalid or will not be infringed.

[63] 21 U.S.C. § 355(j)(2)(B) (2006) (requiring generic applicant to notify the patent owner). Due to data exclusivity, this can happen no earlier than four years after the original patented drug is first approved for sale.

[64] 35 U.S.C. § 271(e)(2) (2006).

[65] 21 U.S.C. § 355 (j)(5)(B)(iii) (2006).

[66] Fed. Trade Comm., Generic Drug Entry, *supra* note 56 at 6.

to do so. In particular, a six-month period of exclusivity is provided to the first generic company establishing that a patent is invalid or not infringed.[67] During this time, no other generic drug can be approved, such that the first generic company can secure a competitive advantage by being the only supplier other than the original patent-owning company. Accordingly, although the United States provides for patent linkage in conjunction with data exclusivity, the two procedures do have mechanisms to minimize potential harm of this linkage to invalid patents.

Although many generic manufacturers have challenged patents (despite the legal expenses of doing so), this does not necessarily result in faster access to low-cost generic. In particular, there have been instances where patent owners and generic manufacturers have reached settlement agreements that result in a payment to generic manufacturers to keep a generic drug off the market until the conclusion of data exclusivity and patent term.[68] This settlement may be financially beneficial to the generic manufacturer, but does nothing to improve access to cheaper generics. In addition, while some might consider such agreements to be in violation of competition laws, a number of courts have sanctioned such agreements.[69]

C. PATENT LINKAGE IN THE GLOBAL ARENA

At the behest of the pharmaceutical industry, the United States has aggressively pursued patent linkage with both developed and developing countries since the conclusion of TRIPS. A handful of countries presently provide patent linkage. Among developed countries, Canada and Australia provide patent linkage, but EU Member States do not. In addition, Canada and Australia only introduced patent linkage recently in compliance with FTAs with the United States. Developing countries that have acceded to demands for patent linkage include China, Jordan, Mexico, UAE, and Singapore.

Patent linkage mandated through FTAs generally require that nations include measures to "prevent" generic companies from marketing a product that is covered by a patent, together with a requirement to notify the original patent owner of any generic applicant that is granted marketing approval during the patent term.[70] Some FTAs preclude

---

[67] 21 U.S.C. § 355(j)(2)(vii) (2006); *see also* Erika King Lietzan, *A Brief History of the 180-Day Exclusivity under the Hatch-Waxman Amendments to the Federal Food, Drug and Cosmetic Act*, 59 FOOD & DRUG L.J. 287, 288 (2004).

[68] *E.g.*, FED. TRADE COMM., GENERIC DRUG ENTRY, *supra* note 56, at 25-34.

[69] *E.g.*, Scott Hemphill, *Paying for Delay: Pharmaceutical Patent Settlement as a Regulatory Design Problem*, 81 N.Y.U. L. Rev. 1553 (2006); Dana A. Elfin, *Official Says FTC Still Seeks to Bring Case on Reverse Payments to End Circuit Split*, 80 PATENT, TRADEMARK & COPYRIGHT J. (BNA) at 788 (Oct. 15, 2010). *See also* Michael Carrier, *A Real-World Analysis of Pharmaceutical Settlements: The Missing Dimension of Product Hopping*, 62 FLA. L. REV. 1009, 1033-36 (2010) (suggesting that settlements in combination with methods used to promote slightly new drugs may in combination create anticompetitive harm).

[70] *E.g.*, U.S.-Australia Free Trade Agreement, art. 17.10.4, U.S.-Australia, May 18, 2004.

approval of any generic drugs unless and until a patent has expired.[71] In other words, during the entire term of the patent, the regulatory authority is prohibited from approving a generic equivalent. This delays the entry of generic drugs and provides an additional de facto period of exclusivity to the originator drug for the time that it takes the regulatory agency to review and approve the generic application once the patent term expires—a period that may easily take several years.

Although patent owners contend that patent linkage is an important complement to patent protection, others suggest it is just another way for patent owners to improperly extend their commercial advantage. For a snapshot of viewpoints, consider the following two views of patent linkage:

> *[L]inkage[] is important. Such a mechanism provides a "procedural gate" or safeguard, because it ensures that drug regulatory entities do not inadvertently contribute to infringement of patent rights granted by another government entity. . . . Legal mechanisms that allow for early resolution of patent disputes before the generic product in question gains marketing approval avoid the need for complex litigation over damages for marketing an infringing product.*
>
> —PhRMA[72]

> *[Patent] [l]inking is an underhanded way of preventing generic competition. . . . Rather than the company having to sue through the courts to enforce its patents, the job is done behind the scenes and without publicity by the [national drug agency].*
>
> —MSF[73]

An important issue with domestic implementation of patent linkages in developing countries that are subject to FTAs is that they may result in overprotection of invalid patents. The linkage requirements imposed by FTAs do not prohibit countries from permitting a generic applicant to challenge a patent as invalid or not infringed (as currently possible under US law). However, they also do not *mandate* that such a procedure exist. There may also not be any strong local generic industry that can lobby to include such provisions.

---

[71] An example of such a requirement lies in the U.S.-Chile Free Trade Agreement, art 17.10.2(c), U.S.-Chile, Jan. 1, 2004 which states: "With respect to pharmaceutical products that are subject to a patent, each Party shall . . . not grant marketing approval to any third party prior to the expiration of the patent term. . . ."

[72] PHARMA, SPECIAL 301 SUBMISSION 2010, *supra* note 32, at 14.

[73] MÉDECINS SANS FRONTIÈRES, MSF BRIEFING NOTE, ACCESS TO MEDICINES AT RISK ACROSS THE GLOVE: WHAT TO WATCH OUT FOR IN FREE TRADE AGREEMENTS WITH THE UNITED STATES 23 (2004), *available at* http://www.twnside.org.sg/title2/FTAs/Intellectual_Property/IP_and_Access_to_Medicines/AccessToMedicinesAtRiskAcrossTheGlobe-WhatToWatchOutForMSF.pdf..

Moreover, incorporating a procedure to sustain challenges by generic companies may be unrealistic in a developing country that has a limited judicial infrastructure. Generic companies may also not have the incentive to challenge patents; notably, generic companies in the United States have an incentive because the first successful challenger of a patent is entitled to a period of market exclusivity that may be extremely valuable in enabling the generic company to gain a commercial advantage over other generic companies. For all these reasons, developing countries that agree to implement patent linkage may end up with the harshest form of linkage that results in providing all patents with total insulation from challenge during their patent term.

1. Mediating the Harm of Patent Linkage

An important question is how harm from patent linkage may be minimized to the extent that nations are pressured into incorporating some type of linkage process. A variety of methods can be pursued, ranging from optional linkage or diluted linkage requirements to additional procedures that protect the public interest.

2. Minimal Patent Linkage

The best way to minimize the harm of patent linkage is not to mandate it at all. This may not seem possible given the United States has been aggressively seeking it. However, there are some US FTAs under which patent linkage is possible, but not required.

The most diluted form of patent linkage is to require only that the patent owner be notified of the application for a generic equivalent, but not mandate that the application be denied. In other words, a generic drug could be approved during the patent term, but the patent owner would be given notice such that it could promptly sue for patent infringement as soon as the generic equivalent is made or sold.

Another diluted form of patent linkage is to require the patent owner be notified of a generic application and make it *optional* for a country to deny approval of the generic, together with additional requirements. As discussed in Chapter 8, this is the approach of the US FTA with Panama and Peru. The additional requirements are that the generic applicant be provided an expeditious procedure to challenge the validity or applicability of the pertinent patent and that an effective reward exist for a successful challenge. In other words, some mechanism to protect the public interest from invalid patents is included. However, whether developing countries will be able to successfully implement these judicial challenges is unclear. Whether the United States may later pressure such countries to adopt mandatory patent linkage also remains to be seen.

3. Mandatory Patent Linkage, but Limited Scope of Patents

Another way patent linkage harm can be mediated is with respect to what patents may bar. Linkage could be limited to only drugs based on new chemical compositions. Patents on new uses of known compositions, or new dosages of existing drugs, would not give rise to patent linkages. Accordingly, although patent linkage would exist, it would cover a smaller group of drugs and thus have less impact on approval of generic drugs.

4. Mandatory Patent Linkage, but with Procedural Safeguards to Protect the Public Interest

Patent linkage harm can also be mediated with a variety of procedural safeguards to balance the public interest in generic drugs and avoid overprotection of invalid patents. As noted earlier, the United States and Canada provide procedural safeguards by giving generic applicants the ability to challenge the validity or applicability of a patent. Variations on such challenges can provide more protection. For example, whereas the United States and Canada stay the approval of a generic application for at least twenty-four months while litigation is pending (unless the patent is found invalid or expires during this time), this stay could be shortened. Another variation that would be even more favorable to the public interest is to have no stay of the approval of the generic application at all during the litigation. In other words, a regulatory agency could approve a generic drug for sale, and the generic drug could be marketed simultaneously with the ongoing litigation. Alternatively, instead of a mandatory stay of regulatory approval, a patent owner could be required to make an additional showing to obtain such a stay.

Additional procedural safeguards may be incorporated to also help protect the public interest. For example, rewards may be provided to generic applicants who successfully challenge a patent; these could be in the form of a period of exclusivity with respect to other generic drugs. Alternatively, a patent owner could be penalized for bringing a lawsuit against a generic applicant if it did not actually have a good faith belief that the patent was valid or infringed. As noted earlier, Australian law takes this approach.[74]

IV. Patents Plus—The Complete Picture

This section builds upon the prior concepts of data exclusivity and patent linkage to show how data exclusivity and patent linkage work together with patent requirements to impede the approval and sale of generic drugs. Understanding how the parts can work together is very important for the many countries that are currently under pressure to

[74] Australia requires the pioneer company to certify that litigation against generic companies is in good faith, has a reasonable prospect of success, and will be conducted without unreasonable delay. *Therapeutic Goods Act 1989* (Cth) § 26C(3)– C(4) (Austl.). If the certification is later discovered to be false or materially misleading, the company may be subject to a fine of up to $10 million. *Id.* § 26(c)(5). The generic applicant may also be subject to a penalty. For a more in-depth discussion of Australia's law, see Thomas Faunce & Joel Lexchin, *'Linkage' Pharmaceutical Evergreening in Canada and Australia*, 4 NEW ZEALAND HEALTH POL'Y, June 1, 2007, at 1.

incorporate data exclusivity, patent linkage, or both into their laws. Accordingly, this section attempts to set forth the primary possibilities that may occur. Before delving into the details, however, here are a few quick clarifications:

Q: If a country provides for data exclusivity, will it necessarily provide for patent linkage?

A: No. Data exclusivity and patent linkage are two separate methods of protecting pioneer drugs. Although both exist in the United States, they do not necessarily always exist together.

Q: Is patent linkage the same as data exclusivity?

A: No. Patent linkage prevents approval of generic versions of drugs that are currently patented whereas data exclusivity may bar approval of unpatented drugs.

Q: Do both data exclusivity and patent linkage impact access to generic drugs?

A: Yes! These are two separate ways that regulatory rules—rather than patent laws—may limit competition from generic drugs. Although each has different requirements, they both pose the same basic problem for the goal of increasing access to generic drugs.

Q: If a firm is issued a compulsory license to make a drug, would either data exclusivity or patent linkage prevent the firm from marketing the drug?

A: Yes! Because patent and regulatory laws operate separately, an exception from one set of laws will not automatically result in an exception from the other. In fact, unless regulatory laws permit an exception in the case of a compulsory license, the firm issued a compulsory license could be effectively precluded from making use of the license as no drugs may be sold without regulatory approval.

Although all WTO nations must provide patents, neither data exclusivity nor patent linkage are required, so there are a variety of possible regulatory schemes that impact when generic drugs can be approved. There are roughly four scenarios:

(1) Patent rights, but no data exclusivity or patent linkage (TRIPS minimum)
(2) Patent rights *plus data exclusivity*, but no patent linkage
(3) Patent rights *plus data exclusivity and patent linkage with challenges.*
(4) Patent rights *plus data exclusivity and patent linkage, with challenges.*

A. PATENTS ONLY

A country that only adopts the minimum requirements of TRIPS would provide for patent rights, but not data exclusivity or patent linkage. In this scenario, a regulatory agency can approve a generic version of a previously approved drug as soon as the generic applicant establishes bioequivalence to the previous drug. In other words, the regulatory agency can approve a generic equivalent without regard to whether the generic might infringe a patent. The patent owner is free to sue for infringement if and when the generic drug is approved for sale. If successful, the patent owner is entitled to recover money damages as set forth in TRIPS. Courts must have the authority to grant an injunction to the prevailing patent owner, but TRIPS does not mandate that courts impose injunctions against losing defendants. In addition, TRIPS permits, but does not require that a losing defendant be penalized with treble damages.

B. PATENTS PLUS DATA EXCLUSIVITY ONLY

The next level of protection for drugs would be a system that accorded patent rights and data exclusivity, but no patent linkage. In this scenario, a regulatory agency would not deny approval of a generic drug because it was covered by a patent. However, it would deny approval of a generic drug during the term of data exclusivity. In addition, the regulatory authority would deny approval of a generic version of previously approved drugs even if the prior drug was unpatented. The sole issue under this system is whether the prior drug had clinical data that the generic applicant wants to rely upon. In addition, to the extent data exclusivity ends before the patent term, a generic drug could be approved during the patent term.

C. PATENTS PLUS DATA EXCLUSIVITY AND PATENT LINKAGE—NO CHALLENGES PERMISSIBLE

The next level of protection for drugs would be patent protection coupled with data exclusivity and patent linkage. In this scenario, a regulatory agency is precluded from approving a generic drug if it is *either* within the period of data exclusivity *or* the patent term. However, this version permits a generic applicant at some point during the data exclusivity to assert that the patent is either invalid or not infringed. So, conceivably a generic drug could be approved during the patent term if the patent was judicially determined to be either invalid or not infringed. This is the basic core of the system adopted in the United States and Canada.

D. PATENTS PLUS DATA EXCLUSIVITY AND PATENT LINKAGE—NO CHALLENGES

The worst-case scenario for generic drugs is one in which patent protection is coupled with data exclusivity and patent linkage, but there are no possibilities for a generic applicant to challenge the patent. This is the scenario that has unfortunately arisen in many countries forced to adopt data exclusivity and patent linkage as a result of FTAs.

V. Concluding Thoughts

Although data exclusivity and patent linkage both impact when generic drugs become available, they do so in different ways. Data exclusivity can limit generic versions of even unpatented drugs. Patent linkage can potentially limit generic versions of patented drugs even when the patent is of suspect validity. When both exist together, the ways in which generics are limited simply multiply. Although there are mechanisms to mediate the harm of each type of regulatory protection, to effectively do so requires not only a firm understanding of the issues, but likely strong political will and resistance to political ressure.

10 Suspensions of In-Transit "Generic" Drugs: A Case Study of Competing Perspectives

This chapter analyzes an important current controversy under TRIPS while simultaneously providing another opportunity to show the impact and influence of competing patent perspectives. In particular, this chapter analyzes a controversial EU Regulation that permits customs officials to detain goods that infringe local patent laws—even though those goods are not intended for local consumption. Pursuant to the Regulation, goods have been detained in a Netherlands port en route to another country. The detentions are permissible under the EU Regulation because Netherlands law considers in-transit goods to infringe patent rights in the Netherlands. These detentions have resulted in delays of goods that violate no patent rights in either the country of origin or destination.

The issue poses difficult interpretive TRIPS questions, yet discussion to date has predominantly suggested the EU law is either clearly consistent or inconsistent with TRIPS. This chapter aims to explain why the TRIPS issue is a nuanced one that is far less clear than most suggest. In addition, this chapter describes how the competing patent perspectives have promoted skewed interpretations of the law and facts. The chapter addresses the following questions:

- Is it clear whether TRIPS permits or precludes a member state from suspending goods for alleged in-transit infringement?
- How do the patent perspectives result in distortions regarding the law and facts involving the in-transit detentions?

I. Introduction 287
II. Background 288
  A. EU Regulation 288
  B. Patent Perspectives Revisited 290
    1. Patent as Mere Privilege 290
    2. Patent as Uber-Right 291
III. Case Study in Competing Perspectives 291
  A. Uber-Right (Predominantly Patent Owner) Perspective 292
  B. Moderated Uber-Right (EU) Perspective 294
  C. Privilege (Health Advocate) Perspective 297
IV. TRIPS Framework 302
  A. TRIPS Border Measures 303
    1. Mandatory 303
    2. Permissible 304
  B. Other TRIPS Provisions 308
    1. Article 41 308
    2. Other TRIPS Context 312
    3. Doha Public Health Declaration 313
V. Impact of Patent Perspectives on Legal Interpretations 314
  A. Explicit TRIPS Interpretations 315
    1. Article 51 315
    2. Article 52 316
    3. Doha Declaration 318
    4. Broader TRIPS Context (Articles 7–8) 319
  B. Other Distorted Legal Interpretations 319
    1. Territoriality 320
    2. Generic 320
    3. Domestic Laws 322
VI. Conclusion 323

I. Introduction

An Indian company specializing in the manufacture of low-cost generic drugs makes a shipment to Brazil. The drug is not protected by a patent in either India or Brazil, so it can be made and sold as a generic, thus reducing drug costs. However, the drugs never reach Brazil because they are suspended en route at a port in Rotterdam, Netherlands, a traditionally popular stopping point for global transport of goods. The drugs are halted pursuant to a EU Regulation that permits customs officials to seize goods that allegedly violate local patent rights—even if they do not infringe patent rights at the country of origin or final destination.[1]

This EU law imposes a TRIPS-Plus standard for enforcement of patent rights. As noted in Chapter 8, any protection of patent rights beyond TRIPS (i.e., a TRIPS-Plus standard) is generally viewed as objectionable to those who consider patent rights to negatively impact access to medicine. The EU law permits drugs to be detained and sometimes even destroyed. This is an important issue for the many countries that rely on imports of generic drugs to lower the cost of medicine.

Since the EU began stopping drugs under its law, this action has prompted heated discussion that often involves competing conclusory allegations concerning TRIPS, as well as name-calling and conflation of issues. The EU asserts that its law is clearly consistent with TRIPS, whereas Brazil and India assert that it is clearly not. Public health advocates have accused patent owners Merck and DuPont of being "corporate criminals . . . acting with impunity to thwart lawful generic competition."[2] Conversely, a EU representative has contended that "countries actually should be grateful to the European Customs" for stopping counterfeit medicines as this action "most likely have saved lives," even though the seized goods were neither counterfeit nor unsafe.[3]

This chapter revisits the competing perspectives of patents first introduced in Chapter 6 to better understand the polarized positions concerning in-transit suspensions of generic drugs. As noted earlier, there is a spectrum of views on patents benchmarked by

---

[1] Council Regulation (EC) No. 1383/2003 of July 20, 2003 Concerning Customs Action Against Goods Suspected of Infringing Certain Intellectual Property Rights and the Measures to Be Taken Against Goods Found to Have Infringed Such Rights, O.J. 2003/L 196/7 [hereinafter EC Regulation 1383/2003]. India and Brazil have sought to challenge this action. Request for Consultations by India, *European Union and Member State - Seizure of Drugs in Transit, WT/DS408/1* (May 19, 2010); Request for Consultations by Brazil, *European Union and Member State - Seizure of Drugs in* Transit, WT/DS409/1 (May 19, 2010).

[2] William New, *International Health Groups Warn WHO, WTO on Medicines Seizures*, INTELL. PROP. WATCH (Feb. 21, 2009), http://www.ip-watch.org/weblog/2009/02/21/international-health-groups-warn-who-wto-on-medicines-seizures/.

[3] Int'l Ctr. for Trade and Sustainable Dev., *Fight over Generic Drug Seizure Takes Centre Stage at TRIPS Council Meeting*, INTELLECTUAL PROP. PROGRAMME (Mar. 11, 2009), http://ictsd.net/i/news/bridgesweekly/42823. As will be explained later, whether a drug is safe is an issue for regulatory agencies to address, but never an issue in assessing patentability or patent infringement.

two distinct and seemingly irreconcilable perspectives. The "privilege view" of patents is that they are a mere privilege granted by a nation, such that patent rights should always be subject to limitations to accommodate other societal goals, such as access to medicine. The alternative perspective views patents as a type of super-property right that should seldom, if ever, be limited, based on the assumption that exclusivity provides an essential incentive to promote the development of new and socially desirable medicines; this is referred to as the "uber-right view".

The impact of these perspectives on suspensions of in-transit goods provides a complimentary illustration to the prior discussion of how perspectives impact views of compulsory licenses under TRIPS. In particular, whereas rules on compulsory licenses are fairly clear under TRIPS, but nonetheless distorted in interpretations by those with an uber-right perspective, the TRIPS rules pertinent to in-transit seizures are more ambiguous, yet distorted in interpretations by those who consider patents to be a privilege.

A better understanding of the competing perspectives that underlie the in-transit seizure controversies should provide an enriched understanding not only of the TRIPS provisions at issue, but ongoing international developments. Specifically, although the EU law has been officially challenged in the WTO as being inconsistent with not only TRIPS, but WTO rules, in-transit suspensions may be an issue that extends beyond the EU.[4] In particular, the EU is attempting through international agreements to require other countries to adopt a similar rule to suspend in–transit goods; for example, this has been a major issue in the negotiations of the Anti-Counterfeiting Trade Agreement (ACTA). If this becomes a new global norm, it could seriously threaten global trade of generic goods as there would be no "safe" country to ship goods through.

II. Background

A. EU REGULATION

Under the EU Regulation, customs officials may act either on behalf of a specific application by a rights-holder or on their own in suspending goods that infringe an intellectual property right based on local right. In other words, although the Regulation applies to all EU member countries, whether there is an intellectual property right being violated is a function of national law. For example, German customs officials should apply German patent law to assess whether in-transit goods infringe patent laws[5]

[4] In addition, there may not be a WTO panel decision to clarify this issue since the EU seems to be moving towards addressing this issue outside the WTO system. *E.g., EU issues note to address drug seizure concerns of India*, Ec. Times, available at http://economictimes.indiatimes.com/news/economy/foreign-trade/EU-issues-note-to-address-drug-seizure-concerns-of-India/articleshow/6869124.cms.

[5] EU Regulation 1383/2003, chap. 3 art. 10.

However, there is a wrinkle in assessing local patent infringement of in-transit goods. In particular, recital 8 of the Regulation suggests that infringement is assessed based on whether the goods would infringe *if* made in the member state.[6] Dutch courts have held that in-transit products can infringe Dutch patents based on the admitted "legal fiction" that the products were made in the Netherlands.[7] Although courts in EU member states have a different view of the Regulation,[8] all the drugs suspended thus far for alleged patent infringement have been in the Netherlands.[9]

Assuming that customs officials can properly consider in transit goods to infringe patents, there are nonetheless additional procedures that must be complied with under the Regulation. For example, the owner of in-transit goods that are seized is provided a mechanism under the EU Regulation to obtain their release. In particular, the owner should be able to obtain release by either objecting to a patent owner's request to destroy the goods or by notifying the customs office that legal procedures have been initiated to evaluate whether patent infringement has occurred.[10]

The EU Regulation permitting suspension of goods that allegedly infringe patents is a relatively recent expansion of a regulation to address other types of intellectual property infringement. The first regulation addressed only counterfeit trademarks, then later was expanded to pirated copyright goods—problems that are considered a major detriment to the EU economy.[11] The expansion to patent infringements was made in 1999 in response to industry interest in including them.[12] However, this inclusion was admitted to be an

---

[6] EC Regulation 1383/2003, recital 8 ("Proceedings initiated to determine whether an intellectual property right has been infringed under national law will be conducted with reference to the criteria used to establish whether goods produced in that Member State infringe intellectual property rights.").

[7] *E.g.*, Sosecal v. Sisvel, IER 2008/83, Case No. 311378, Roll No. KG ZA 08-617 [2008] (The Netherlands, The Hague District Court), English translation available at http://www.eplawpatentblog.com/PDF_December09/The%20Hague%20DC%20Sisvel%20v%20Sosecal%20EN.pdf.

[8] Nokia v. HMRC, [2009] EWHC (Ch) 1903. In addition, some also suggest that an earlier case, Montex/Diesel, a trademark case that precluded seizure of in-transit goods without evidence of likely diversion into the EU markets suggests that there can be no infringement of in-transit goods without evidence of likely diversion. Case C-281/05, Montex Holdings Ltd. v. Diesel SpA, 2006 E.C.R. I-10881.

[9] There are two cases pending before the European Court of Justice regarding the present Regulation, or its predecessor. Case C-495/09, HMRC v. Nokia, 2009, *available at* http://www.eplawpatentblog.com/2010/January/Court%20of%20Appeal%20Order%20Nokia.pdf; Case C-446/09, Phillips v. Lucheng Meijing Indus. Co., *available at* http://eur-lex.europa.eu/LexUriServ/LexUriServ.do?uri=OJ:C:2010:024:0029:0030:EN:PDF. However, neither of these cases are patent cases.

[10] EC Regulation 1383/2003, arts. 13–14.

[11] 1986 O.J. (L 357) 1; *see also Commission Combating Counterfeiting and Piracy in the Single Market—Green Paper*, at 98, COM (1998) 569 final (Oct. 15, 1998).

[12] The expansion to incorporate patents was officially justified as important to "safeguarding innovation," although the focus of the regulation has always been—and remains—on counterfeiting and pirating, which are considered a major detriment to the economy. *See Commission Proposal for a Council Regulation Amending Regulation (EC) No 3295/94 Laying Down Measures to Prohibit the Release for Free Circulation, Export, Re-Export or Entry for a Suspensive Procedure of Counterfeit and Pirated Goods*, at ¶ 6.3, COM (1998) 25 final (Jan. 28, 1998).

"experiment" in that identifying patent infringement is more difficult than identifying counterfeit trademarked or pirated copyrighted goods.[13]

Although EU customs officials were permitted to suspend alleged patent infringements in 1999, actual suspension of goods did not happen for another decade. Beginning in 2008, EU customs officials began using their authority to detain drugs that were alleged to infringe patent rights. Over a period of about eighteen months, roughly twenty shipments were detained, with many delayed for months and some even destroyed.[14] All but one shipment originated from India, where the drugs were made legally as generics. The shipments were intended for developing countries where the drugs would also not violate patent rights. The seized drugs treated a variety of conditions including HIV, heart disease, dementia, and schizophrenia.[15]

B. PATENT PERSPECTIVES REVISITED

This section briefly recaps the two theories of patent perspectives first introduced in Chapter 6 and provides additional insight into how they apply to the issue of in-transit seizures.

1. Patent as Mere Privilege

One view of patents is that they are a tool to promote innovation that is inherently subject to limitations regarding promoting other societal goals. Patent rights would be seen as inherently posing a conflict with the goal of promoting immediate access to medicine as patents limit the ability of low-cost generics to enter the market.

The privilege view would be totally opposed to in-transit patent rights. After all, the privilege view of patents wants to generally limit traditionally recognized patent rights and create generous exceptions to them. Especially as in-transit patent rights are not universally recognized, this would be considered a new and improper expansion of patent rights.

In-transit patent infringement would be viewed as particularly onerous to those with a privilege view because it is a type of patent right to which no exception can be created. For example, if a country that viewed patents as a privilege were nonetheless required to provide them, it can still accommodate its privilege view by limiting what is patentable and by utilizing exceptions to patent rights, such as compulsory licenses. However, a

[13] *Id.* ¶¶ 6.4–6.5.

[14] *E.g.*, Request for Consultations by India, WT/DS/408/1 (May 19, 2010).

[15] Letter from J. van der Vlist on behalf of Netherlands State Secretary for Finance to Ms. S. Bloemen, Health Action International Europe (HAI) (May 7, 2009), ***available at*** http://www.haiweb.org/19062009/7%20May%202009%20Dutch%20government%20response%20to%20Freedom%20of%20Information%20request%20(EN).pdf [hereinafter FOI Letter from Netherlands] (English translation).

country that views patents as a privilege is powerless to control whether another country expands patent rights to include in-transit goods (except a WTO challenge).

Holders of the privilege perspective may view the national right to determine patent scope as inviolate so as to disregard the ability of other countries to consider in-transit goods as infringing patent rights. In particular, the privilege perspective would consider a drug that was generic (without patent rights) at its point of export and final destination to never be subject to patent rights while the drug was en route. Although the privilege view may hold the right of a nation to make its own assessment of patent rights as being supreme, this is not necessarily consistent with the country where the goods are in transit also being able to make its own assessment.

2. Patent as Uber-Right

The alternative perspective of patents views them as a privileged property right that is stronger than other property rights. In contrast to the privilege perspective that views patents as inherently conflicting with access to medicine, holders of the uber-right view see no conflict. Under this view, patents are consistent with promotion of public health because they promote innovation that results in improved treatments benefiting all.

The uber-right view of patents may play a role in considering in-transit drugs to be patent infringement. Although patent rights have not traditionally extended to in-transit goods, an uber-right view would be consistent with promotion of stronger patent rights. Alternatively, holders of the uber-right view may believe so strongly in the sanctity of patent rights that if an invention is granted a patent in one country, a patent should be granted in other jurisdictions as well.

Those espousing the uber-right view of patents would not see any conflict between in-transit infringement and legitimate trade in generic drugs as under an uber-right view, drugs that infringe in transit would not be *legitimate* trade. Accordingly, those with an uber-right perspective could claim to endorse legitimate trade in generic goods while neglecting to say they do not actually consider goods to be generic if they infringe a local patent while in transit.

III. Case Study in Competing Perspectives

This section illustrates through a series of three arguments the competing perspectives of patents regarding the seizure of drugs in the EU. Each argument provides a view of the seizures through the lens of competing patent perspectives; none represent the views of the author. The arguments are distilled from publicly available sources such as news stories and press releases. The first two reflect a view of patents as an uber-right, although the two arguments are slightly different. The first presents the most extreme uber-right view that would likely be embraced by patent-owning pharmaceutical companies and those who support the interests of patent owners. The second presents a view that is consistent with the uber-right perspective, but slightly more moderate; this view reflects public positions taken by the EU. The last reflects a privilege view that would likely be embraced by developing countries and public health advocates who support their interests.

A. UBER-RIGHT (PREDOMINANTLY PATENT OWNER) PERSPECTIVE

Holders of the more extreme uber-right perspective might view the in-transit drug suspensions in this way:

There has been great confusion and many overstatements made concerning a few shipments of drugs that were temporarily detained.[16] There is a critical need to put these limited situations into their proper perspective rather than have them continue to be misconstrued.[17]

The impact of limited detentions of a few shipments of goods has been totally overstated. The number of drugs detained is negligible compared to the overall trade in goods, as well as drugs that actually reach the patients who need them. Moreover, most of the detained drugs are not essential ones, so claims the detainments created life-threatening delays to treatment are simply unwarranted. The detained drugs include treatments for dementia and schizophrenia, which are not life-threatening conditions. In addition, nearly 20 percent of the detained drugs are for lifestyle conditions.[18] Although some of the drugs were for heart treatment, unpatented and affordable alternatives existed. For example, aspirin has been used as a substitute for the patented heart drug sold as Plavix.

Although there are two instances of HIV treatments being temporarily detained, they comprise a mere 10 percent of all goods detained in the EU, and they were promptly released with only negligible delay. Moreover, one of the HIV shipments was destined for Brazil, which has its own generic drug industry such that it could have easily made the desired drug.

Indian companies and organizations have loudly protested detention of drugs from India, yet failed to note their own complicity in any unnecessary delay of drugs to developing countries. In at least one case after a temporarily detained shipment was released, the Indian company returned the shipment to India rather than continuing it on the

[16] *See* Press Release, European Federation of Pharmaceutical Industries and Associations (EFPIA), Customs Seizures of In-Transit Medicines, Mar. 13, 2009, *available at* http://www.efpia.eu/content/default.asp?PageID=559&DocID=6574 [hereinafter EFPIA Press Release] (acknowledging "a certain amount of media coverage" concerning in-transit drugs detained).

[17] *Id.* (noting "it is important that these events are not misconstrued").

[18] FOI Letter from Netherlands, *supra* note 15, Appendix.

original route.[19] In other cases, Indian companies have abandoned their shipments after those shipments have been released.

Patent owners are acting in the interest of patients in ensuring they receive the highest quality medicines. Innovative companies that patent new drugs are in the best position to provide quality ones. Companies that create new (and usually patented) drugs must always provide extensive documentation of the safety and efficacy of their innovative drugs before they are marketed. Conversely, generic medicines are approved for sale based on a mere showing that they are bioequivalent to the innovative drug; regulators *assume* the generic is likely to be as safe and effective as the original. However, patients do not always find the generic version to have the same effect, and in some instances, they may even have unexpected negative side effects. This is not entirely surprising as approval of generic drugs is not based on the same type of rigorous clinical testing as new drugs. Accordingly, there is no guarantee that drugs are safe and effective when they are not made by the patent owner.[20]

The pharmaceutical industry is dedicated to improving the health and safety of citizens across the world. The industry is the primary source of funding for research into developing newer and better treatments that benefit all. The industry also has a long-standing commitment to improving access to medicines in developing countries.[21]

Moreover, patent owners have fully complied with EU laws. The EU regulation properly recognizes that patent owners can and should protect consumers from unauthorized drugs that infringe patent rights. Patent owners are well within their legal rights to ask customs officials to verify whether generic drugs in transit are counterfeit to ensure that the drugs do not violate patent laws.[22] Moreover, to help protect consumers, pursuant to the EU law a patent owner may request surrender of detained goods for destruction. It is up to the shipper to decide whether to contest this action, and in no case have goods been destroyed without the shipper being given an opportunity to do so. Patent owners have consistently notified those shipping unauthorized drugs of any suspensions to permit them to exercise their legal rights. Patent owners have even taken the additional step of attempting to avoid unnecessary detentions by requesting that unauthorized companies

---

[19] Dr. Reddy's returned the shipment of the hypertension drug losartan potassium to India rather than continuing on to Brazil.

[20] Letter from B.E.M. van Kessel, Attorney, Eli Lilly, to Cipla Ltd. (Dec. 9, 2008), http://online.wsj.com/public/resources/documents/eudrugs2009letter3.pdf [hereinafter Eli Lilly Letter to Cipla] stating that Cipla's generic tablets "may not be safe or effective" because they are "not genuine Eli Lilly products. While this statement reflects an uber-right view, it ignores the fact that all drugs legally sold are evaluated for safety and efficacy, as explained in Chapter 1.

[21] EFPIA Press Release, *supra* note 16 (asserting the "pharmaceutical industry has demonstrated a long-standing commitment to improving access to medicines in developing countries").

[22] *See id.* (recognizing the right of countries to "stop products that they suspect may be counterfeit from entering the supply chain").

not send other shipments that would violate intellectual property rights.[23] If unauthorized companies decline to grant this request, they should assume the risk that their unauthorized goods may be detained.

Last, but not least, European patent owners have clarified that although they recognize they have the legal authority to stop goods under EU law, it is not their intent to restrict access to medicines in developing countries.[24] In other words, companies should not be criticized when it is not their policy or practice to encourage detention of legitimate generic drugs intended for shipment to customers beyond the EU.[25]

B. MODERATED UBER-RIGHT (EU) PERSPECTIVE

Holders of a more moderate uber-right perspective might view the in-transit drug suspensions in this way:

The EU has an interest in safeguarding the health of its own citizens, as well as all the citizens of the world. Counterfeit drugs pose serious risks to health and constitute a major problem. The EU has no intent of hampering legitimate trade in generic drugs, and its laws do not have this effect.[26] The EU supports the goal of promoting low-cost drugs to developing countries.[27] However, counterfeit drugs are not beneficial to anyone. EU customs authorities play an important role in addressing the global trade in counterfeit products, especially fake medicines whose effects disproportionately impact developing countries.[28] In fact, EU customs officials have played an important role in protecting consumers from dangerous counterfeit medicines such that consumers should be grateful to the EU for its laws and policies.[29]

---

23 Patent owner Eli Lilly requested generic drug manufacturer Cipla to sign a declaration agreeing it would "not send other consignments containing products which infringe upon Eli Lilly's intellectual property rights, including infringing tablets, or otherwise infringing products." Eli Lilly Letter to Cipla, *supra* note 20.

24 EFPIA Press Release, *supra* note 16 (stating "it is neither the policy nor practice of [EFPIA] members to encourage Member States to use the powers of detention available to them to prevent the flow of legitimate generic products from manufacturer to customer outside the EU. This applies even where goods transit through EU countries where intellectual property legislation could be applied.").

25 *Id.* (noting it is not "the intention of EFPIA members to restrict access to medicines in developing countries").

26 *E.g.*, EC Intervention, WTO General Council, Feb, 3, 2009, *available at* http://keionline.org/blogs/2009/02/05/ec-intervention-at-wto [hereinafter EC Intervention, Feb. 2009].

27 *E.g.*, Resolution of July 12, 2007 on the TRIPS Agreement and Access to Medicines, EUR. PARL. DOC. P6_TA(2007)0353.

28 EC Statement to TRIPS Council, Agenda Item M—Other Business, June 8–9, 2009, http://www.ipeg.eu/blog/wp-content/uploads/EU-comments-on-use-of-BDO-against-seized-medicines.pdf [hereinafter EC Statement, June 2009] (noting "it is important to continue to allow the Customs Authorities to control goods in transit and ensure that measures can be taken against global trade in counterfeit products, and in particular fake medicines whose effects mainly hit developing countries").

29 *E.g.*, EFPIA Press Release, *supra* note 16.

The EU has been improperly and incorrectly criticized by overly zealous healthcare advocates and misguided news reporters as "seizing" drugs. The EU does not seize goods.[30] Rather, small shipments of goods are merely temporarily detained. It is incorrect and unfair to characterize the EU as seizing generic drugs when it is only temporarily detaining them for further investigation. The number of shipments from India that have been detained is miniscule compared to the number of counterfeit drugs that have been seized;[31] the detained drugs are likely only a "nanopercentage" of medicine passing through the EU.[32] Moreover, Indian companies share the blame for detained goods that do not reach their final destination. In some cases, after shipments are released to the original Indian companies, the company elects to either send the goods back to India or abandons them entirely.[33]

There is no systemic problem that needs to be corrected. Only a few isolated cases are at issue out of the thousands of shipments that travel through Europe.[34] Customs officials are acting properly when they detain drugs that may be counterfeit (and thus violate patent rules).[35] Objections to the temporary delays are an "over-exploitation" of limited instances.[36] If there were an actual problem with the EU Regulation permitting suspensions, medicines would be continuously being stopped in all EU Member States since the Regulation was enacted. To the contrary, goods have only been detained in a handful of countries, and the few instances are all past history; there is no current problem.[37] Moreover, there is no need for concern as the Regulation at issue is already under review.

---

[30] In an official statement to the WTO, the EU stated "Dutch authorities *temporarily* detained (which does *not* mean *seize, confiscate or destroy*) a small shipment of drugs" in a Dutch airport, but underscored that this was consistent with TRIPS. EC Intervention, Feb. 2009, *supra* note 26, ¶ 4 (emphasis in original).

[31] *E.g.*, EFPIA Press Release, *supra* note 16 (characterizing the number of incidents as "minuscule relative to the massive flow of medicines . . . transiting through the EU").

[32] Catherine Saez, *WTO Forum: Bypassing International Agreements May Hamper Medicines Access*, INTELL. PROP. WATCH (Oct. 11, 2009), http://www.ip-watch.org/weblog/2009/10/11/wto-forum-bypassing-international-agreements-may-hamper-medicines-access/ (quoting Luc Devigne of the European Commission trade directorate).

[33] *See generally* EC Intervention, Feb. 2009, *supra* note 26 (noting that goods were eventually released to the Indian owner that sent goods back to the country of origin even though there was no legal obligation to do so).

[34] *E.g.*, Kaitlin Mara, *Medicines Access Again Captures Attention at WTO as Progress Urged in Round*, INTELL. PROP. WATCH (Oct. 30, 2009), http://www.ip-watch.org/weblog/2009/10/30/medicines-access-again-captures-attention-at-wto-as-progress-in-round-urged/.

[35] *See Nigeria: Seizure of Drugs Threatens HIV Access*, IRIN, Mar. 13, 2009, http://www.irinnews.org/Report.aspx?ReportID=83459.

[36] *E.g.*, Jonathan Lynn, *Developing States Attack EU on Generic Drug Seizure*, REUTERS, Feb. 4, 2009 at http://in.reuters.com/article/idINIndia-37821420090203, (noting that EU ambassador Eckart Guth warned WTO members not to blow the case out of proportion).

[37] *See* Saez, *supra* note 32 (Luc Devigne of the European Commission trade directorate noted that there were only a few cases of detention in 2008, so there was "much noise about nothing").

Even some of the examples advocates put forth as being illustrative of a problem suggest otherwise. For example, public health advocates strongly protested the temporary detention by German customs of a shipment of antibiotics from India destined for Vanuatu. However, the shipment was detained for suspected infringement of *trademark* and not patent rights, so this event has no bearing on the issue of trade in generic drugs.[38] Moreover, once customs received information that there was no trademark infringement, the goods were promptly released and reached their final destination.

The EU remains committed to the policy of promoting access to medicine. The Regulation at issue is designed to prevent intellectual property infringement and trade of counterfeit goods. Contrary to repeated allegations, the EU is not confused concerning the distinction between generic medicines and fake medicines. However, many dangerous goods are being shipped through European ports, with the trade in fake medicines having increased substantially.[39] Although generics may be distinct from fakes, "EU customs probably saved lives around the world by stopping "fakes" pursuant to the Regulation.[40]

The EU acknowledges that detentions on the ground of suspected infringement of intellectual property rights might result in some delay, but border enforcement by customs is expressly contemplated by both the WTO and the World Customs Organization.[41]

The EU law has built-in safeguards to protect public health. EU customs officials do not make any final decisions whether goods infringe any intellectual property rights, including patents.[42] Although customs authorities may instigate a detention, it is up to a rights-holder to pursue litigation. In the unlikely event the detention is later discovered to be unsubstantiated, the owner of the shipment is entitled to compensation. In addition, the EU has clarified that customs authorities should avoid actions that would delay or unnecessarily disrupt legitimate trade in generic drugs.[43]

[38] EC Statement, June 2009, *supra* note 28.

[39] *Id.*

[40] Saez, *supra* note 32 (quoting Luc Devigne).

[41] EC Statement, June 2009, *supra* note 28.

[42] *Id.* ("Under EU legislation, customs do not make a final decision on whether goods are infringing IPR").

[43] Explanatory Note on the Application of Council Regulation (EC) No 1383/2003 Concerning Customs Action Against Goods Suspected of Infringing Certain Intellectual Property Rights and the Measures to be Taken Against Goods Found to Have Infringed Such Rights, July 13, 2009 [hereinafter EU Explanatory Note, July 2009], *available at* http://www.mszh.hu/hirek/kapcsolodo/Explan_note_1383_2003_EN.pdf (noting that "customs authorities . . . are invited to pay particular attention when controlling pharmaceutical products in transit in order to avoid actions that would delay or cause unnecessary disruption of legitimate trade in generic drugs").

The EU customs regulation does not constitute extraterritorial enforcement of patent rights—and all claims to the contrary are erroneous.[44] Any goods seized according to the EU customs law must be in violation of local rights.[45] Moreover, the EU Regulation does not require widespread detention of drugs. It is up to each Member State to decide whether in-transit goods violate patents. Notably, not all Member States consider in-transit goods to violate patent laws.[46]

C. PRIVILEGE (HEALTH ADVOCATE) PERSPECTIVE

Holders of the privilege perspective might view the in-transit drug suspensions in this way:

The EU is putting public health at risk by its aggressive and abusive implementation of its EU Regulation. Any law that permits seizure of in-transit generic drugs presents a serious risk to public health as treatment delays are not merely undesirable, but potentially life-threatening. Indeed, the UN Special Rappoteur recently expressed concern that the seizures jeopardize achievement of human rights norms.[47] Accordingly, the seizures must be condemned as unacceptable and promptly reviewed and altered.[48]

The EU law creates barriers to the export of quality, low-cost generic drugs. Patients in the developing world are deprived of access to affordable life-saving drugs due to the EU law.[49] Generic drugs from India have been a "lifeline for countries" that cannot afford

[44] *Contra* Intervention by Brazil, WTO General Council, Feb, 3, 2009, *available at* http://keionline.org/blogs/2009/02/03/intervention-by-brazil-at-wto-general-council-on-seizure-of-500-kilos-of-generic-medicines-by-dutch-customs-aut [hereinafter Brazil Intervention, Feb., 2009].

[45] EC Statement, June 2009, *supra* note 28 ("regarding the principle of territoriality let me reassure you that the EU Customs Regulation has no extra-territorial effect").

[46] EU Explanatory Note, *supra* note 43 (noting that in several member states, patent rights do not extend to goods in transit).

[47] Intervention by India, TRIPS Council, Agenda Item M—OTHER BUSINESS—Seizure of generic drug consignments at EC ports, June 8–9, 2009, http://www.ip-watch.org/weblog/wp-content/uploads/2009/06/intervention-by-india-seizure-of-generic-drug-consignments-at-ec-ports.pdf [hereinafter India Intervention, June 2009].

[48] *E.g.*, Press Release, HAI et al., Seizure of UNITAID/Clinton Foundation Anti-Retroviral Medicines by Dutch Customs Authorities "Unacceptable" (Mar. 6, 2009), *available at* http://www.haiweb.org/06032009/6%20Mar%202009%20Press%20release%20More%20generic%20medicines%20intercepted%20in%20the%20Netherlands%20(English).pdf [hereinafter HAI et al. Press Release, Mar. 6, 2009] ("condemn the unacceptable seizure of legitimate generic antiretroviral medicines in transit from India to Nigeria by Dutch customs officials, and call on the European Commission to immediately review and modify its regulations . . . to allow lawful trade of generic medicines.").

[49] *E.g.* Radhieka Pandeya, *Dr. Reddy's Consignment of Drugs to Brazil Seized*, LIVEMINT.COM (Jan. 15, 2009, 12:09am), http://www.livemint.com/2009/01/14220926/dr-reddy8217s-consignment-o.html (quoting attorney Leena Menghaney, for MSF, stating "the fallout will be on patients' lives in the developing world who will not be able to access affordable life-saving drugs from India").

patented drugs—a lifeline that is now in jeopardy.[50] Trade in legitimate medicines is critical to ensuring access to medicines for millions.[51]

The EU is repeatedly placing embargos on medicines at the behest of drug company "bullies" under a law they masterminded to give them authority to harass generic drug companies.[52] The EU law allows "corporate criminals"[53] to act with impunity when making frivolous claims concerning goods regarding intellectual property rights.[54] These claims have no legal basis. If the goods were intended for actual sale in any EU Member State, there might be grounds for patent infringement. However, all the seized goods were merely part of the regular and legitimate flow of products in transit that do not infringe.[55]

The EU seizures are not only unacceptable, but they set a dangerous precedent.[56] Even if the number of seizures relative to global trade is modest, the fact that there is a law . . . permitting seizures of generic drugs is a continuing threat to legitimate trade in generic medicines. While the EU law is in effect, generic companies must either contend with unexpected delays or increased costs. For example, some companies whose goods are seized have had to abandon them even after those goods were released because the cost of contesting the EU law is more than that of the drugs.[57] In addition, some companies have had to use alternate shipping routes to avoid seizures in the EU, which substantially raises costs.[58]

The type of drugs seized is irrelevant to the fact that *all* generic drugs are threatened. "It's hypertension medicine today, but tomorrow its AIDS medicines, cancer medicines,

---

[50] John Miller & Geeta Anand, *India Prepares EU Trade Complaint*, Wall St. J. Online, Aug. 6, 2009 (quoting Sophie Bloemen of Health Action International).

[51] Priyanka Golikeri, *EU Seizes Another Generic Package*, Daily News & Analysis (Nov. 3, 2009, 2:13 IST), http://www.dnaindia.com/money/report_eu-seizes-another-generic-package_1306411 (quoting an official from an Indian pharmaceutical company that "trade in legitimate medicines between countries is fundamental to ensuring access to medicines for millions").

[52] Brook Baker, *Pointing the Finger at Big Pharma—Dutch Seizure of Generic Medicines*, Program on Information Justice and Intellectual Property (Feb. 10, 2009), http://www.wcl.american.edu/pijip/go/blog-post/pointing-the-finger-at-big-pharma-dutch-seizure-of-generic-medicines.

[53] *Id.* ("too little attention has been directed at these corporate criminals who are acting with impunity to thwart lawful generic competition even in countries where their patents and marketing rights have no effect").

[54] *Id.* (noting the EU Regulation "gives impunity to Big Pharma to make frivolous claims of its 'suspicion' that the products 'might'violate intellectual property rights in the Netherlands").

[55] *See id.* (noting that shipment would infringe if for sale in the Netherlands, but neglecting any mention of Dutch law that would consider in-transit goods to infringe).

[56] *See* Brazil Intervention, Feb. 2009, *supra* note 44, para. 4 (noting the Dutch seizure was unacceptable and sets a dangerous precedent).

[57] For example, Cipla was stated to abandon a consignment of drugs confiscated in the Netherlands. P.T. Jyothi Datta, *Cipla Export Consignment Too Seized at Amsterdam*, Bus. Line, Feb. 4, 2009, http://www.thehindubusinessline.com/2009/02/04/stories/2009020451610200.htm (noting the "cost of litigation was disproportionate to the value of the consignment").

[58] Miller & Anand, *supra* note 50 (noting that Ind-Swift no longer transports through Netherlands and the alternative route through Singapore/Malaysia costs twice as much).

and pediatric formulations."[59] Indeed, AIDS medicines *have* been stopped, including one shipment of HIV medication sent by UNITAID, a Geneva-based organization. Ironically, the EU funds the important work of this organization—and this monetary support was undermined by the seizures. Although the EU has stated that the goods were "promptly" shipped, there was a delay of several weeks; and any delay in treatment of HIV is not tolerable. Moreover, this example underscores that the EU law threatens the work of the WHO, UNAIDS, Global Fund, and other organizations who provide low-cost generic drugs to countries that need them.

The EU repeatedly conflates generic drugs with counterfeits and suggests that it is somehow protecting the public against dangerous counterfeits.[60] However, the EU is confusing multiple issues. Counterfeit drugs are ones that improperly use the trademark of another; they may pose health risks if they are substandard drugs, but the word *counterfeit* by itself does not mean that the drug is of poor quality. Moreover, generic drugs are by definition legal drugs that have been properly evaluated and certified as being equivalent in safety and efficacy to the original/brand name version. Improperly using the term *counterfeit* to refer to both substandard drugs and those that may violate intellectual property rights misleadingly creates an overtone of righteousness.

EU claims about saving lives are a transparent attempt to misrepresent the facts; in none of the cases of seized drugs was quality an issue.[61] To the contrary, the EU is engaged in overzealous enforcement of EU patents.[62] The EU's law is a thinly disguised trade barrier that protects the European pharmaceutical industry[63] while undermining the Indian generics industry.[64]

---

[59] Baker, *Pointing the Finger at Big Pharma*, *supra* note 52.

[60] Mara, *supra* note 34 (quoting India's statement that "underlying the drug seizures is also a deliberate mixing up of the issue of spurious/sub-standard drugs with IPRs."); Sanjay Suri, *Health: EU Blocking Medicines for the Poor*, IPSNEWS, Oct. 20, 2009, http://ipsnews.net/news.asp?idnews=48935 (Sophie Bloemen, from Health Action International quoted as stating "the EU has argued that it needed to check for counterfeits as these are dangerous for public health . . . but counterfeits actually relate to trademark infringement, not a patent infringement").

[61] Brazil, Intervention, TRIPS Council, June 8–9, 2009, *available at* http://www.ip-watch.org/weblog/wp-content/uploads/2009/06/brazil-statement-trips-june-09.pdf [hereinafter Brazil Intervention, June 2009] ("This is a blatant attempt to confound the issue . . . in none of the cases . . . was there any issue with the quality of the drugs."); *see also* India Intervention, June 2009, *supra* note 47, at 2 ("It seems that it has been ingrained very deeply within the EC authorities *that IP violative products are synonymous with potentially dangerous substances.* This is clearly an untenable logic.") (emphasis in original).

[62] Brazil Intervention, June 2009, *supra* note 61, at 2 ("EC customs authorities did not worry about quality—their sole concern was the EC patents").

[63] *See* Pallavi Aiyar, *No Cure in Sight for India-EU Drug Seizure Controversy*, BUS. STAND., Nov. 14, 2009, http://bsl.co.in/india/news/no-cure-in-sight-for-india-eu-drug-seizure-controversy/18/58/376436/ (suggesting India "maintains that European countries are creating trade barriers against Indian drug companies to protect the interests of their firms").

[64] *See* Miller & Anand, *supra* note 50 (quoting Rajeev Kher, Joint Secretary of India as stating that the country views the in-transit seizures "as an attack on the Indian generic industry").

The EU is evidencing a deep—yet wrongly held—belief that violation of intellectual property rights (including patent rights) must necessarily involve potentially dangerous substances.[65] Such a linkage is not only overly simplistic, but simply wrong.[66] Generic medicines neither infringe nor are dangerous. [67]

Even if the EU were genuinely concerned about counterfeit drugs, its procedure is seriously flawed as EU customs officials seem unable to distinguish legitimate generic drugs from counterfeit products.[68] There was at least one shipment of drugs stopped not for any patent infringement, but on the grounds of a purported counterfeit trademark. The supposed counterfeit trademark—amoxicillin—was in fact the internationally recognized nonproprietary (i.e., *non-trademark*) name of the drug. Notably, counterfeit trademarks *should* be easy to identify, as that is one of the reasons they have been considered appropriate for customs officials to police. In contrast, patent infringement is much harder to assess, as even the EU admitted when it first added patent infringements to the scope of what customs officials should police (the inclusion of patents was referred to as an "experiment"). The inability of EU customs officials to distinguish between a generic drug and one with a counterfeit trademark—a supposed "easy" task—is not reassuring to generic drug manufacturers. To the contrary, this suggests that EU customs officials are seriously misguided in their enforcement attempts.

The fact that some suspended drugs have been released (after protracted delay) simply underscores that the EU allegations concerning counterfeit products and patent allegations are baseless.[69]

The EU's claim that there is no "systemic problem" or need to be concerned because the Regulation is already under review is also untenable. EU law permitting seizures of generic drugs constitutes a systemic problem as it means generic drugs can be detained at any time. This problem exists regardless of the number of drugs seized as the law allows unacceptable unpredictability and delays in treatments as well as increased expenses as generic companies seek to avoid transport through the EU. In addition, the EU's "review" of the Regulation is of no help to the issue of in- transit drugs. This "review" is just a periodic one required by the Regulation—not one instigated in response to the problem of

---

65 India Intervention, June 2009, *supra* note 47, at 2 ("it seems that it has been ingrained very deeply within the EU authorities that IP violative products are synonymous with potentially dangerous substances").

66 *Id.* ("This clearly is an untenable logic").

67 *Id.* ("[W]e are talking about generic medicines, which neither infringe IPRs nor are they 'potentially dangerous'").

68 Brazil Intervention, June 2009, *supra* note 61 (noting that the "losartan" and other episodes suggest that EC customs authorities may mistake legitimate generic medicines for fake products).

69 *See* India Intervention, June 2009, *supra* note 47 at 1 (stating that India takes "*serious exception to such unsubstantiated and wild allegations* [that seized drugs were counterfeits, fake drugs, patent violations, etc]. *The fact that the drugs were subsequently released are a proof that the allegations were baseless*") (emphasis in original).

seized generic drugs.[70] In fact, the EU's own documents indicate the review is not focused on access to medicine. To the contrary, the review focuses on enhanced enforcement and increased education efforts.[71]

The EU seizure of goods, as well as its broader international efforts to associate generics with counterfeit medicines, go beyond TRIPS and negate steps that the international community—including the EU—has taken or promise to take to promote access to medicine.[72] For example, the EU Regulation that authorizes seizures of generic goods threatens a global accord the EU has implemented to permit compulsory licensing of goods for export to developing countries.[73] In addition, the seizures are inconsistent with a recent resolution adopted by members of the World Health Assembly—again including the EU—suggesting that public health be considered when adopting TRIPS-Plus rules.[74] Moreover, the same resolution declared a commitment to improving access to health products and overcoming access barriers: commitments that are clearly undermined by the EU seizures.[75]

The broader picture of the EU's actions in the global arena is deeply troubling. Not only does the EU seem bent on disrupting legitimate trade in generic drugs from India, but it wants to impose its improper law on other countries. Patent holders and those who support them are promoting a coordinated and global approach toward maximizing intellectual property rights while simultaneously threatening the delicate balance

[70] *See* Mara, *supra* note 34 (commenting that an Indian representative noted that the review mechanism started in September 2008 before the in-transit seizure was raised as a problem and that the mechanism is focused on reviewing the Regulation in relation to other EU laws, rather than its consistency with TRIPS or the Doha Declaration, such that it is not an adequate response to the concerns of developing countries).

[71] EU Council Resolution to Review EU Regulation on In-Transit Seizures, 1999 O.J. (C 71) 01, Annex, Mar. 25, 2009 (suggesting campaign to inform consumers of dangerous counterfeit goods, problem of organized crime, and global counterfeiting).

[72] *See, e.g.*, India, Intervention under Agenda Item "Other Business," WTO General Council Meeting, Feb. 3, 2009, *available at* http://www.ip-watch.org/weblog/2009/02/03/concern-erupts-over-wto-system-and-medicines-shipments-trips-talks-rekindling/ (click on "statement by India to the General Council is available here") [hereinafter India Intervention, Feb. 2009] (noting that the intellectual property maximalist trend, including efforts to link generics with counterfeit medicines, circumscribes TRIPS flexibilities and is "counter to the spirit of the TRIPS Agreement which is a minimum standards agreement").

[73] *See, e.g.* India Intervention, June 2009, *supra* note 47, at 2 ("It is ironical that while on one hand WTO has taken steps to promote access to affordable medicines and remove obstacles to proper use of TRIPS flexibilities, on the other hand *some members seek to negate the same by seizing drug consignments in transit and creating barriers to legitimate trade*") (emphasis in original).

[74] *E.g.*, Letter from Christian Wagner-Ahlfs, BUKO Pharma-Kampagne et al to Margaret Chan, Director-General, WHO, Feb. 18, 2009, at 3, http://www.haiweb.org/07032009/18%20Feb%202009%20Open%20Letter%20to%20WHO%20on%20Dutch%20seizure%20of%20generics.pdf [hereinafter NGO Letter to Chan, Feb. 2009] (citing WHA 61.21).

[75] *See* WHO, Access to Medicine, Mar. 13, 2009, http://www.who.int/mediacentre/news/statements/2009/access-medicines-20090313/en/index.html.

under TRIPS.[76] The in-transit seizures in the EU are a mere symptom of a much larger and more dangerous phenomenon that involves creating new laws to control generic drugs as well as unduly confusing low-cost generics with counterfeit medicines—one that is totally inconsistent with TRIPS norms.[77]

IV. TRIPS Framework

This section addresses whether in-transit drugs may be properly detained by customs officials for patent infringement under TRIPS. This question necessarily contemplates the EU law, but the analysis here will not be limited to its specifics as there are other existing or potential laws that similarly aim to permit suspension of in-transit goods for alleged patent infringement. This section focuses on a few TRIPS provisions[78] that are most pertinent to explaining the crux of the present controversy.[79] Overall, it attempts to illustrate that the TRIPS issues are not as clear as most countries and commentators appear to view them—an issue that will be explored in detail in the next section about the impact of competing patent perspectives. This is admittedly a somewhat ambitious endeavor because there are a number of provisions involved and a comprehensive interpretation according to customary rules of international law (which apply) would likely be very lengthy. However, given that the goal is to demonstrate that the TRIPS issues are not easily resolved (and subject to the influence of patent perspectives), rather than to arrive at a definitive single meaning, a slightly abbreviated interpretation hopefully will suffice.

---

[76] *See* Intervention by India, TRIPS Council, Agenda Item M—OTHER BUSINESS—Public Health Dimensions of TRIPS Agreement, March 3, 2009, http://keionline.org/node/309 [hereinafter India Intervention, Mar. 2009] (noting trend "to implement the protection and enforcement of IPRs in a maximalist manner and thereby upset the delicate balance between rights of IPR holders and the public policy objectives under the TRIPS Agreement,").

[77] *Id.* ("We . . . note with dismay efforts by some Members to link safe and efficacious but low cost generics with counterfeit medicines, which is essentially an IPR issue."); *see also* Lison Joseph, *Shipments Seizure: India's Drug Makers May Avoid EU Route*, LIVEMINT.COM (Dec 12, 2008, 12:00 AM), http://www.livemint.com/2008/12/12000018/Shipments-seizure-India8217.html (quoting Greg Perry, Director General of the European Generic Medicine, that there is "growing confusion" and conflation of patent issues with counterfeit and piracy in areas such as ACTA and the World Custom Organization's draft SECURE treaty).

[78] There are many TRIPS provisions of possible pertinence to in-transit seizures. This is underscored by Brazil's initial request for consultations under the Dispute Settlement Procedure. Brazil cited fifteen separate articles of TRIPS (not including cited subsections of TRIPS articles) as potentially being violated. *See* Request for Consultation by Brazil, May 19, 2010, WT/DS409/1, at 4.

[79] The seizures also raise issues under the General Agreement on Trade and Tariffs (GATT), which applies to trade in goods for the same member states that are obligated to comply with TRIPS. General Agreement on Tariffs and Trade art. V, Oct. 30, 1947, 61 Stat. A-11, 55 U.N.T.S. 194; s*ee also* Frederick M. Abbott, Comment, *Worst Fears Realised: The Dutch Confiscation of Medicines Bound from India to Brazil*, 1 BRIDGES, at 13 (Feb.–Mar. 2009); T.B. Simi, Briefing Paper, *Patent Rights on Goods in Transit: A Threat to Access Affordable Medicine*, CUTS Center for International Trade, Economics & Environment (CUTS-CITEE) (Apr. 2009). However, the GATT issue is beyond the scope of this book.

This section proceeds in two parts. First, the section explains that the basic TRIPS provisions concerning border measures may arguably permit a country to suspend in-transit goods that violate patent rights. Then, it demonstrates how such suspension may nonetheless be impermissible under TRIPS because of incompatibility with other TRIPS provisions.

A. TRIPS BORDER MEASURES

This section focuses on TRIPS provisions that explicitly address border measures, including an explanation of what is required versus what is permitted. Although the ultimate question is what is *permissible*, it is helpful to start with what is mandatory as that is clearly articulated and also helps set the stage for later explanations of misconceptions influenced by patent perspectives.

No TRIPS provision *requires* customs to suspend in-transit goods for violating patent rights. However, TRIPS articles 51–52 address the fundamental prerequisites concerning when member states must enforce intellectual property rights at their borders.[80] Article 51 discusses what types of goods must be policed, whereas article 52 specifies what a rights-holder must show before customs officials are required to suspend goods.

This section establishes what these provisions require members to provide at their borders before addressing the more ambiguous issue of what member states could potentially do under these provisions. Ultimately this section concludes that under these articles, transit of goods that allegedly infringe patents could be suspended under TRIPS article 51, but that it is unclear how to assess infringement because of a difficult interpretive question concerning the relevant "country of importation" pursuant to TRIPS article 52.

1. Mandatory

a. Article 51–Goods That Must be Policed

Customs officials in member states must police goods that meet two separate criteria. First, the goods must be intended for importation into the member state for free circulation: the goods are intended for importation and are not in transit to a different member state. Second, customs officials only need suspend imported goods that violate a small subcategory of intellectual rights violations. In particular, they only need to police goods that involve either counterfeit trademarks or pirated copyrighted material. To consumers,

---

[80] There are other TRIPS provisions that relate to the procedures for seizures, as well as possible destruction of seized goods. *See* TRIPS, arts. 53–59; *see also* Henning Grosse Ruse-Khan & Thomas Jaeger, *Policing Patents Worldwide?—EC Border Measures Against Transiting Generic Drugs under EC and WTO Intellectual Property Regimes*, 40 INT'L. REV. INTELL. PROP. & COMP. L. 502 (2009).

counterfeits are commonly referred to as "knock-offs";[81] for example, a counterfeit trademarked good would be a sneaker marked "Nike" that was not made by the major shoe company Nike.

b. Article 52—Prima Facie Violation

Customs officials are only required to suspend counterfeit trademark and pirated copyrighted goods when the rights-holder has established prima facie infringement (and also provided a detailed description of the pertinent good to enable custom officials to recognize and seize the infringing import).[82] Prima facie infringement is established with respect to the laws of the country of importation. In other words, if Nike wanted to suspend imports of counterfeit Nike sneakers into Germany, it would need to establish that such imports would violate its trademark pursuant to German law.

2. Permissible

Now that the TRIPS requirements are established, this section explains what is permissible, although not required. In particular, this section addresses whether customs officials *may*—but need not—suspend in-transit goods suspected of infringing local patent rights under articles 51 and 52.

a. Article 51—Goods That May be Policed

Although member states are only required to police counterfeit trademark and pirated copyrighted goods, TRIPS *permits* other types of intellectual property rights to be policed at the borders. TRIPS article 51 states that members may apply the same suspension procedures to "other intellectual property rights," and patents are undeniably a type of intellectual property right.[83] The word *may* means that it is permissible, but not required. So, based on this provision, extending protection at the border to imports of infringing patents is not a problem.

The next issue is whether the border measures can be applied to patent infringements of in-transit goods. In other words, can member states also police goods destined for other markets on the ground that the in-transit good infringes a patent? Article 51 refers to "importation," without defining this term. The word import is defined by the Oxford online dictionary as to "bring (goods or services) into a country from abroad for sale."[84]

[81] Officially, a counterfeit trademark is one that is identical or otherwise indistinguishable from the trademark, although the good is often sold under circumstances that suggest it is not legitimate, such as on street corners. *See, e.g.*, TRIPS art. 51, fn 14(b) (defining counterfeit trademark goods).

[82] TRIPS art. 52.

[83] TRIPS art. 51 (stating "Members may enable such an application to be made in respect of goods which involve other *infringements of intellectual property rights*").

[84] Oxford Online Dictionary.

This definition is broad enough to cover importation of in-transit goods that have a different final destination. In addition, there is additional context that supports such an interpretation. In particular, a footnote to article 51 states there shall be "no obligation" to extend border measures to goods "in transit,"[85] with the phrase "*no obligation*" suggesting that it is specifically contemplated as permissible, even if not required.

Some have stated that article 51 may not apply to in-transit goods that arguably infringe patents.[86] One proffered reason with a contextual basis is that because the language of article 51 speaks to customs authorities suspending goods from release into free circulation, it cannot apply to in-transit goods that are never released into free circulation—unless the provision was drafted incorrectly.[87] Although it is true that the text of article 51 releasing goods into free circulation, general principles of treaty interpretation require promoting interpretations that do not reduce any clauses to a nullity. If this principle is applied to article 51 and its corresponding footnote—which clearly contemplates in-transit goods—a possible interpretation would be that although the required goods are released into free circulation, optional (i.e., in-transit) goods would be released for continued transit. An additional reason offered for why article 51 should not apply to in-transit patent infringement is that the practice was not known at the time TRIPS was enacted.[88] However, even if there was no practice, this does not mean it would be inconsistent with TRIPS. In addition, both of these arguments seem to be arguments that article 51 does not *require* customs officials to suspend in-transit goods that may infringe patents; however, neither argument necessarily precludes member states from taking such action. They may also suggest that article 51 does not apply at all, in which case there would be no need to analyze article 52 compliance. However, none of the arguments presented definitively conclude that member states are *precluded* from taking action against in-transit goods that might infringe patents.

In conclusion, it is at least plausible to read article 51 to permit customs officials to police imported goods that infringe patents, as well as in-transit goods that do so. This conclusion is supported by the text of article 51 as well as its clarifying footnote. In addition, although policing in-transit goods for possible patent infringement is enforcement

---

[85] TRIPS art. 51, n.13 ("It is understood that there shall be no obligation to apply such procedures to imports of goods put on the market in another country by or with the consent of the right holder, or to goods in transit").

[86] For example, Seuba states "TRIPS only *enables* border measures taking place regarding goods that are imported, exported and re-exported, but not border measures against goods that do not need their release by customs agencies." Xavier Seuba, Free Trade of Pharmaceutical Products: The Limits of Intellectual Property Enforcement at the Border, ICTSD Issue Paper No. 27, at 5.2 (2009), http://ictsd.org/downloads/2010/04/seuba_web_10.pdf.

[87] *Id.*

[88] *Id.* (quoting Abbott).

that exceeds the minimum required in article 51, member states are permitted to exceed the minimums so long as they do not contravene other parts of TRIPS.[89]

b. Article 52—Relevant Country of Importation

TRIPS article 52 requires the rights-holder initiating border procedures "under article 51" to establish prima facie infringement in the country of importation. A procedure "under article 51" refers to not only mandatory procedures for imports of counterfeit trademarked and pirated copyrighted goods, but procedures regarding other goods that countries have the option to police. As noted above, a member state *may* (but need not) apply border procedures to in-transit goods that allegedly infringe patents. However, article 52 requires that before such procedures are applied, the patent owner must establish a prima facie case of infringement.

A critical issue is what country's laws are relevant. TRIPS article 52 states that a prima facie case of infringement must be established "under the laws of the country of importation." The country of importation is easy to identify and apply where goods are intended for importation into a country for free circulation. For example, where a good is intended for import into Germany, German law would clearly be relevant to whether the good infringed a patent (or other intellectual property right). However, what country's law is relevant when the goods are in transit and destined for another country? If the good is only temporarily in Germany, with a final destination of Brazil, should the laws of Germany or Brazil apply to whether any patent right is being violated? Unlike imported goods intended for free circulation, in-transit goods are intended for another destination. Some may consider the goods to never be imported under article 51, such that article 52 would not apply to in-transit goods. However, that would produce a curious result where in-transit goods that allegedly infringe patents could be stopped pursuant to article 51, but there would be no proof required pursuant to article 52.[90]

One possible way to interpret "country of importation" is to consider it to be the "in transit" country. Although in-transit goods move on to another destination, they could arguably be interpreted as being at least temporarily imported, then subsequently moved. This reading of importation is consistent with the dictionary definition of import discussed above. Also, applying local law would be consistent with traditional patent laws that vary by nation. Patent rights and patent laws are known to be territorially restricted. Nothing in TRIPS changes this; in fact, the flexibility of member states to adopt different interpretations of the TRIPS standards is premised on this territoriality, which also

[89] TRIPS art. 1.1.

[90] An alternative view would be that this arguably confirms that in-transit goods may not be policed. However, given that the prior section concluded that in-transit goods may be suspended consistent with article 51, this view is rejected.

means nations generally apply their *own* domestic law.[91] Interpreting the country of importation to be the one where goods are in transit would be consistent with this tradition. However, this is not the only possible interpretation.

"Country of importation" could also be interpreted as being the final destination where the goods are imported for free circulation. This would also be consistent with the dictionary definition of import. The broader context of TRIPS also lends supports to this conclusion. For example, one part of the TRIPS preamble states that enforcement of intellectual property rights should not be a barrier to "legitimate trade." In addition, both articles 7 and 8 (respectively labeled "objectives" and "principles") speak to the importance of protecting public health as well as ensuring that intellectual property rights are conducive to social and economic welfare. These provisions could suggest that a proper TRIPS interpretation should promote global trade in generic goods since the generics are legitimate trade that support public health. In addition, the Doha Public Health Declaration's suggestion that TRIPS "can and should" be interpreted in a manner consistent with public health concerns would also support an interpretation that considered in-transit goods to not infringe so as to enable their delivery to their final destination where they will be noninfringing.

Another relevant part of the interpretive context for devining what is meant by "country of importation" is the official Waiver of the TRIPS compulsory licensing provision that usually requires the license to be for domestic use. As explained in Chapter 7, the Waiver permits member states to issue compulsory licenses on drugs destined for exportation to countries that lack the ability to manufacture generic drugs. The Waiver is intended to promote the shipment of drugs that are generic both as to country of origin and destination. The generic drugs made and shipped pursuant to the Waiver are a subcategory of generics as they are all made under a compulsory license. Nonetheless, if drugs made pursuant to the Waiver were suspended in transit for alleged patent infringement, that would arguably hinder the intended purpose of the Waiver in providing low-cost drugs to poor countries.[92] The existence of the Waiver and its intended purpose of promoting trade destined for poor countries might be grounds to interpret the term "country of importation" in a manner that would more likely result in drugs more quickly reaching their intended destination.

There are thus two possible ways of interpreting the application of article 52 with respect to in-transit goods. In particular, the relevant "country of importation" could refer either to the country of final destination or the in-transit country. Although multiple provisions of TRIPS may seem to lend support for considering the "country of importation" to be the final destination, that would lead to the in-transit country assessing

[91] One possible caveat would be if a nation had a choice of law issue with respect to what country's law to apply.

[92] However, as suspension delays, but does not necessarily completely prevent drugs from reaching the final destination, there may not be a problem.

infringement based on another's country laws. This approach would seem at odds with the general rule that patents are territorial rights, with each nation enforcing its *own* rights. In contrast, if the country of importation is considered the in-transit country, suspension would seem consistent with TRIPS. However, as discussed in the next section, even if suspending in-transit goods as patent infringement is arguably consistent with the TRIPS border provisions, they may ultimately be impermissible as going beyond the minimum obligations of TRIPS.

B. OTHER TRIPS PROVISIONS

The section addresses whether suspension of in-transit goods might be impermissible under TRIPS—even if suspension is permissible under the usual border provisions. As noted earlier, although nations are generally permitted to provide stronger rights than the minimums required under TRIPS, they may only do so to the extent the stronger rights do not "contravene" other provisions of TRIPS.[93] This section considers possible other provisions of TRIPS that may be contravened.[94] Although TRIPS does not define *contravene*, its ordinary meaning as set forth in dictionaries is something that is inconsistent with a law, right, or principle.[95] A provision that is inconsistent with a law would seem to suggest there is an explicit inconsistency with a TRIPS provision that sets forth an affirmative obligation.

This section addresses various aspects of TRIPS that could possibly be contravened. Technically, only one provision of TRIPS needs to be contravened for the in-transit suspensions to be inconsistent with TRIPS. However, exploring other possible provisions establishes a basis for a later discussion of the distorted perspectives of TRIPS requirements. This section considers several parts of TRIPS, beginning with the most likely provision to be contravened and concluding with the least likely. The section concludes that in-transit suspensions may violate an article of TRIPS concerning general enforcement of rights, but not the more contextual parts of TRIPS that have been invoked.

1. Article 41

In looking for TRIPS articles that may be inconsistent with suspension of in-transit goods for alleged patent infringement, the most promising one may be article 41, the

---

[93] TRIPS art. 1.1.

[94] For the sake of brevity, this section does not include an extensive analysis of what constitute "provisions" under TRIPS—especially as article 1.1 requires any provision to also be contravened. Although a WTO panel would likely consult dictionary definitions to clarify the ordinary meaning of the undefined term "provision," this is not necessary to conclude what parts of TRIPS are contravened.

[95] 3 THE OXFORD ENGLISH DICTIONARY 847 (2d ed. 1989).

general TRIPS requirement concerning all enforcement procedures.[96] Under article 41, member states must ensure that all enforcement procedures avoid the creation of barriers to "legitimate trade."[97] The issue thus is whether suspending in-transit goods for alleged patent infringement creates a barrier to legitimate trade. TRIPS does not define this key phrase, but it is clearly an important foundation to TRIPS as the first sentence of the TRIPS preamble—which is part of the interpretive context for all provisions of TRIPS, including article 41—similarly states that there is a need to "ensure that measures and procedures to enforce intellectual property rights do not themselves become barriers to legitimate trade."[98]

Whether in-transit drugs are considered "legitimate trade" seems to be a function of whether generics are considered legitimate products or suspect goods that are more akin to illegitimate goods, such as counterfeits. Although generic manufacturers and patent owners are likely to have two contrary views on what is meant by "legitimate," a proper interpretation of TRIPS should not necessarily adopt the view of any single perspective.[99] Even though "legitimate trade" is not defined anywhere under TRIPS, some have asserted that "the validity of generics in international trade is not contested at all,"[100] or that it "will easily qualify" as legitimate trade.[101] A contrary position was offered in a blog post by Professor Bryan Mercurio, a specialist in international trade, who suggested that he was not persuaded by an interpretation that would permit in-transit seizures under article 51, yet be impermissible under article 41 (as not being legitimate trade).[102]

An important question is whether legitimate trade should be defined by the member state where goods are in transit or whether there should be a global definition. A prior WTO panel interpreting the term "legitimate *interests*" suggested that the word *legitimate* is *not* subject to member state interpretation—unlike other undefined terms such as what constitutes a "new" invention for patentability. Rather, the panel defined "legitimate" based on dictionary entries and the general understanding of what consti-

---

[96] This article easily meets the definition of a "provision" as it imposes affirmative obligations on member states regarding the nature of enforcement procedures.

[97] TRIPS art. 41(1). There is also a requirement that enforcement procedures not result in "unreasonable" time limits or "unwarranted delays," although those issues are not addressed here.

[98] TRIPS pmbl.

[99] To generic manufacturers, international transport of their goods is seen as obviously legitimate trade. However, to a patent owner, an in-transit good might be reasonably considered illegitimate.

[100] Seuba, *supra* note 86, at 4.

[101] Shamand Basheer, *India's "TRIPS" Case against the EU: How Strong is It?*, SpicyIP Blog (Jan. 30, 2009, 11:39 pm), http://spicyipindia.blogspot.com/2009/01/indias-trips-case-against-eu-how-strong.html.

[102] Bryan Mercurio, *Generic Pharmaceuticals, Patent Infringement and Freedom of Transit*, Int'l Econ. L. Pol'y Blog (Jan. 30, 2009, 10:53 PM), http://worldtradelaw.typepad.com/ielpblog/2009/01/generic-pharmaceuticals-patent-infringement-and-freedom-of-transit-.html.

tutes a legitimate interest in common legal discourse.[103] In particular, the panel suggested that legitimate interests referred to protection of interests that are justifiable in that they are supported by relevant public policies or other social norms,[104] as reflected in common patent law doctrines.[105] In addition, the panel suggested that a legitimate interest must be one that is universally recognized rather than only recognized by a few countries.[106] Although panel decisions are technically not binding on subsequent panels, in practice, WTO panels generally defer to prior panels, making pertinent this prior interpretation of "legitimate interests." However, although their ruling may be of some interpretive value, the prior panel was notably interpreting legitimate *interests* in the context of a different provision that had other clauses, rather than legitimate *trade*. Nonetheless, some of the rationale concerning the meaning of "legitimate" may still be applicable

This definition of "legitimate" raises the question of whether goods that infringe only in transit but not at the point of departure or destination would be considered legitimate in legal discourse or social norms. One long-standing norm in the context of patent rights is that patents are territorial rights, such that each member state can decide what constitutes patent infringement. This notion of territoriality would arguably support a domestic law that considers in-transit goods to infringe so long as they are considered to be within the territory of that nation. In other words, based on the principle of territoriality, a country could consider in-transit goods that infringe its laws to not be legitimate because they violate patent laws in transit. Although some might suggest that in-transit goods should never infringe because they do not result in competition with the patent owner, that is a policy argument that does not directly bear on whether a country can define what it considers to be infringement.

Moving beyond patent specific norms, there is evidence that many believe at least some trade in generic goods would be considered legitimate trade so long as it did not infringe at the point of departure and final destination. This position could be supported by the formal Waiver of the usual TRIPS requirement that compulsory licenses be issued predominantly for domestic use.[107] In particular, the Waiver is intended to enable a country to issue a compulsory license to make a generic drug to export to a country without adequate resources to effectively make its own generic drugs. Thus, the Waiver necessarily contemplates that generic goods in one country could be shipped to another. However,

---

[103] Panel Report, *Canada—Patent Protection of Pharmaceutical Products*, WT/DS114/R, ¶¶ 7.68–.70 (Mar. 17, 2000) [hereinafter WTO Panel Report, *Canada Generics*].

[104] *Id.* ¶ 7.69.

[105] *Id.*

[106] *Id.* ¶¶ 7.79–.82 (rejecting argument that there was a legitimate interest in providing patent owners with the same effective patent term as all other patent owners because not all countries followed this reasoning).

[107] If the broader WTO context (including its genesis in free trade in goods were considered), that could lend additional support to considering the generic goods to be legitimate goods.

some might believe that only generic goods produced under this specific provision would be considered legitimate trade.

The next issue is whether a law that permits suspension of allegedly infringing goods in transit is a *barrier* to legitimate trade. Proper interpretation of this term should consider the broader TRIPS context, which repeatedly emphasizes that intellectual property enforcement should not be a barrier to legitimate trade. In addition, because TRIPS is part of the WTO framework (which is premised on free trade in goods), this may suggest that promoting trade is relevant to interpretation of this provision. Promoting free trade is also supported by the TRIPS preamble which explicitly states that enforcement of intellectual property rights should not become a barrier to free trade; articles 7 and 8, suggest the need to consider intellectual property rights in a broader context that includes the interests of users.[108] However, a significant question remains concerning what might be a barrier to legitimate trade. As *barrier* is another important, yet undefined term, resort to a dictionary definition is appropriate. A barrier may be defined as a physical obstacle, or something that promotes separation as an "obstacle," or something that makes it "difficult" or "impossible" for something to happen.[109] A provision permitting suspension of in-transit drugs would seem to make trade more difficult, although not necessarily impossible. Of course, if the broader context of the WTO and TRIPS were considered, a temporary suspension could possibly be considered a barrier to free trade.

In conclusion, it is not clear whether TRIPS article 41 prohibits a national law permitting suspension of in-transit goods for alleged infringement. A nation desiring to suspend such goods could argue that infringing goods are not legitimate trade and that each nation should be entitled to assess what constitutes legitimate trade because that term is not defined under TRIPS. However, the broader context of TRIPS, especially considering the genesis of the WTO in promoting free trade, suggests there is a reasonable argument that the suspension of in-transit goods may pose a barrier to trade. But, this is not crystal clear. This conclusion depends on interpreting generic drugs as "legitimate trade" and considering temporary suspensions as a "barrier"—even when goods ultimately reach their final destination.

---

[108] TRIPS pmbl.; *see also* TRIPS arts. 7-8 (providing support for recognizing interests beyond those of the owner of an intellectual property right).

[109] *E.g.*, 1 THE OXFORD ENGLISH DICTIONARY 972 (2d ed. 1989) ("a fence or material obstruction of any kind . . . anything immaterial that stops advance hostile or friendly, that defends from attack, prevents intercourse or union, or keeps separate or apart"); THE RANDOM HOUSE DICTIONARY OF THE ENGLISH LANGUAGE 122 (1973) ("anything that restrains or obstructs progress, access, etc . . . a limit or boundary or any kind").

2. Other TRIPS Context

This section considers whether the broader context of TRIPS contains provisions that are "contravened" by the in-transit seizure of drugs.[110] In particular, this section considers the preamble,[111] as well as articles 7 and 8 (which are labelled "objects" and "purposes" collectively) as related provisions that could be considered contravened. All of these parts of TRIPS have been invoked to suggest that in-transit seizures are inconsistent with TRIPS. However, as will be explained in this section, in–transit seizures do not clearly violate any of these provisions.

It may be difficult for these provisions to be "contravened" under article 41. After all, to contravene a provision requires something that is inconsistent with a law, right, or principle. Generally, to assess whether something is inconsistent, there must be a clear understanding of what is required. However, all of these contextual provisions seem more aspirational rather than concrete obligations whose violations would be more easily seen.

Article 7 may seem initially promising as a provision that may be contravened as it actually refers to "obligations." However, a closer look reveals that it may be difficult to contravene this provision. In particular, article 7 states "the protection and enforcement of intellectual property rights *should* promote innovation to the mutual advantage of producers and users . . . in a manner conducive to social and economic welfare, and to a balance of rights and obligations."[112] An initial problem is that because the clause simply says that rights *should* promote certain things—in contrast to other provisions that state what members *shall* be obliged to do—it is difficult to imagine how the clause can be clearly contravened. In addition, even overlooking this important wording, it remains difficult to see how the desired promotion would be contravened. After all, producers and users of intellectual property would inevitably have differing views on whether their interests were adequately promoted and there is no clear objective benchmark that would indicate how to balance these interests. Accordingly, it would seem difficult to contravene this provision.

Similarly, article 8 may also be difficult to contravene. Specifically, although article 8 promisingly states that "members *may* . . . adopt measures necessary to protect public

[110] These are arguably provisions of TRIPS in that they set forth conditions imposed on parties to the agreement, albeit much weaker ones than in article 41.

[111] Indeed, the preamble was in part relied upon by some groups to suggest that the in-transit seizures were improper. *See, e.g.*, Letter from Christian Wagner-Ahlfs, BUKO Pharma-Kampagne et al. to Pascal Lamy, Director General, WTO (Feb. 18, 2009), http://www.haiweb.org/07032009/18%20Feb%202009%20Open%20Letter%20to%20WTO%20on%20Dutch%20seizure%20of%20generics.pdf [hereinafter NGO Letter to Lamy, Feb. 2009].

[112] TRIPS art. 7 (emphasis added).

health," the very same sentence qualifies that statement: permissible measures are only those that "are consistent with the provisions of this Agreement."[113]

The last part of the broader TRIPS context that has been repeatedly cited to suggest that in-transit seizures are impermissible is the TRIPS preamble. One part of the preamble that on its face may seem relevant is the recognition of "the underlying public policy objectives of national systems for the protection of intellectual property, including development and technological objectives."[114] This clause could support the development objective of India to have patent laws that promote its development goals, including benefiting from global trade in generic drugs. However, this clause cannot be examined in isolation from the rest of TRIPS, which clearly requires India to provide patents on some drugs—even if that is not considered consistent with India's development objectives. In addition, it is difficult to consider this provision as being contravened as the clause seems primarily to be an aspirational one about how intellectual property *should* promote developmental objectives. India might allege that the EU law permitting suspension of drugs does not promote any development objective. However, given that each country has the right to decide on its own patent laws so long as they comply with TRIPS, India's views on whether the EU law promotes *India's* development do not seem relevant.

3. Doha Public Health Declaration

In-transit suspensions have been repeatedly suggested as being inconsistent with the Doha Public Health Declaration and thus impermissible. Although the Doha Public Health Declaration is technically a subsequent agreement between the parties to TRIPS and not part of TRIPS itself, it will be considered consistent with this section's broad approach to TRIPS provisions. The two sentences that some have suggested are violated by in-transit seizures are as follows:

> We agree that the TRIPS Agreement does not and should not prevent members from taking measures to protect public health. Accordingly, while reiterating our commitment to the TRIPS Agreement, we reaffirm that the Agreement can and should be interpreted and implemented in a manner supportive of WTO members right to protect public health and in particular, to promote access to medicine for all.[115]

Although the last clause about promoting access to medicine for all may be tempting to focus on, neither this clause nor the sentence in its entirety sets forth a condition that can be easily violated. In particular, this sentence is about how TRIPS "can and should be

---

[113] TRIPS art. 8.

[114] TRIPS pmbl.

[115] Doha Public Health Declaration, para. 4

interpreted," but does not set forth an affirmative obligation. Even if "contravene" is broadly interpreted to include contravention of principles rather than affirmative obligations, the Doha Declaration is still not clearly contravened. Importantly, the Declaration does not provide an unqualified permission to take any action that promotes access to medicine. To the contrary, the clause concerning access to medicine is importantly limited at the start by a reiteration of the TRIPS requirements. In other words, the goal or principle of promoting access to medicine must be couched in the context of existing TRIPS provisions. This is very important because although immediate access to medicine would be maximally promoted with no patent rights, that would be in complete contravention of TRIPS requirements.

This section hopefully clarifies that assessing whether in-transit seizures of drugs is permissible under TRIPS poses interpretive challenges; it also provides some guiding principles. Before examining how patent perspectives may distort appropriate interpretations of TRIPS, some core concepts of pertinent TRIPS provisions are recapped below:

Q: Does TRIPS require member states to suspend in-transit goods that allegedly infringe local patent rights in transit?

A: No. This should be an easy answer as the issue is not whether TRIPS requires members to take such action, but whether members *may* take such action. TRIPS only requires member states to suspend imports of counterfeit trademark or pirated copyrighted goods intended for free circulation in the country of import. TRIPS does not require suspension of either in-transit goods or goods that violate patent rights.

Q: If TRIPS does not require member states to suspend in-transit goods that allegedly infringe patent rights, may member states nonetheless do so?

A: Maybe. Although member states can probably suspend in-transit goods under the border provisions, such a suspension may be impermissible under other TRIPS provisions, such as the provision requiring enforcements to not constitute a legitimate barrier to trade.

Q: If a drug is made in a country without violating patent rights, is it a generic drug not only in that country, but in any country to which it is exported?

A: No! Whether a drug violates patent rights or is generic depends on the national laws of each country. Just because a drug is generic where made does not mean it cannot infringe a patent in another country.

V. Impact of Patent Perspectives on Legal Interpretations

There are a variety of ways in which patent perspectives seem to color legal interpretations of TRIPS relating to in-transit seizures of drugs. In some cases positions have been asserted that belie the text of TRIPS. In others, positions are advocated purportedly on the basis of TRIPS, but not actually supported by any TRIPS provision. In yet other cases, perspectives may be reflected in positions that do not necessarily contradict the explicit text of TRIPS, but where the reality is more ambiguous. The patent perspectives also seem at play in misinterpretations of applicable Dutch law, as well as invoking irrelevant national laws to address the TRIPS issue.

A. EXPLICIT TRIPS INTERPRETATIONS

1. Article 51

This section first provides a brief review of the proper interpretation of TRIPS before discussing distorted interpretations. As noted earlier, although TRIPS article 51 does not require members to act on infringements by in-transit goods, it also does not expressly preclude them from doing so. To the contrary, a clarifying footnote provides that there is "no obligation" to extend measures to goods in transit, which suggests that although not required, suspension of in-transit goods would be permissible.

Although the important issue is whether in-transit products may be suspended, the view of patents as privilege may overemphasize that TRIPS does not *require* suspension of in-transit goods while ignoring the general rule that TRIPS permits member states to provide more protection than is required. For example, in a letter to the WTO director-general, several health groups stated "TRIPS does not require Article 51 to apply to patented goods, and the exception for goods in transit is discretionary rather than mandatory."[116] In a separate letter to the WHO, the same group of health advocates said that "consistent with WTO rules, goods in transit are exempt from normal restrictions associated with patent . . . when in route to a market where the use is legitimate," citing TRIPS article 51, footnote 13.[117] However, it is not entirely clear whether the view is that in-transit goods cannot be policed, or that they do not normally need to be policed.[118] If the latter, the statement is not necessarily inaccurate, but nonetheless seems off the focus of the central query of whether in-transit goods may be policed at all.

Sometimes there is an erroneous conflation between what is not required under TRIPS article 51 versus what is permissible. For example, Daya Shanker, a scholar in the area of international business, stated that the footnote to article 51 "specifically mention[s] that there shall be no obligation to apply such provision [to suspend] to the goods in transit. The use of the term 'shall be' does not leave any doubt that the provision of Section 51 is

[116] NGO Letter to Lamy, Feb. 2009, *supra* note 111, at 1.

[117] NGO Letter to Chan, Feb. 2009, *supra* note 74, at 1.

[118] The very next sentence states "[t]his approach is not uniform, however, as illustrated recently by several seizures of medicines by Dutch customs officials." *Id.* It is unclear whether this sentence means that the Dutch law breaks from legal tradition or permissible action under the WTO/TRIPS framework.

not to be applied to the case of goods in transit."[119] Shanker is correct in initially stating that there is no obligation under TRIPS to suspend goods in transit; the footnote states that "it is understood that there shall be *no* obligation to apply such procedures to . . . goods in transit."[120] However, his emphasis on the words "shall be" seems to distort the broader context. After all, just because there is no obligation to suspend goods in transit does not mean that such suspension is prohibited. This is especially true because TRIPS only provides minimum standards that members can exceed unless they contravene specific provisions of TRIPS. Shanker's assertion seems to overlook the basic minimum standard framework; it also does not suggest that contravention of other provisions of TRIPS are at issue. Similarly, as discussed earlier, Professor Seuba concludes that TRIPS article 51 "cannot apply to goods in transit," but his discussion seems to focus on the fact that nothing in this article *requires* application to goods in transit, rather than whether there is anything that would prohibit such an extension.[121]

2. Article 52

The view of patents as privilege may also be very influential in interpreting TRIPS article 52, which requires a rights-holder to establish prima facie infringement under the law of the "country of importation." As noted earlier, there is a reasonable basis for concluding the pertinent country of importation *could* be the in-transit country. However, as every public health advocate knows, if the in-transit country is considered the relevant "country of importation," seizures of in-transit drugs in the EU would be permissible under TRIPS—at least according to this provision. Knowledge of this undesirable result may lead to awkward interpretations concerning what constitutes the relevant country of importation.

Those in favor of promoting public health interests often assume that the country of importation *must* necessarily be the country of final destination.[122] For example, patent

---

[119] Posting of Daya Shanker, daya.shanker@deakin.edu.au, to IP-Health (Feb. 7, 2009), *available at* http://lists.essential.org/pipermail/ip-health/2009-February/013452.html. Daya Shanker is a lecturer at Deakin University.

[120] TRIPS art. 51, n.13.

[121] Seuba, *supra* note 86, at 14. Nirmalya Syam, a program officer with the South Centre, a self-proclaimed intergovernmental think tank of developing countries (that would therefore be expected to hold a privilege view of patents) also asserts that the article 51 footnote "clearly obliges WTO members not to apply border measures to goods in transit"—although just one paragraph later he seems to realize that member states *could* apply such measures when he states that members "are *not expected* to apply border measures to the movement of goods in transit." Nirmalya Syam, *Seizures of Drugs in Transit: Why Europe's Law and Actions are Wrong*, South Centre, 41 S. BULLETIN, Sept. 22, 2009, http://www.southcentre.org/index.php?option=com_content&task=view&id=1073&Itemid=279.

[122] Although there are other examples of distorted interpretations, this section focuses primarily on the views of scholars who should have an excellent understanding of the law and are also known to generally favor access to medicine and thus likely to embrace a view of patents as privilege. Distorted interpretations in articles written by

scholar Carlos Correa has suggested that the EU lacks competence to determine patent infringement at the country of importation, which necessarily assumes the relevant country is the final destination, rather than an EU country where goods are in transit.[123] Although Professor Correa is correct in opining that it would be difficult for EU customs officials to apply the law of a different country, interpretation of TRIPS provisions does not turn on what would be most practical. In a blog post, scholar Grosse Ruse-Khan similarly cited the difficulty EU customs officials would have in applying the law of another nation as being relevant in the context of assuming that country of importation should be the country of final destination,[124] although he candidly admits at one point that his "desire to limit TRIPS plus . . . may play a role"—perhaps an open admission of the relevance and impact of patent perspectives.[125]

Professor Seuba takes a more nuanced position, but seems to nonetheless be influenced by a view of patents as privilege. He correctly notes that given the lack of a definition of "importation," any good that crossed a border could "in principle" be considered.[126] However, he almost immediately disregards this possible interpretation based on the "territoriality principle, the lack of exploitation of the patent in the national territoriality, as well as the non-violation of any of the rights granted by a patent."[127] Although he does not clearly explain why considering in-transit goods to infringe a patent violates territoriality, his assumptions concerning what is necessary to violate a patent actually contradict traditional notions of territoriality that permit countries to decide whether to grant patents, as well as the scope of patent rights. He may be suggesting that to consider in-transit goods to infringe violates the territorial expectations of other member states. However, this is only one possible view. As he rightly notes, importation can be more broadly defined.

---

students are not included as those distortions could equally likely be a function of an incomplete understanding. However, some of their conclusions seem similar. *E.g.*, Shashank Kumar, *Freedom of Transit and Trade in Generic Pharmaceuticals*, 31 Eur. Intell. Prop. Rev. 791 (2010) (article by LLM candidate that asserts there is a "trend" of interpreting the phrase "country of interpretation" to be the country of final destination based on a review of cases in a variety of jurisdictions that interpret the word *import* in cases not only beyond the patent arena, but also outside of the intellectual property context—such as tax law—while giving virtually no weight to the Indian case that clearly concluded that in-transit goods could infringe copyright laws).

[123] *See, e.g.,* Saez, *supra* note 32.

[124] Henning Grosse Ruse-Khan, Comment to *Generic Pharmaceuticals, Patent Infringement and Freedom of Transit*, International Economic Law and Policy Blog (Jan. 25, 2009, 10:44 PM), http://worldtradelaw.typepad.com/ielpblog/2009/01/generic-pharmaceuticals-patent-infringement-and-freedom-of-transit-.html.

[125] *Id.* Of course, a blog post necessarily reflects an informal and possibly preliminary view. In addition, his view, as explained in more detail in a subsequent article, could alternatively reflect a view that the GATT/WTO framework should play a substantial role even in interpreting TRIPS provisions. *See* Grosse Ruse-Kahn & Jaeger, *supra* note 80 at 534-55.

[126] Seuba, *supra* note 86, at 5.1.

[127] *Id.*

3. Doha Declaration

The Doha Public Health Declaration is repeatedly invoked and often relied upon improperly to assert that in-transit seizures are clearly impermissible. Some have stated that the EU regulations permitting seizures of alleged patent infringement are "clearly inconsistent"[128] with or a "direct violation" of Doha.[129] Similarly, a representative from Oxfam characterized the seizures as a "nonsensical encroachment . . . on the WTO's Doha Declaration, which *ensures* that intellectual property rules should not interfere with the ability of developing countries to protect and promote public health."[130] However, as noted earlier, although the Doha Declaration does mention that intellectual property rules should not interfere with public health, it does so in the same breath as a clear affirmation of TRIPS rules. In other words, the Doha language concerning promoting public health cannot override TRIPS requirements.

A less radical proposition is the assertion that the Doha Public Health Declaration supports an interpretation of TRIPS (article 51) that would prohibit goods in transit from being infringing patent claims based on the Declaration's "central objective" of providing "access to medicine for all."[131] For example, Professor Frederick Abbott, a noted scholar in international intellectual property law, accurately commented that TRIPS article 51 does not require suspension of in-transit goods that violate patent rights, but nonetheless argued that Doha "should preclude" such action because the Doha Public Health Declaration is a relevant interpretive agreement.[132] Although Professor Abbott is correct that the Doha Declaration is an interpretive agreement, his subsequent assertion that the EU Regulation conflicts with Brazil's "right . . . to protect the public health of its citizens" borders on an overbroad interpretation of what the Doha Declaration provides. As noted above, the Doha Declaration does not guarantee nations the ability to take any action to promote the public health of their citizens. Professor Abbott may also realize that his reliance on the Declaration is overbroad as he later states in the same document that the in-transit seizures may present an opportunity for Brazil and India to claim that their benefits under TRIPS are nullified or impaired because they had a legitimate expectation that products unpatented in their territories could be freely traded.[133] Although it is technically possible for a country to assert that benefits under TRIPS have

---

[128] NGO letter to Chan, Feb. 2009, *supra* note 74, at 2; *see also* Brazil Intervention, Feb. 2009, *supra* note 44, ¶. 8 (asserting "extraterritorial enforcement of patent rights cannot be reconciled with the terms of the Doha Declaration" that TRIPS "can and should be interpreted in a manner supportive of the right to protect public health").

[129] Baker, *supra* note 52.

[130] HAI et al. Press Release, Mar. 6, 2009, *supra* note 48, at 1 (quoting Rohit Malpani from Oxfam International) (emphasis added).

[131] NGO Letter to Lamy, Feb. 2009, *supra* note 111, at 2.

[132] Abbott, *supra* note 79, at 13.

[133] *Id.* at 14.

been nullified even if no actual provision of TRIPS is violated (so-called "non-violation" claims), such claims are currently under a moratorium; in other words, it is not a valid claim at this time. Also, the fact Professor Abbott mentions a non-violation claim suggests that he knows there is no *direct* violation of TRIPS.

4. Broader TRIPS Context (Articles 7–8)

The broader TRIPS context has also been invoked as support for the proposition that in-transit goods cannot be policed under TRIPS.

Brazil has asserted "the TRIPS Agreement *does not allow* the detention of goods in transit" because violating IP rights in the country of transit violates the principle of territoriality and is counter to the objectives and purposes of TRIPS.[134] However, as noted earlier, TRIPS nowhere states that it does not allow detention of in-transit goods; to the contrary, TRIPS expressly *permits* such action in the footnote to article 51. In addition, Brazil's reliance on articles 7 and 8 is likely overstated. Brazil asserts that any seizure of such drugs "offends articles 7 and 8," which suggests a belief that these provisions can be affirmatively violated.[135] As noted earlier, it is difficult to find a clear violation of these contextual provisions—unless only limited parts of them are quoted. For example, Brazil cites article 7 as stating that enforcement of rights must be done "in a manner conducive to social and economic welfare" while article 8 "upholds members' rights to protect public health and nutrition."[136] However, article 7 actually states that enforcement of rights *should* contribute to social and economic welfare. Similarly, article 8 does not provide member states an unfettered right to take any action that promotes public health; to the contrary, an essential caveat to article 8 is that measures to protect public health must be "consistent" with TRIPS. In other words, article 8 does not provide a complete loophole from TRIPS obligations. What Brazil is suggesting is that its desire that in-transit goods be considered not infringing should *trump* the EU's ability to frame its own intellectual property rights. At a minimum, articles 7 and 8 are not as clearly breached as Brazil's statement seems to suggest.

B. OTHER DISTORTED LEGAL INTERPRETATIONS

This section discusses other skewed interpretations that are not always clearly tied to specific TRIPS provisions. In particular, some have suggested that in-transit suspensions are inconsistent with a TRIPS principle of "territoriality," without citing any provision of TRIPS. Similarly, others have suggested that the suspensions are inconsistent with what constitutes a generic drug that may—or may not—be related to any specific TRIPS

---

[134] Brazil Intervention, June, 2009, *supra* note 61, at 3.

[135] *Id.*

[136] *Id.*

provision. Last, there are also skewed interpretations of the Dutch law applicable to the EU seizures as well as irrelevant invocations of other domestic laws.

1. Territoriality

A view of patents as privilege may lead to some overstatements concerning the role of territoriality with respect to in-transit seizures. For example, India has said that the in-transit seizures violate territoriality principles that are the "key stone in the edifice" of TRIPS, without citing a specific provision of TRIPS or explaining the territorial violation[137] Similarly, Brazil has asserted that "TRIPS does not allow the detention of goods in transit" because it "violates the principle of territoriality, a keystone of the international IP system"—again without citing a specific provision of TRIPS that is being violated.[138] In both cases, the desire to consider the in-transit goods as not infringing may result in overstatements concerning TRIPS provisions. Indeed, the fact that no article of TRIPS was cited suggests there is no violation of any actual provision.

Although it is true that TRIPS recognizes the rights of individual member states to establish patent laws that are consistent with TRIPS, it is a leap in logic to suggest the suspension of in-transit goods violates territoriality. To the contrary, permitting member states to stop in-transit goods on the basis of patent infringements (or to not do so) would seem consistent with the notion that each nation determines the scope of its own patent laws. In addition, this distortion contradicts the actual text of TRIPS article 51, note 13 that suggests countries may in fact control goods in transit.

2. Generic

Another misstatement (or at least red herring likely put forth by those with a perspective of patents as privilege) is the misconception that drugs are deemed generic based on where they are commercialized. For example, Brazil stated that "[u]nder TRIPS, the medicines seized are generic under the law of the market in which they were meant to be commercialized. . . . [w]hether or not the medicines were generic under the law of the country of transit is an irrelevant question."[139] However, Brazil cites no authority for this

[137] India Intervention, March 2009, *supra* note 76, at 1; *see also* Minister of State Preneet Kaur Address to High-Level Segment of the Economic and Social Council Special Event on Africa and the Least Developed Countries: Partnerships and Health (July 8, 2009), portion *available at* http://www.keionline.org/blogs/2009/07/08/india-ecosoc-seizures (noting the "seizures contravene the concept of territoriality enshrined in the TRIPS Agreement").

[138] Brazil Intervention, June, 2009, *supra* note 61, at 3; *see also* Brazil Intervention, Feb. 2009, *supra* note 44, ¶ 8 (stating "[e]xtraterritorial enforcement of patent rights can not be reconciled with the terms of the Doha Declaration on TRIPS and Public Health").

[139] Brazil Intervention, Feb. 2009, *supra* note 44, ¶ 6.

supposed proposition of TRIPS and misstates the relevant issue: whether a good infringing a patent in the country of transit is in fact an open issue.[140]

In a later statement, Brazil asserted that "trade in generic medicines is perfectly legal" without acknowledging that what is considered generic (and thus not in violation of a patent) may be subject to local laws.[141] Not only does the statement not seem to have any relevance as to whether in-transit drugs may be permissibly controlled under TRIPS, but it is also confusing. It is unclear whether the word *generic* is used here to refer to unpatented goods or goods deemed generic and thus bioequivalent by a national regulatory authority. If the statement is meant to refer to unpatented goods, it has no support in TRIPS. Nowhere does TRIPS state that whether a good is patented (or not) is determined by where the drug is sold.[142] In addition, if the word *generic* is used to refer to goods approved for sale as bioequivalents, that would simply have no relevance to the in-transit situation as there is no dispute that the goods were approved by the national regulatory authority in the country of final destination.

More likely, the statement is a poorly articulated attempt to refer to the "country of importation" under TRIPS article 52 as necessarily being the final destination as that is where goods are commercialized. As noted above, there are some flaws to assuming the country of importation must necessarily be the country of final destination. Moreover, the assertion that drugs are unpatented (and thus generic) based on where they are commercialized is a further distortion as TRIPS does not make any reference or suggestion that commercialization is relevant to assessing the country of importation.[143] Although it might be possible to make an argument that in-transit goods should be excluded as they do not affect the patent owner's commercial market in the nation where they are briefly present, this would be a mere policy argument that does not impact TRIPS requirements. At most, the policy argument might be relevant in discussions of what domestic law should or should not be considered to be infringement.

---

140 *Id.*; *see also* Priyanka Golikeri, *EU Seizes Another Generic Package*, Daily News & Analysis, Nov. 3, 2009, *available at* http://www.dnaindia.com/money/report_eu-seizes-another-generic-package_1306411 (in the context of discussing seizure of clopidogrel, citing an unnamed Indian "patent expert" that claimed the seizures were impermissible because although the seized drugs were patented in the EU markets, they "are not meant for the EU population").

141 *E.g.*, Brazil Intervention, June 2009, *supra* note 61, at 1 (affirmatively stating "trade in generic medicines is perfectly legal from the intellectual property point of view"). However, this statement may be intended to indicate that generic drugs are not counterfeit.

142 This section assumes the term *generic* is used here to refer to unpatented drugs. Technically, the term refers to a drug's being approved by a national regulatory agency as the bioequivalent of a previous drug. However, it has been used more broadly to refer to unpatented drugs that are bioequivalent.

143 The only TRIPS provision that may refer to some aspects of commercialization relate to the basic patent rights that nations must provide. TRIPS, art. 28. However, control of in-transit goods is clearly a TRIPS-Plus requirement.

3. Domestic Laws

This last part looks at the misinterpretation or improper invocation of national laws in the context of discussing the in-transit suspensions. First, there are misstatements concerning whether in-transit drugs can infringe a patent under Dutch law—the relevant law at issue for the EU suspensions. Second, there is also an overstatement of existing European law that seems to overlook there being contrary Dutch law at issue. Last, other laws and "traditions" are sometimes discussed in the context of whether the in-transit suspensions comply with TRIPS without any explanation for why they might be relevant to a proper interpretation of TRIPS.

a. Dutch Law

The most extreme distortion of existing law is an assertion that drugs that are in-transit in the Netherlands with no intent to enter Netherlands for free circulation cannot infringe. Although there may be reasonable policy arguments for such a law, this statement does not reflect existing case law that clearly holds that in-transit products can be deemed patent infringement. But this has not prevented erroneous statements from being made. For example, Brook Baker, an active leader with Health GAP, stated that if a shipment was "intended for sale and use in Netherlands, its importation would violate patent holders rights in the Netherlands," but seems to assume that in-transit goods simply cannot infringe.[144] Notably, he fails to acknowledge the existing Dutch case law that provides otherwise. In addition, he suggests that the EU Regulation permits "Big Pharma to make frivolous claims"—presumably based on his (erroneous) belief that in-transit drugs cannot infringe patent rights.[145]

b. Other Laws and Traditions

The view of patents as a privilege may also lead some to invoke irrelevant laws and traditions in the context of discussing whether the EU law is consistent with TRIPS. For example, a group of health advocates asserted that there are "long standing traditions to provide exceptions to patent rights for aircraft and marine vessels in route to markets, and for goods in transport." [146] However, tradition does not always control; after all, there is also a long-standing tradition for countries to not provide any patents on drugs, which TRIPS has clearly overturned. In addition, even the supposed tradition providing for

---

[144] Baker, *supra* note 52.

[145] *Id.* Baker suggests that there can be no basis for stopping products unless they might be diverted, which suggests that he thinks the ECJ law regarding trademark law (rather than specific patent law cases) applies.

[146] Along similar lines, a group of health advocates asserted in a letter to the WTO director general that there are "long standing traditions to provide exceptions to patent rights for aircraft and marine vessels in route to markets, and for goods in transport," although the very next paragraph recognizes that control over in-transit goods is discretionary. NGO Letter to Lamy, Feb. 2009, *supra* note 111, at 1.

exceptions for goods in transit are often overstated. For example, Daya Shanker invoked a 1913 case from the United States as purported support for his assertion that goods in transit cannot infringe patent laws because they have not entered the customs area.[147] However, it is a basic principle of patent law that laws of individual nations do not control other nations. Moreover, even if the case was binding, it does not actually address in-transit goods; rather, the issue there was whether a patented part of a *vessel* that was temporarily in a US port would infringe.[148]

VI. Conclusion

In-transit suspensions of generic medicines are one of the latest challenges to providing access to medicine. Analysis of the TRIPS consistency is complicated, despite polarized views that seem to suggest otherwise. This issue thus presents another opportunity to show the impact of patent perspectives in distorting facts and law. Regardless of any possible resolution of the issue at the WTO, the same issues continue to linger in the international arena.

Although this chapter focused on a single issue where competing patent perspectives are at work, the same perspectives play a role in many situation. As discussed in the next chapter, both patent proponents as well as public health advocates have keenly moved their arguments to different fora (whether legal or public) so as to best achieve their goals. Although the next chapter is not framed as a case study, the different patent perspectives nonetheless are likely at work in prompting divergent actions that promote different interests—and also make it difficult for participants to find common ground.

---

[147] Shanker, *supra* note 118.

[148] Brown v. Duschene, 60 U.S. 183 (1913).

11 A History of Access to Medicine through the Lens of Patent Perspectives

This chapter provides a broader view of the development of international laws impacting the protection of drugs. Whereas prior chapters focused primarily on how patent-owning companies impact these laws, this chapter shows that such companies do not necessarily always dominate the conversation. In particular, just as companies have utilized tools such as forum shifting and trade pressures to achieve their goals, access to medicine proponents have used forum shifting and public pressure to promote their interests. This chapter demonstrates that the development of laws and policies impacting access to medicine is a dynamic process in which competing patent perspectives (as well as their confluence) play an important role.

I. Introduction 326
II. Pre-Doha 327
A. South Africa 327
1. Patents as Privilege 328
2. Patents as Uber-Right 328
3. Confluence of Perspectives 330
B. Brazil 333
III. Doha Public Health Declaration 335
A. Competing Proposed Declarations 336
1. Patents as Privilege 336
2. Patents as Uber-Right 337
B. Confluence of Perspectives 339
IV. Post-Doha 339
A. Uber-Right 340
1. Uber-Right Views and the Waiver 340
2. Uber-Right View of TRIPS and Doha 341
3. Uber-Right View of Health Impacts 342
4. TRIPS-Plus Activities 343
5. Uber-Right Outreach Efforts 345
6. Uber-Right View of a Middle Ground 347
B. Privilege 348
1. Privilege View of TRIPS, Doha, and The Waiver 348
2. Solutions Steeped in a View of Patents as Privilege 349
3. WHO 351
4. WIPO 352
V. Conclusion 353

I. Introduction

This chapter provides a broader view of how competing patent perspectives relate to issues involving access to medicine under TRIPS and TRIPS-Plus measures. The use of patent perspectives in this chapter contrasts with their use in earlier chapters where each chapter focused on an in-depth analysis of an isolated issue. This chapter addresses a wide range of controversies through the lens of competing perspectives to show how the perspectives have shaped and continue to shape the development of international law and policy.

Each section of this chapter illustrates the interplay of patent perspectives in developing international law and policy. As discussed earlier, controversies concerning the impact of patents on access to medicine can be better understood as a competition between two polarized views of patents. One view is that patents are a privilege granted by the state and inherently subject to limitation—especially when they conflict with other social priorities, such as access to medicine. An alternative view of patents is that they are a special type of property right that deserves stronger protection than traditional property rights during a limited time to promote innovation; as the rights are stronger, they are an "uber-right." Under the uber-right view, patents are consistent with promotion of public health because patents promote innovation that results in improved treatments benefiting all.

This chapter is divided into three sections. The first section evaluates the clash of perspectives in two controversies that arose shortly after TRIPS was concluded, but before the Doha Public Health Declaration. The second section considers the impact of the perspectives, as well as other fortuitous factors that led to the Declaration. The third section highlights the continued existence of perspectives post-Doha, not only in how the Declaration is viewed, but also in terms of other TRIPS-Plus activities.

II. Pre-Doha

A. SOUTH AFRICA

The confluence of competing patent perspectives is illustrated by a controversy over a South African law aimed at promoting access to medicine. The controversial law was adopted in December 1997 to provide more affordable medicines by permitting broad use of compulsory licenses and parallel imports.[1] Under the law, the South African minister was authorized to prescribe compulsory licenses for any patented drug that was initially marketed by the patent owner; however, this law did not provide any specific guidelines or conditions for when a compulsory license would issue beyond that it would be to protect the public health.[2] The South African Minister was also authorized to prescribe conditions to permit importation of a drug originally made by the patent owner abroad, but not imported with her consent.[3] In other words, the law permitted parallel imports of drugs.

South Africa adopted the law despite strong objections from both the pharmaceutical industry and foreign governments. South Africa considered the law necessary to address an HIV epidemic. At the time, South Africa had the most people with HIV/AIDS, yet most could not be treated with antiretrovirals, which then cost about $1000/month per patient. Moreover, treatment was considered essential in light of a newly minted constitutional mandate to provide access to health care for all.[4]

---

[1] Medicines and Related Substances Control Amendment Act 90 of 1997 (S. Afr.)

[2] *Id.* § 15C(a).

[3] *Id.* § 15C(b).

[4] S. Afr. Const., 1996 (arts. 27(1)(a)–27(2)). The constitutional mandate was one of the initial actions after President Mandela was elected in the country's first democratic elections. *See, e.g.*, William W. Fisher & Cyrill

1. Patents as Privilege

The South African law is easily consistent with the view of patents as privilege. In particular, given a healthcare crisis that involved an HIV epidemic, the law was designed to ensure that patents did not stand in the way of public health.

In addition, holders of the view of patents as privilege would find the law to be in complete compliance with TRIPS. First, they would consider the law consistent with TRIPS compulsory license requirements. Even though the law includes none of the procedural requirements of TRIPS,[5] holders of a privilege view would find it adequate because it *could* be applied consistent with these requirements. Second, they would believe parallel imports are clearly permitted under TRIPS even though the text of TRIPS is more nuanced in only eliminating the issue from a formal dispute before the WTO.

2. Patents as Uber-Right

Holders of an uber-right view would take a dim view of the broad powers of the South African minister to impose compulsory licenses without specific limits in the laws that guarantee compliance with TRIPS compulsory license requirements.[6] Although the law at issue here was particularly broad, those with an uber-right view have contested compulsory licenses without regard for TRIPS compliance, as discussed more extensively in Chapter 6. In addition to the compulsory license objection, those with an uber-right view would consider parallel imports to violate TRIPS patent rights, as argued by PhRMA.[7]

Rigamonti, *The South Africa AIDS Controversy: A Case Study in Patent Law and Policy* 2–3 (2005), *available at* http://cyber.law.harvard.edu/people/tfisher/South%20Africa.pdf.

[5] For example, it did not state that patent owners would be entitled to remuneration, or that they would have the right to challenge the issuance of a compulsory license, as required by TRIPS. Medicines and Related Substances Control Amendment Act 90 of 1997 § 15C(a) (S. Afr.) ("[N]otwithstanding anything to the contrary contained in the Patents Act, 1978, [the Minister may] determine that the rights with regard to any medicine under a patent granted in the Republic shall not extend to acts in respect of such medicine which has been put onto the market by the owner of the medicine, or with his or her consent . . . .").

[6] For example, the 1998 Special 301 report asserted that the Minister of Health was empowered to "abrogate patent rights for pharmaceuticals." *See* Press Release, USTR, USTR Announces Summary of Special 301 Annual Review, at 21 (May 1, 1998), *available at* http://keionline.org/sites/default/files/ustr_special301_1998.pdf [hereinafter 1998 Special 301 Report Summary]. It is unclear whether this statement is meant to indicate that the law abrogates rights, or whether any compulsory license would abrogate patent rights.

[7] Pharmaceutical Research and Manufacturers of America, *Submission of the Pharmaceutical Research and Manufacturers of America for the 1999 National Trade Estimate Report on Foreign Trade Barriers* (Dec. 4, 1998), *available at* http://www.cptech.org/ip/health/phrma/nte-98/nte-98.html. The US government took a more moderate approach—the United States recognized that there can be no WTO disputes concerning parallel imports, but nonetheless asserted that TRIPS does not authorize parallel imports. World Trade Organization, Statement by the US Delegation, *Minutes of the Council for TRIPS Special Discussions on Intellectual Property and Access to Medicines*, IP/C/M/31 (July 10, 2001), at 40. A decade since the South Africa dispute, PhRMA no longer suggests that parallel imports clearly violate TRIPS, but instead focuses on drug safety and counterfeits.

Although pharmaceutical companies and some governments repeatedly suggested that the South African law was inconsistent with TRIPS, South Africa was never formally challenged in an official action before the WTO. On some level, those espousing these views may have recognized that their interpretations were not consistent with TRIPS and thus could not form the basis of an official WTO dispute.[8] In particular, they may have implicitly recognized that their proposed interpretations reflected not TRIPS itself, but their desired views of patents.

The uber-right view of patents is also reflected in criticisms that have no basis in TRIPS and in fact seem inconsistent with what TRIPS permits. For example, a representative from PhRMA stated that the compulsory licensing authorized by the South African law is "a form of patent piracy" or that "it's stealing."[9] Although this is consistent with an uber-right view that opposes compulsory licenses, TRIPS explicitly permits them. Along similar lines, a representative of the major multinational pharmaceutical company Bristol-Myers Squibb stated, "Patents are the lifeblood of our industry. Compulsory licensing and parallel imports expropriate our patent rights."[10]

Parallel imports were also falsely conflated with counterfeit drugs. For example, one spokesman from the pharmaceutical industry asserted, "A lot of parallel imports come from places like India, and half the time there are no active ingredients."[11] As discussed in Chapter 2, parallel imports are legitimate goods made (or authorized to be made) by patent owners and thus are not counterfeits. However, holders of an uber-right view may view the patent right as requiring global control, such that even products originally made by the patent owner may be considered illegitimate and thus somewhat similar to counterfeit goods.

The uber-right view is also reflected in statements that suggest there is no conflict between patents and public health. For example, one representative of Bristol Myers Squibb suggested "[t]here is a need not to fight the firefighters."[12] This characterization of patent-owning companies as heroic is consistent with the uber-right view that strong patent rights (including the ability to charge patent-based profits) are essential to

---

See, e.g., PHARMACEUTICAL RESEARCH AND MANUFACTURERS OF AMERICA, SPECIAL 301 SUBMISSION 2010 at 6-13, 65, 146, http://keionline.org/sites/default/files/USTR-2010-0003-0245.1.pdf. However, this may not reflect an actual change in perspective. After all, a country may be listed in a Special 301 Report without violating TRIPS.

[8] Another possible explanation is that those with an uber-right view made a calculated political decision to forgo an official WTO dispute because the same result could be achieved more predictably without the WTO.

[9] Merrill Goozner, *Third World Battles for AIDS Drugs*, CHI. TRIB., Apr. 28, 1999.

[10] Sabin Russell, *New Crusade to Lower AIDS Drug Costs*, S.F. CHRON., May 24, 1999, at A1. For a deeper discussion of why parallel imports and compulsory licenses are opposed by pharmaceutical companies, see Chapter 2 (parallel imports) and Chapter 6 (compulsory license case study).

[11] Sabin Russell, *World Trade Showdown*, S.F. CHRON., Nov. 24, 1999, at A1.

[12] Russell, *supra* note 10, at A1.

promote more research. Such a view might help explain why South Africa was listed on the US Special 301 list (of countries with inadequate intellectual property protection) for improperly taking a leadership role at the WHO in arguing for reduced global protection of drugs.[13] In addition, consistent with the view that patents do not conflict with access, companies suggested that factors beyond patents were the actual problem in addressing HIV, with poverty and a lack of infrastructure being the real culprits in HIV epidemics.[14]

3. Confluence of Perspectives

This section shows how the competing patent perspectives can create an important dynamic in shaping the ultimate outcome of disputes. Not only do competing perspectives play a role in how laws are interpreted, but they may have an important role in the outcome of international disputes. The dispute over the South African law highlighted the existence of differing patent perspectives for the global community as each side sought to influence public opinion. The ultimate outcome was thus influenced not only by the existence of the differing perspectives, but how such perspectives were perceived by the public.

Patent-owning pharmaceutical companies took a multipronged and sustained approach against the legislation. Just two months after the law was enacted, PhRMA both directly sued the South African government to try to block implementation of the law[15] and requested that the US government designate South Africa a "priority foreign country"—a designation that refers to a country with the most "onerous and egregious policies" with respect to intellectual property and requires immediate action by the US Trade Representative (USTR).[16] The USTR declined to designate South Africa a "priority watch" country, but did list it on the "watch list."[17] The United States then imposed trade sanctions on South Africa, withdrawing preferential tariff treatment for several

---

[13] *See* Press Release, USTR, USTR Announces Results of Special 301 Annual Review (April 30, 1999), *available at* http://keionline.org/ustr/1999special301 [hereinafter 1999 Special 301 Report Summary].

[14] Moreover, PhRMA suggested that providing broader access to antiretrovirals might actually lead to more problems because patients would be unlikely to be able to follow the then-complex daily regime, which actually could create drug-resistant strains of HIV.

[15] Notice of Motion in the High Court of South Africa, Case No. 4183/98, *available at* http://www.cptech.org/ip/health/sa/pharmasuit.html). They argued that the law was an impermissible delegation of power, that patent owners would be improperly deprived of property without due compensation, and that it was a violation of TRIPS.

[16] 19 U.S.C. § 2242(a)-(b) (2000); Fisher & Rigamonti, *supra* note 4, at 7.

[17] *See* 1998 Special 301 Report Summary, *supra* note 6, at 21. The "watch list" refers to countries the United States considers to have inadequate intellectual property protections, but no legal action must be immediately taken. Nonetheless, countries placed on a watch list typically modify their laws to avoid any prospect of sanctions.

South African goods pending "progress" on intellectual property.[18] Several months after imposing these sanctions, the United States eliminated financial aid to South Africa pending a report from the Secretary of State indicating steps taken by the US government in conjunction with South Africa to repeal, suspend, or terminate the provision.[19]

Although PhRMA attempted to preempt negative publicity [20] and had the support of many government actors, there were countervailing forces. For example, although the US Trade Representative's Office condemned South Africa's law, some US lawmakers disagreed with this action.[21] In 1998, the Executive Board of the World Health Organization adopted a resolution explicitly suggesting that member states should "ensure that public health rather than commercial interests have primacy" and to consider options under TRIPS to safeguard access to essential medicines.[22] The following year, the World Health Assembly enacted a resolution that called upon member states to ensure equitable access to essential drugs and to review options to address such access under international agreements.[23] The UN Commission on Human Rights also issued a resolution that recognized access to medicine in the context of pandemics (such as HIV/AIDS) as fundamental to the human right to health.[24]

Nongovernmental consumer groups also played an important role in establishing global support for South Africa's law (and also for paving the way for the Doha Public Health Declaration). For example, the groups Health Action International and Consumer Project for Technology helped to draft language that led to the 1998 WHO Executive Board resolution.[25] In 1999, several groups held international meetings to address not

[18] *E.g.*, Fisher & Rigamonti, *supra* note 4, at 7.

[19] The withdrawal of aid was accomplished as an amendment to an appropriations bill. Omnibus Consolidated and Emergency Appropriations Act, Pub. L. No. 105-277, 112 Stat. 2681-155 (Oct. 1998).

[20] Even before the controversial legislation was signed, PhRMA had brought suit claiming that certain South African officials had made improper statements in the news to create a "perception in the minds of the general public that medicines in South Africa are unreasonably expensive." PUBLIC PROTECTOR OF THE REPUBLIC OF SOUTH AFRICA, REPORT ON THE PROPRIETY OF THE CONDUCT OF MEMBERS OF THE MINISTRY AND DEPARTMENT OF HEALTH RELATING TO STATEMENTS IN CONNECTION WITH THE PRICES OF MEDICINES AND UTILIZATION OF GENERIC MEDICINES IN SOUTH AFRICA, SPECIAL REPORT NO. 6 (1997), *available at* http://www.polity.org.za/polity/govt/pubprot/report6.html.

[21] For example, Rep. Jesse Jackson, Jr. introduced the HOPE for Africa Bill that requested the US government to "not . . . seek, through negotiation or otherwise, the revocation or revisions of any sub-Saharan African intellectual property or competition law or policy that is designed to promote access to pharmaceuticals" if it complies with TRIPS. *See* HOPE for Africa Act, H.R. 772, 106th Cong. § 601 (1999).

[22] WHO Executive Board, Revised Drug Strategy EB 101.R24, ¶ 1(2) (Jan. 1998), *available at* http://keionline.org/who/EB101.R24. Notably, the United States was not part of the rotating WHO Executive Board the year this resolution passed. SUSAN K. SELL, PRIVATE POWER, PUBLIC LAW 148 (2003).

[23] World Health Organization [WHO], *Revised Drug Strategy*, World Health Assembly Res. 52.19, ¶ 1 (Mar. 24, 1999); *see also* WHA Res. 54.10 (May 21, 2001); WHA Res. 54.11 (May 21, 2001).

[24] United Nations High Commissioner for Human Rights, Access to Medication in the Context of Pandemics such as HIV/AIDS, Res. 2001/33, ¶ 1 (Apr. 23, 2001).

[25] SELL, *supra* note 22, at 146–50.

only the use of compulsory licensing to increase access to AIDS drugs, but other issues under TRIPS that posed hurdles to accessing lower-cost medicine.[26] Also, Doctors Without Borders (also known as MSF—the acronym for their French name) launched an Access to Medicine Campaign in 1999 and has since played a key role in providing low-cost HIV drugs to patients worldwide, as well as promoting laws and policies to encourage greater use of generic medicine.[27] The ultimate successful resolution of the South African dispute is often credited to the coordinated efforts of nongovernmental organizations.[28]

Vice-President Gore initially played a role in promoting the interests of PhRMA.[29] As a result, he was subject to increasing public criticism by those eager to use his public presidential campaign as an opportunity to promote their interests.[30] One of the most publicly visible acts was a disruption of Gore's campaign in the summer of 1999 with signs that declared "Gore's Greed Kills."[31] Later that summer, activists locked Gore out of his office in response to a proposed US compromise that would permit South Africa to use parallel imports and compulsory licenses for HIV drugs, but no other medications.[32] Gore subsequently met privately with the South African president, and the United States removed South Africa from its watch list of countries for trade sanctions.[33] At the November 1999 Seattle WTO Ministerial Meeting, President Clinton indicated that the United States would alter its trade policy to promote more access to medicine in poor countries.[34] Then, in May 2000, roughly three years after South Africa enacted its controversial law, President Clinton signed an executive order that barred further trade pressure

[26] *See, e.g.*, ELLEN' T HOEN, THE GLOBAL POLITICS OF PHARMACEUTICAL MONOPOLY POWER: DRUG PATENTS, ACCESS, INNOVATION AND THE APPLICATION OF THE WTO DOHA DECLARATION ON TRIPS AND PUBLIC HEALTH 30–31 (2009). The HIV advocacy group ACT UP also played a major role in the South African controversy.

[27] *See, e.g.*, MSF, Access to Medicines, http://www.doctorswithoutborders.org/news/issue.cfm?id=2379.

[28] *E.g.*,' T HOEN, *supra* note 26, at 30–31; Thomas Olesen, *'In the Court of Public Opinion': Transnational Problem Construction in the HIV/AIDS Medicine Access Campaign, 1998–2001*, 21 INT'L SOCIOLOGY 5, 20-21 (2006).

[29] Al Gore apparently pressured South Africa in his role as head of the US-South Africa Binational Commission. *E.g.*, PETER DRAHOS WITH JOHN BRAITHWAITE, INFORMATION FEUDALISM: WHO OWNS THE KNOWLEDGE ECONOMY 7 (2002).

[30] *See, e.g.*, Debora Halbert, *Moralized Discourses: South Africa's Intellectual Property Fight for Access to AIDS Drugs*, 1 SEATTLE J. SOC. JUST. 257, 272–74 (2002); Charles R. Babcock & Ceci Connolly, *AIDS Activists Badger Gore Again*, WASH. POST, June 18, 1999, at A12.

[31] Tony Reichhardt, *Gore Under Fire in Controversy over South Africa AIDS Drug* Law, 399 NATURE 717 (1999); William Dowell, *Ethics and AIDS Drugs*, TIME, July 12, 1999; Neil A. Lewis, *U.S. Industry to Drop AIDS Drug Lawsuit against South Africa*, N.Y. TIMES, Sept. 10, 1999.

[32] Halbert, *supra* note 30, at 273–74 (citing Press Release, Treatment-Access, Activists Lock Gore Out of His Office, Aug. 25, 1999).

[33] *E.g.*, Karl Vick, *African AIDS Victims Losers of a Drug War*, WASH. POST, Dec. 4, 1999, at A1.

[34] *E.g.*, SELL, *supra* note 22, at 153.

against Sub-Saharan African countries.[35] At the same time, several multinational companies announced price reductions for AIDS drugs to developing countries.[36]

Although the US government stopped exerting pressure on South Africa, pharmaceutical companies did not immediately follow suit. Rather, it took roughly another year of negative public press as well as a fortuitous synchronicity of events involving an announcement by Bristol Myers Squibb (BMS) that it would not enforce a critical patent in Sub-Saharan Africa on part of a three-drug HIV cocktail. Although the patent was owned by Yale, BMS did the clinical trials necessary to bring the drug to market and had an exclusive license from Yale to sell the drug. The BMS decision to forego its rights in Sub-Saharan Africa was in response to negative widespread publicity concerning Yale's refusal to license the drug to MSF (on grounds that it needed to respect the exclusive license). This result likely could not have been reached without the assistance of then-first-year law student (and now law professor) Amy Kapczinski, who happened to meet a key MSF player at a conference and agreed to help MSF's licensing effort.[37] She played an important role in coordinating the pressure on Yale that eventually led to international negative press,[38] including a statement by one of the coinventors that he wished BMS "would provide the drug free."[39] Just a few days after this story was published, BMS announced that it would not enforce its license in South Africa.[40]

B. BRAZIL

The controversy over Brazil's patent law that permitted compulsory licenses on the grounds of "lack of local working" provides an interesting companion to the South Africa

---

[35] Access to HIV/AIDS Pharmaceutical and Medical Technologies, Exec. Order No. 13,155, 65 Fed. Reg. 30521 (May 12, 2000).

[36] *E.g.*, SELL, *supra* note 22, at 153–54. The price drops were part of the "Accelerating Access Initiative" that involved Beohering, BMS, GSM, Merck, Hoffman, and Abbott. *Id. But see* Barton Gellman, *An Unequal Calculus of Life and Death*, WASH. POST, Dec. 27, 2000 (finding that promises by six of the largest pharmaceutical companies have amounted to little help for AIDS patients in South Africa).

[37] Daryl Lindsey, *Amy and Goliath*, SALON.COM (May 1, 2001), http://www.salon.com/news/feature/2001/05/01/aids.

[38] Julian Borger & Sarah Boseley, *Campus Revolt Challenges Yale over $40 Million AIDS Drug Deal: University Claims Its Hands are Tied by Deal with Drug Firm*, GUARDIAN (London), Mar. 13, 2001; Donald McNeil, *Yale Pressed to Help Cut Drug Costs in Africa*, N.Y. TIMES, Mar. 12, 2001.

[39] McNeil, *supra* note 38. The coinventor alternatively suggested that BMS permit India or Brazil to produce the drug cheaply for underdeveloped organizations, but seemed skeptical that this would happen as he noted "the problem is, the big drug houses are not altruistic organizations. Their only purpose is to make money." *Id.*

[40] In its announcement, BMS stated "[w]e seek no profits on AIDS drugs in Africa, and we will not let our patents be an obstacle." Melody Petersen & Donald G. McNeil, Jr., *Maker Yielding Patent in Africa for AIDS Drug*, N.Y. TIMES, Mar. 15, 2001; *see also* Andrew Clark & Julian Borger, *Cheaper Drugs for Africa: Manufacturer to Relax Its Patent on Two AIDS Remedies*, GUARDIAN (London), Mar. 15, 2001.

case in evaluating the intersection of patent perspectives and access to medicine.[41] In both cases, there was an HIV epidemic, and patent laws were modified to attempt to provide better access to medicine. However, the modifications of the patent laws as well as the outcomes are different, as explained below.

Around the time that South Africa enacted its controversial law, Brazil enacted its own controversial one. Brazil was also facing an HIV epidemic, and in November 1996, it began offering free ARVs to treat HIV,[42] consistent with its constitutionally guaranteed universal access to healthcare. In 1996, Brazil amended its patent laws to establish a "local working" requirement; this required patent owners to make patented inventions in Brazil within three years of the grant of the patent or else be subject to a compulsory license.[43] From this local working requirement international controversy arose.

The Brazilian dispute reflects a long-term controversy concerning local working that reflects the two polar views of patents. Those who consider patents as a privilege view compulsory licensing as reasonable if the patented invention is not being made by the patent owner locally. After all, pursuant to the privilege view, a patent should benefit society through technology transfer—and a local working requirement ensures that patent rights are used in the public interest and not abused. However, those holding an uber-right view would not require a patent owner to use the invention in the country where it was patented. Those who hold an uber-right perspective . . . perspective consider patents as only instrumental to promoting *future* development of drugs; current access to technology is not seen as essential. This dichotomy mirrors the same views on local working held when TRIPS was negotiated, suggesting that despite the conclusion of TRIPS, views have not changed.

Patent-owning companies successfully lobbied the United States to take action against Brazil's law. In February 2001, the United States brought a formal WTO dispute against Brazil alleging that its local working provision impermissibly discriminated against US patent owners.[44] However, before a WTO panel could address the legal issue, public opinion appeared to affect the dispute. Controversy surrounding Yale's exclusive license to Bristol Myers Squibb publicly broke a month after the WTO complaint. In addition, Brazil's success in stemming its AIDS epidemic was internationally recognized. In June

[41] For the sake of brevity, this section discusses only these two controversies. Other controversies existed, including one concerning Brazil's repeated threats of compulsory licenses used to obtain deeper discounts from drug companies. However, as competing perspectives on compulsory licenses are extensively discussed elsewhere, this section will not specifically address this aspect of Brazil's compulsory licenses.

[42] Lei No. 9.313, de 13 de Novembro de 1996, Diário Oficial Da União [D.O.U.] de 14.11.1996 (Braz.), *available at* http://www.planalto.gov.br/ccivil_03/Leis/L9313.htm.

[43] Lei No. 9.279, art. 68(1)(I), de 14 de Maio de 1996 (Brazil), *available at* http://www6.senado.gov.br/legislacao/ListaPublicacoes.action?id=102474&tipoDocumento=LEI&tipoTexto=PUB (Portuguese); http://www.wipo.int/wipolex/en/details.jsp?id=515 (English translation).

[44] Request for the Establishment of a Panel by the United States, *Brazil—Measures Affecting Patent Protection*, WT/DS199/3 (Jan.8, 2001).

2001, on the opening day of a UN AIDS summit, the USTR announced the case would be dropped and asserted that the United States wanted to "protect intellectual property rights without compromising . . . efforts to combat HIV/AIDS."[45] However, it was not a complete victory for Brazil because part of the negotiated compromise was that Brazil would give advance notice if it intended to use the local working provision—and the United States reserved its rights to bring a new WTO action.[46]

The controversy over Brazil's local working basis for compulsory license presents an interesting contrast to the South African controversy. In each case, public opinion (including a strong spotlight on the view of patents as privilege) seemed to play a role in moderating the outcome. In addition, whereas the South African controversy resolved itself clearly in favor of South Africa, the outcome in Brazil is not a clear win for any side.

III. Doha Public Health Declaration

The genesis of the Doha Public Declaration also illustrates how the confluence of competing patent perspectives may impact the development of law. Although this section focuses primarily on competing perspectives reflected in proposed versions of the Declaration and how the confluence of those perspectives culminated in the final version, it is important to note how that version came to fruition. The path towards the eventual Declaration was paved in part by the display of divergent perspectives concerning TRIPS requirements, as seen in the South African and Brazil disputes. In addition, those who viewed patents as privilege were also concerned about political pressure to adopt laws that reflected an uber-right view.[47] This likely prompted a group of African countries to seek a special session with the TRIPS Council for the first-ever discussion concerning the impact of TRIPS on access to medicine. After the TRIPS Council agreed to convene this session, papers were submitted reflecting a range of competing perspectives; the

[45] Press Release, USTR, US and Brazil Agree to Use Newly Created Consultative Mechanism to Promote Cooperation on HIV/AIDS to Address WTO Patent Dispute (June 25, 2001).

[46] *See Brazil—Measures Affecting Patent Protection—Notification of Mutually Agreed Solution*, WT/DS199/4 (July 19, 2001).

[47] *E.g.*, World Trade Organization Council for Trade-Related Aspects of Intellectual Property Rights, *Proposal from a Group of Developing Countries: Draft Ministerial Declaration on the TRIPS Agreement and Public Health*, IP/C/W/312, WT/GC/W/450 (Oct. 4, 2001) [hereinafter Developing Country Draft Ministerial Declaration, IP/C/W/312] (suggesting that members adopt the following language: "acknowledging the vulnerability of developing and least developed country members to the imposition or the threat of imposition of sanctions and to the prospect of being deprived of incentives or other benefits . . . beyond the framework of the WTO").

themes of these papers are largely reflected in the subsequent draft declarations that are the subject of discussion here.[48]

A. COMPETING PROPOSED DECLARATIONS

A pair of proposed declarations that circulated in advance of the Doha Ministerial Conference provides an excellent example of competing perspectives, along with important background to the subsequent discussion of how the confluence of perspectives resulted in the Doha Public Health Declaration. This section provides highlights from each draft that showcase a view of patents either as a privilege or as an uber-right. The full text of each draft declaration, as well as the final Declaration are printed in the Appendix of this book.

1. Patents as Privilege

The proposed draft declaration by developing countries evidences a strong view of patents as a privilege throughout the four-page document. The langugage includes both extensive preambular language and specific text aimed to clarify TRIPS.

The proposed preamble underscores a privilege view in several ways. It specifically articulates the privilege view that patents may conflict with access to medicine.[49] This conflict is placed within the context of broad international support for promoting public health as a social priority.[50] In addition, the proposal suggests that patent protection should not be aimed solely at promoting innovation, but should also encourage technology transfer to enable local production of medicine, consistent with the objectives and principles of TRIPS.[51]

---

[48] *See* World Trade Organization Council for Trade-Related Aspects of Intellectual Property Rights, *EU's Paper: Communication from the European Communities and Their Member States*, IP/C/W/280 (Jun. 12, 2001) [hereinafter EU Paper, IP/C/W/280]; WTO Council for Trade-Related Aspects of Intellectual Property Rights, *Draft Ministerial Declaration: Proposal by the African Group, Barbados, Bolivia, Brazil, Cuba, Dominican Republic, Ecuador, Honduras, India, Indonesia, Jamaica, Pakistan, Paraguay, Philippines, Peru, Sri Lanka, Thailand, and Venezuela*, IP/C/W/296 (June 29, 2001) [hereinafter African Group et al. Paper, IP/C/W/296].

[49] Developing Country Draft Ministerial Declaration, IP/C/W/312 ("Cognizant of the concerns expressed by non-governmental organizations, public health advocates and the worldwide public regarding potential implications of . . . the TRIPS Agreement[] on the availability and affordability of needed medicines and other healthcare products").

[50] *Id.* (noting the obligation of members to promote public health, as well as the international right to health); *see also* African Group et al. Paper, IP/C/W/296, ¶¶ 7–14 (June 19, 2001) (invoking multiple UN resolutions, a WHO medicines resolution, and a document by the heads of state of the "Group of 15" that all supported the need to promote access to drugs, especially with respect to HIV).

[51] Developing Country Draft Ministerial Declaration, IP/C/W/312; *see also* African Group et al. Paper, IP/C/W/296, ¶ 19 (stating that TRIPS article 7 "stems from the recognition by Members that the mere existence and

The proposed text of the declaration repeatedly reflects a view of patents as a privilege that should never hinder access to medicine. The strongest example is the very first proposed statement, which boldly asserts "nothing in the TRIPS Agreement shall prevent Members from taking measures to protect public health."[52] If taken literally and in accordance with a privilege view, this would permit members to deny granting any patents if that might hinder public health. Similarly, the belief that patents will conflict with public health is implicit in a request that the TRIPS Council monitor the impact of TRIPS on health on an ongoing basis.

The proposal also aimed to provide specific clarification of TRIPS consistent with a privilege view. For example, the proposal aimed to clarify one issue involved in the South African controversy: that member states have the right to recognize international exhaustion of patent rights.[53] The proposal also attempted to clarify both basic requirements of compulsory licenses that were subject to controversy in the South Africa case, as well as issues that had not yet been internationally addressed. For example, the proposal included clarifications on when member states could grant compulsory licenses.[54] In addition, the proposal suggested interpretations to the existing TRIPS articles to ensure that developing countries without manufacturing capacity (and thus unable to effectively use compulsory licenses) could still access low-cost drugs.[55]

2. Patents as Uber-Right

Somewhat predictably, the proposed declaration by some developed countries primarily reflected an uber-right perspective. After all, some of the countries involved are home to multinational pharmaceutical companies, such as the United States and Switzerland.[56]

---

the exercise of IPRS, such as patents, do not necessarily result in the fulfillment of the objectives of the Agreement" with respect to providing benefits for both patent owners and users of patented medicines).

[52] Developing Country Draft Ministerial Declaration, IP/C/W/312, ¶ 1. Similarly, the proposal suggests that members refrain from bringing official actions against developing countries who are taking steps to promote and protect public health (seemingly without regard to whether those steps are consistent with TRIPS), a suggestion that is consistent with the view that patent rights should not impede public health. *Id.* ¶ 11.

[53] *Id.* ¶ 2 (each member retains the right to establish its own policy and rules regarding the exhaustion of intellectual property rights). The suggested language also attempted to clarify that TRIPS does not require data exclusivity. *Id.* ¶¶ 7–8.

[54] *Id.* ¶ 3 (clarifying that each member state has the right to determine the grounds for issuing compulsory licenses); *Id.* ¶ 4 (clarifying that in cases of national emergency, extreme urgency, or public noncommercial use, there is no need to first request permission from the patent owner).

[55] *Id.* ¶¶ 5, 9. Notably, these suggested interpretations require far less logistics than the Waiver eventually adopted by the TRIPS Council.

[56] WTO Council for Trade-Related Aspects of Intellectual Property Rights, *Preambular Language for Ministerial Declaration, Contribution from Australia, Canada, Japan, Switzerland, and the United States: Access to Medicines for HIV/AIDS and Other Pandemics*, IP/C/W/313 (Oct. 4, 2001) [hereinafter Developed Country Proposed Preambular Language, IP/C/W/313]. Notably, the EU did not participate in this proposal although it had submitted a paper at the earlier WTO session.

The circulated document contained only preambular language, which may reflect that these countries believed TRIPS already struck a proper balance such that no further clarifications were necessary.[57] Rather than clarify TRIPS terms, the proposal suggested a pledge to "work with" both the private sector and affected countries to "facilitate the broadest possible provision of drugs in an affordable, medically effective and WTO-consistent manner."[58]

This preambular language is consistent with the general premise of the uber-right that patents are not inherently in conflict with public health. Rather, as stated in the proposed declaration, "strong" protection is a necessary incentive for research that "contributes to public health objectives globally." Furthermore, the declaration affirms that TRIPS in fact "contributes to the availability of medicines"[59]—presumably by providing an incentive for development of drugs consistent with the uber-right focus on future development rather than current access. Higher costs for patented drugs seem implicitly sanctioned by the statement that "strong" protection is necessary, as strong protection generally means there will be no or only minimal exceptions to patent rights.

The uber-right view of patents is also reflected in the limited number of specific TRIPS provisions discussed, as well as narrow interpretations of TRIPS. For example, the draft language stated that members "reaffirm the appropriateness of Members using the flexibility afforded by the Agreement to ensure that medicines for treatment of HIV/AIDS and other pandemics are available to their citizens . . . ."[60] Although this statement acknowledges HIV as a problem, it also seems to suggest that TRIPS flexibilities only are relevant to pandemics—even though there are no such limits in TRIPS. Similarly, the draft does not directly address the contentious issue of whether members have a right to recognize international exhaustion. Instead, the draft language of the developed countries focuses more on the sanctity of the model for segregating markets and thus maximizing profits. In particular, the proposed language states that "whatever the exhaustion regime that they may have chosen," members are encouraged to "take measures to prevent pharmaceuticals provided to the poorest populations of the globe under discounted pricing schemes or supplied under aid schemes from being diverted . . . to markets for which they were not intended; . . . ."[61]

[57] Although some draft text was later circulated, it is not available for analysis. In addition, regardless of the wording of this proposed text, it is still notable that the originally circulated text contained no explicit provisions.

[58] Developed Country Proposed Preambular Language, IP/C/W/313, at 2.

[59] *Id.* at 1. In addition, the last statement of the proposal reaffirms the importance of countries carrying out their TRIPS obligations "while taking into account health concerns." *Id.* at 2.

[60] *Id.* at 2.

[61] *Id.*

B. CONFLUENCE OF PERSPECTIVES

The unanimously concluded Doha Public Health Declaration likely reflects a series of unrelated and unexpected events, rather than a true convergence of opinion. A few months before the Doha meeting, developed countries and pharmaceutical companies continued to insist that issues besides patents were the true hurdles to access to medicine. This uber-right view was set forth in an article in a leading medical journal.[62] However, the impact of that article was short-circuited by unexpected events. In particular, in the wake of the Sept. 11, 2001 terrorist attacks on the United States, there was a bioterrorism threat involving mailed anthrax. Canada initially stated it would issue a compulsory license on the patented antibiotic marketed as Cipro to ensure adequate supplies to treat anthrax inhalation, and a US official stated the United States was contemplating a similar move.[63] Although no compulsory licenses were ultimately issued, the United States was widely mocked for its hypocrisy. The United States had condemned the use of compulsory licenses to treat HIV pandemics, but considered issuing a compulsory license for its own use to address an unrealized threat of limited scope.[64]

Although the Declaration was unanimously concluded, competing views of patents persist in the form of divergent interpretations of the Declaration. Different views on the issues of compulsory licenses and in-transit patent infringement have already been explored in Chapters 6 and 10. However, these are not the only instances. The different perspectives have continued to play a role in a number of other situations, some of which are further discussed in the following section.

IV. Post-Doha

This section provides an overview of some activity since the Declaration (beyond what has been addressed in other Chapters) that reflects each perspective, as well as how those competing perspectives may sometimes help to create more moderate approaches. The discussion of each perspective includes not only a discussion of commonly held views, but how holders of each perspective have attempted to use a variety of forums to achieve immediate goals and influence public opinion.

---

[62] Amir Attaran & Lee Gillespie-White, *Do Patents for Antiretroviral Drugs Constrain Access to AIDS Treatment in Africa?*, 286 J. AM. MED. ASSOC. 1886 (2001); *see also* Donald G. McNeil, *Patents or Poverty? A New Debate over poor AIDS Care in Africa*, N.Y. TIMES, Nov. 5, 2001.

[63] *E.g.*, Dan Ackman, *A New Deal on Cipro*, FORBES, Oct. 24, 2001, http://www.forbes.com/2001/10/24/1024topnews_print.html.

[64] *E.g.*, Geoff Dyer & Adrian Michaels, *A Bitter Pill for Drug Makers*, FIN. TIMES (London), Oct. 23, 2001, at 27; Sarah Boseley, *Drug Dealing*, GUARDIAN, Oct. 24, 2001, *available at* http://education.guardian.co.uk/businessofresearch/comment/0,579977,00.html.

A. UBER-RIGHT

This section shows how an uber-right perspective of patents influences the interpretation of existing laws, the perceived impact of laws, and the desire to modify laws. In particular, this section first shows how holders of the uber-right view TRIPS influenced the development of the Waiver (of the usual TRIPS requirement that compulsory licenses be predominantly for domestic use). Then, this section discusses illustrates the impact of such views on the development and perception of free trade agreements (FTAs) on public health. Last, this section describes activities that reflect an uber-right view of patents.

1. Uber Right Views and the Waiver

The uber-right view of Doha was often reflected in discussions concerning what was first labelled the "Paragraph 6" problem (with reference to paragraph 6 of the Doha Public Health Declaration). As noted earlier, developing countries had suggested a number of possible solutions to ensure that poor countries with inadequate resources to make their own generic drugs under a compulsory license could access low-cost drugs. However, all the proposed solutions were inconsistent with an uber-right view and thus rejected. Paragraph six of the Declaration is notable for reflecting an international consensus that TRIPS posed a barrier to providing low-cost drugs to some poor countries such that the TRIPS Council should take action. Nonetheless, the suggestion that the TRIPS Council find a solution implicitly rejected the privileged-based ones.

The uber-right view of patents was reflected in negotiations that occurred after the TRIPS Council provided a "solution" in the form of the current Waiver. For example, proposals to limit the scope of the Waiver to only a narrow range of diseases or to eligible countries reflects the uber-right view that patent rights should only be restricted in very limited circumstances.[65] Proposals to limit a solution to only cases of epidemics reflects a narrow reading of the Doha Public Health Declaration; although epidemics are certainly noted (as discussed above), the Declaration language is inclusive in indicating that there are problems *including* epidemics.

The Waiver also reflects some of the uber-right interest. As discussed in Chapter 7, the Waiver is extraordinarily complex. This complexity favors the interests of the uber-right

[65] *See, e.g.*, World Trade Organization Council for Trade-Related Aspects of Intellectual Property Rights, *Second Communication from the United States: Paragraph 6 of the Doha Declaration on the TRIPS Agreement and Public Health*, IP/C/W/358 (July 1992), at 15-16 (suggesting a moratorium on WTO disputes for LDCs only in the case of infectious epidemic diseases). A more moderate view that still tilts more to the uber-right perspective was offered in the EU's suggestion that the Waiver not be limited to epidemics, but to a set of diseases that could be subject to expansion in specific cases, depending on WHO's determination of the severity of the problem. *See* INTERNATIONAL TRADE CENTER, NEGOTIATIONS ON PARAGRAPH 6 OF THE DOHA MINISTERIAL DECLARATION ON THE TRIPS AGREEMENT AND PUBLIC HEALTH : PROGRESS REPORT 14 (2003) (*citing* Communication from the European Communities and their Member States on Paragraph 6 of the Doha Declaration on the TRIPS Agreement and Public Health, JOB(03)/9 (Jan. 24, 2003), *available at* http://www.intracen.org/btp/issues/trips/publications/para6_august2003.pdf.

in enabling their owners to maximize profits (including through market segregation); to protect such market segmentation, drugs made for export to poor countries need special marking and special notifications under the Waiver to try to minimize parallel imports.

The opposition of holders of the uber-right view to patent exceptions even in the case of a recognized need is also reflected in the continuing controversy concerning the Waiver. In particular, when the Waiver was proposed as a formal amendment to TRIPS, some countries wanted to limit developing countries' use of it to emergencies.[66] In addition, the uber-right view is also reflected in resistance to revisiting the Waiver as potentially not realizing its goal of promoting access to medicine. Although developing countries have contended that the Waiver is ineffective because it has only been used once in five years, the holder of an uber-right view predictably may suggest that there is no real problem if the Waiver is not used.

2. Uber Right View of TRIPS-Plus and Doha

An uber-right view is often seen in discussions concerning FTAs that impose TRIPS-Plus obligations. In particular, the holders of the uber-right view seem to interpret TRIPS and the Doha Public Health Declaration very restrictively. For example, they have repeatedly suggested that TRIPS requires data exclusivity, even though an accurate interpretation of TRIPS would permit but not require it (as explained in Chapter 3).[67] In addition, an uber-right view of existing flexibilities may overlook the ability to take a narrow view of patentability[68] and instead focus only on the complicated provision to enable compulsory licenses for export as a method to promote low-cost access to drugs.[69] An uber-right view may also ignore the clear language in the Declaration concerning the ability of countries to determine themselves the appropriate grounds for compulsory licenses. For example, the United States has negotiated a number of FTAs that limit these grounds, yet insists that its trade agreements comply with the Declaration. In fact, some of the "side letters" to its FTAs purport that the TRIPS-Plus provisions "do not affect a Party's ability

---

[66] This was reflected in controversy over whether a statement read by the WTO Council should be part of a formal amendment when the Waiver was proposed as an official amendment to TRIPS. *E.g.*, Ian Swanson, *WTO Members Re-Open Fight over Substance of TRIPS-Health Agreement*, INSIDE US TRADE, Mar. 12, 2004, *available at* http://www.cptech.org/ip/wto/p6/insideustrade03122004.html.

[67] *E.g*, INT'L FED'N PHARM. MFRS. & ASS'NS, ENCOURAGEMENT OF NEW CLINICAL DRUG DEVELOPMENT: THE ROLE OF DATA EXCLUSIVITY paras. 3–4 (2000); EU, *EU-India FTA Negotiations and Access to Medicines: Questions and Answers*, *available at* http://trade.ec.europa.eu/doclib/docs/2010/may/tradoc_146191.pdf [hereinafter *EU-India FTA Q&A*] (characterizing data exclusivity as "required by Article 39.3 TRIPS").

[68] As discussed in Chapter 3, because TRIPS does not define key terms (such as what is an "invention," or what is "new"), member states have some flexibility to limit the scope of patentability.

[69] *E,g*, *EU-India FTA Q&A*, *supra* note 67, question 1 (in responding to a question concerning how the proposed FTA will limit access to affordable medicines in developing countries, the EU focuses solely on the complex procedure to issue a compulsory license to export drugs while ignoring the fact that India's patent provisions provide a simpler alternative to creation of generic drugs by limiting patentability).

to take necessary measures to protect public health by promoting access to medicines for all"—even though the FTAs do in fact limit flexibility under TRIPS to restrict patentability, and to more widely use compulsory licenses and parallel imports.[70]

3. Uber-Right View of Health Impacts

The uber-right view that patents and increased patent rights do not compromise public health is seen in many statements made in conjunction with FTAs that impose TRIPS-Plus requirements. First of all, the FTAs are not seen as compromising public health even though they require more drugs to be patented and for longer patent terms. This will lead to reduced access to drugs in the short term while such drugs are patented. However, holders of an uber-right view do not consider this a problem because patent rights are seen to promote innovation that aids public health. Similarly, the uber-right view that patent rights increase availability of drugs overlooks the fact that drugs are not truly available to consumers if they are priced out of reach.[71]

An uber-right view may view patents as beneficial even in the face of contrary evidence. For example, US officials have repeatedly stated that the US FTA with Jordan has encouraged foreign direct investment and stimulated local research,[72] consistent with the uber-right view that enhanced patent protection promotes social benefits.[73] However,

---

[70] *E.g.*, CAFTA Understanding Regarding Certain Public Health Measures, U.S.-Costa Rica-Dom. Rep.-El. Sal.-Guat.-Hond.-Nicar., Aug. 5, 2004; Side Letter on Public Health Letter to Ambassador Zoellik from Taib Fihhri, Minister of Morocco, June 15, 2004, *available at* http://www.ustr.gov/sites/default/files/uploads/agreements/fta/morocco/asset_upload_file258_3852.pdf; United States-Peru Trade Promotion Agreement, art. 16.13: Understandings Regarding Certain Public Health Measures, US-Peru, April 12, 2006, *available at* http://www.ustr.gov/webfm_send/1031.

[71] For example, whereas the USTR has suggested that stronger patent protection will make companies willing to release innovative drugs in the markets of free trade partners, recent reports show that newly launched medicines in Jordan were virtually unaffordable to all patients. OXFAM INT'L, ALL COSTS, NO BENEFITS: HOW TRIPS-PLUS INTELLECTUAL PROPERTY RULES IN THE US-JORDAN FTA AFFECT ACCESS TO MEDICINES 17 (2007), *available at* http://www.oxfam.org.uk/resources/issues/health/downloads/bp102_trips.pdf; Hamed El-Said & Mohammed El-Said, *TRIPS-Plus Implications for Access to Medicines in Developing Countries: Lessons from Jordan–United States Free Trade Agreement*, 10 J. WORLD. INTELL. PROP. 438 (2007).

[72] MICHAEL P. RYAN & JILLIAN SHANEBROOK, ESTABLISHING GLOBALLY COMPETITIVE PHARMACEUTICAL AND BIO-MEDICAL TECHNOLOGY INDUSTRIES IN JORDAN, INTERNATIONAL INTELLECTUAL PROPERTY INSTITUTE (2004); Marilyn Chase & Sarah Lueck, *In New Trade Pacts, U.S. Seeks to Limit Reach of Generic Drugs*, WALL ST. J., Jul. 6, 2004, at A1 (USTR spokesperson stated that Jordan had a blossoming of its pharmaceutical industry after implementation of the FTR).

[73] OFFICE OF THE U.S. TRADE REPRESENTATIVE, POLICY BRIEF: CAFTA AND ACCESS TO MEDICINES (2005), http://ustraderep.gov/assets/Trade_Agreements/Regional/CAFTA/Briefing_Book/asset_upload_file433_7198.pdf; *see also US-Bahrain FTA: Fact Sheet on Access to Medicines*, OFFICE OF THE U.S. TRADE REPRESENTATIVE (Sept. 2004), http://www.ustr.gov/about-us/press-office/fact-sheets/archives/2004/september/us-bahrain-fta-fact-sheet-access-medicines (asserting that Jordan has developed its own medicines since enactment of the FTA as evidence that stronger patent protection helps developing countries); Jeanne Holden, *Jordan Benefits from Intellectual Property Reforms*, *in* FOCUS ON INTELLECTUAL PROPERTY

although Jordan's economic growth improved after the FTA, there has been no increase in pharmaceutical investment, and especially no increase in the number of multinational firms that have established production facilities in Jordan or licensed patents to Jordanian firms.[74] In addition, Jordanian companies have not become more innovative—less than 0.1 percent of sales are invested into R&D, and since the FTA no local manufacturers have filed a patent application on a new medicine.[75]

4. TRIPS-Plus Activities

The uber-right view is reflected in a number of activities in the international arena. First, a US law that requires a listing of countries that fail to provide "adequate" intellectual property protection (together with the US ability to impose sanctions on those countries) is consistent with an uber-right view that more protection is socially desirable.[76] Second, the consistent use of political and economic pressure to coerce countries to adopt stronger rights also reflects an uber-right view that more protection is better and does not have negative impacts for public health; this activity can be seen in the consistent negotiation of TRIPS-Plus agreements that generally create ever-higher standards of protection. Third, holders of an uber-right view of patents may view "complementary" ways of protecting patented drugs as part and parcel of patent rights to be protected. For example, the uber-right view may be reflected in FTAs that require regulatory protection of drugs (through data exclusivity and patent linkage),[77] as well as control over a nation's usual discretion to control drug prices to enable patented drugs to be sold for a maximum price.[78]

RIGHTS 28 (U.S. State Department 2008), *available at* http://www.america.gov/media/pdf/books/iprbook.pdf#popup.

[74] Hamed El-Said & Mohammed El-Said, *TRIPS-Plus Implications for Access to Medicines in Developing Countries: Lessons from Jordan-United States Free Trade Agreement*, 10 J.WORLD INTELL. PROP. 438, 453–55 (2007).

[75] *Id.* at 453-57; OXFAM INT'L, *supra* note 69, at 17.

[76] In addition, the fact that compliance with TRIPS is stated as not necessarily being adequate also reflects an uber-right view that increased protection is positive.

[77] For example, the pharmaceutical industry has stated "[t]he very concept of intellectual property rights breaks down if a patent holder loses the ability to sell his or her product at a market-determined price. . . . Such a scheme takes value away from the patent and is the equivalent of expropriating intellectual property." PHARMACEUTICAL RESEARCH AND MANUFACTURERS OF AMERICA, SPECIAL 301 SUBMISSION 2009, at 18, *available at* http://www.phrma.org/files/attachments/PhRMA%20Special%20301%20Submission%202009%5B2%5D.pdf.

[78] *E.g.*, PHARMACEUTICAL RESEARCH AND MANUFACTURERS OF AMERICA, SPECIAL 301 SUBMISSION 2010, at 19, *available at* http://keionline.org/sites/default/files/USTR-2010-0003-0245.1.pdf (suggesting the Australia and Korea FTAs "set an important precedent for mechanisms that should be adopted in other countries that place pricing and reimbursement constrictions on pharmaceuticals").

A major current international initiative that reflects an uber-right view is the pending Anti-Counterfeiting Trade Agreement (ACTA).[79] While a detailed discussion of the development of ACTA is beyond the scope of this chapter, its uber-right emphasis can be quickly highlighted. As just one example, there was a proposal to mandate all members to adopt a law similar to the current EU Regulation that permits countries to consider in-transit goods to potentially infringe patent laws; as noted in Chapter 10, this reflects an uber-right view of patents. Although there was no consensus on that issue, more modest language currently proposed would nonetheless *permit* countries to adopt such measures. In addition, as TRIPS-Plus agreements and activities repeatedly demonstrate, whenever an uber-right view is permissible, there may be political pressure to modify laws to mirror that perspective. Also, other provisions may also reflect an uber-right view that patented drugs are necessarily superior over generics, although the provisions actually regulate trademark. In particular, because trademarks are usually sold by an international non-proprietary name that may bear some similarity to the brand name, a generic could be improperly considered trademark infringement. This is not a hypothetical fear, as demonstrated by the improper seizure in Germany of a shipment of generic antibiotics goods that were properly using this common name.[80] However, ACTA raises the stakes by imposing higher sanctions for trademark infringement than TRIPS and thus posing an additional threat to access to medicine.

In addition, the uber-right view that patented (and branded) drugs are superior may also be seen in a variety of national laws that would seem to legalize conflation of the terms *counterfeit* and *generic*. As discussed in Chapter 10, generics are legitimate goods and not counterfeit. However, there are some proposed laws that bar "counterfeiting," yet define it sufficiently broadly that it could extend beyond the traditional trademark connotation of counterfeit to include generic drugs because it refers to a violation of "*any* intellectual property right" that involves goods that are "identical or substantially similar" to the protected goods.[81] Given that generic drugs are legally required to be equiva-

---

[79] Anti-Counterfeiting Trade Agreement, Final Text, Dec. 3, 2010, available at http://trade.ec.europa.eu/doclib/html/147079.htm; *see also* Washington College of Law, Program on Information Justice and Intellectual Property (PIJIP), http://sites.google.com/site/iipenforcement/home (providing links to additional resources and drafts). ACTA is being negotiated by the US, EU, Japan, Switzerland, Australia, South Korea, Singapore, and Morocco.

[80] *E.g.*, Press Release, Health Action International, Another Seizure of Generic Medicines Destined for a Developing Country, this time in Frankfurt (June 5, 2009), available at http://www.haiweb.org/19062009/5%20Jun%202009%20Press%20release%20Seizure%20of%20generic%20medicines%20in%20Frankfurt.pdf.

[81] *E.g.*, Kenya Anti-Counterfeit Act, section 34 (2008), ***available at*** http://sites.google.com/site/iipenforcement/home/anti-counterfeiting-laws; *see also* Nicholas Wadhams, *Kenya Pressured to Implement Anti-Counterfeit Law Despite Access Fears,* INTELL. PROP. WATCH (July 2, 2009), http://www.ip-watch.org/weblog/2009/07/02/kenya-pressured-to-implement-anti-counterfeit-law-despite-access-fears/; Joe Mathew, *Ugandan Move on Patented Drugs' Import Worries Indian Companies*, BUS. STAND., Mar. 6, 2009.

lent from a bioequivalent standpoint, this definition could embrace generic drugs rather than the more traditional counterfeit trademark.

Along similar lines, there is controversy concerning whether the WHO should use the word *counterfeit* in referring to fake medicines. Some countries oppose use of this term on the grounds that it refers to trademark issues that distract from the public health goals of the WHO.[82] Also, some suggest that use of the word in discussions of adulterated drugs improperly creates the impression that there are intellectual property issues at stake. Although this issue has been percolating for several years, the in-transit suspensions of generic drugs in Europe have likely heightened this controversy, as the EU has repeatedly used the term "counterfeit" to improperly refer to generic drugs that are legitimate and not in violation of trademark laws.[83] Additional discussion of the term counterfeit in conjunction with substandard drugs may further instill the incorrect view that generics are similarly substandard.

5. Uber-Right Outreach Efforts

An uber-right view is also evident beyond official activities of governments. This section briefly highlights a few examples of how the uber-right view has been promoted by rightsholders in collaboration with different groups, including the US Patent and Trademark Office (USPTO), a US university, and an academic.

Multinational company Pfizer teamed up with the USPTO[84] in September 2009 to present a seminar in India that focused on "misconceptions" of evergreening, as well as

---

[82] *E.g.*, Draft Intervention by India on Agenda Item No 11.20, Counterfeit Medical Products, 63rd WHA, para ¶ 3, http://www.ip-watch.org/weblog/wp-content/uploads/2010/05/India-Statement-on-Counteifeit-medical-products.pdf; Third World Network, WHO's Counterfeit Programme: Legitimises IP Enforcement Agenda, Undermines Public Health, Briefing Paper, 63rd World Health Assembly, May, 2010, *available at* http://www.twnside.org.sg/title2/briefing_papers/nontwn/Briefing.paper.on.WHO.Counterfeits.pdf; *see also* KM Gopakumar & Sangeeta Shashikant, Unpacking the Issue of Counterfeit Medicines (2010).

[83] *See, e.g.*, Kaitlin Mara, *WHO: No Evidence of Misleading on Counterfeit Medicines*, Intell. Prop. Watch, (May, 20, 2010), http://www.ip-watch.org/weblog/2010/05/20/who-no-evidence-of-misleading-on-counterfeit-medicines/; *E.g.*, Kaitlin Mara, *Debate Erupts over WHO Police Work aAgainst "Counterfeit" Drugs Trade*, Intell. Prop. Watch (May 20, 2010), http://www.ip-watch.org/weblog/2010/05/20/debate-erupts-over-who-police-work-against-counterfeit-drugs-trade/; Kaitlin Mara, *Pointed Exchange of Views at WHO Briefing on Counterfeit Drugs*, Intell. Prop. Watch (Mar. 31, 2010), http://www.ip-watch.org/weblog/2010/03/31/flare-up-in-who-briefing-on-counterfeit/.

[84] Although the USPTO has an official program for providing assistance to countries, it later acknowledged that it was a mistake to sponsor a program with a stakeholder. *E.g.*, http://keionline.org/node/796. The seminars are part of a broader program by the USPTO called the Global Intellectual Property Academy. This program was established in 2006 and is designed to advance IPR policies through training of foreign officials as well as stakeholders and the general public. USPTO, Training and Education: Global Intellectual Property Academy, http://www.uspto.gov/ip/training/index.jsp.

the importance of regulatory data protection and patent linkage.[85] Notably, these are all topics where Indian law does not mirror the desires of pharmaceutical companies. Prior to the seminar, the USPTO suggested it would "provide some insights into how India can realise its full potential through the appropriate legal and regulatory measures to reward innovation."[86] This USPTO statement reflects the uber-right view that more protection necessarily promotes more innovation without compromising public health. Similarly, the substance of the meeting also reflected an uber-right view—at least as reflected in notes made public from an anonymous member of civil society who was present. For example, the US ambassador to India asserted that US negotiators present during the TRIPS negotiations did not recall that data exclusivity was rejected and that there was no record that language on data exclusivity was specifically rejected, such that TRIPS must require data exclusivity.[87] Although there may be no record that data exclusivity was rejected, that is not necessary. The record shows that prior language proposing data exclusivity was proposed, but not adopted.[88] Under the usual canons of treaty interpretation, proposed language that is not adopted is not binding and should not be read as a restriction on the final text. Similarly, the US ambassador suggested that he was "not impressed" with WHO's statements cautioning developing countries to avoid adopting TRIPS-Plus measures. Moreover, he asserted that WHO had no role to play regarding intellectual property.[89] Holders of an uber-right view may discount the role of those who disagree with the idea that patents (and related rights, such as data exclusivity) never present a conflict with access to medicine.

An uber-right view has been consistently promoted to high-level Indian government officials and judges through a major US university in conjunction with a variety of patent-owning companies. In particular, the "GWU India Project" focuses on promoting the "benefits of intellectual property protection."[90] The project was launched in 2004—one

---

[85] The seminar was targeted at media and civil society organizations.

[86] *USPTO-Pfizer Workshop on IP and Innovation in Mumbai*, INTELL. PROP. WATCH (Sept. 4, 2009), http://www.ip-watch.org/weblog/2009/09/04/uspto-pfizer-workshop-on-ip-innovation-in-mumbai/.

[87] Meeting Notes, Delhi USPTO-Pfizer, at 4 (Oct. 9, 2009), http://www.centad.org/Delhi-USPTO-Pfizer-Meeting-Notes.pdf (discussing Session 4 presentation by Dominic Keating, USPTO/US Embassy, New Delhi).

[88] In particular, a widely available prior draft of the eventual TRIPS provision had language that stated "data may not be relied upon for the approval of competing products for a reasonable time." Brussels Draft, reprinted in DANIEL GERVAIS, THE TRIPS AGREEMENT: DRAFTING HISTORY AND ANALYSIS 421 (3d ed. 2008). This language is nowhere present in the final TRIPS provision regarding protection of data. *See* TRIPS art. 39(3).

[89] Meeting Notes, *supra* note 87, at 6, http://www.centad.org/Delhi-USPTO-Pfizer-Meeting-Notes.pdf (discussing WHO in "Other Questions and Comments").

[90] *Overview: India Project*, GEORGE WASHINGTON UNIVERSITY, http://www.law.gwu.edu/Academics/research_centers/india/Pages/Overview.aspx. The conferences, as well as the sponsoring university, have been criticized. *E.g.*, Judit Rius, *Indian NGOs Confront GWU Law School Efforts to Push Maximalist IPR Norms in India*, JUDIT RIUS'S BLOG, KNOWLEDGE ECOLOGY INT'L (Mar. 4, 2010, 17:10), http://keionline.org/node/793; *Stop George Washington U from Being a Pharma Front*, UNIVERSITIES ALLIED FOR ESSENTIAL MEDICINES, http://essentialmedicine.org/stop-george-washington-u-being-pharma-front; *see also* Kaitlan

year before India was required to begin providing patents on drugs and drug compounds under TRIPS.[91] Since then, the project has offered annual conferences with the financial support and involvement of major corporations such as Novartis and groups such as PhRMA.[92] The presentations seem to put forward an uber-right view.[93] For example, Novartis argued that section 3d of India's patent laws—the provision that was used to deny patent protection to Gleevec, a Novartis cancer treatment—was inconsistent with TRIPS; the company reportedly suggested a more "appropriate" interpretation of section 3d.[94]

6. Uber-Right View of a Middle Ground

The uber-right view of patents is also seen in a 2008 letter co-authored by Pfizer and a law professor to the US Finance Committee that proposes a "practical vision" for addressing health care in the developing world that allegedly acknowledges the need for access, but also stresses innovation—primarily through patents.[95] The proposal advocates that all developed nations "ensure that pricing and reimbursement policies recognize and reward innovation."[96] As noted earlier, holders of an uber-right view of patents would consider things that help protect patented drugs as consistent and complementary to patent protection. In addition, the proposal relies on the uber-right assumption that more protection necessarily promotes more research—even for diseases that impact developing countries with insufficient resources to pay a patent premium. There is some acknowledgement that such countries would lack resources to pay a premium, as the proposal suggests that those who purchase drugs for poor countries should be sure to cover the costs of developing drugs for poor countries.[97] This again reflects an uber-right view that

---

Mara, *Indian Civil Society Raises Concerns over US Industry-Sponsored IP Summits*, INTELL. PROP. WATCH (Mar. 11, 2010), http://www.ip-watch.org/weblog/2010/03/11/indian-civil-society-raises-concerns-over-us-industry-sponsored-ip-summits/.

[91] India Project, *supra* note 90; *see also* Latha Jishnu, *Insidious India Project*, BUS. STAND., Aug. 15, 2010.

[92] *See, e.g.*, Mara, *supra* note 90.

[93] This would be consistent with the mission of the Center established at GWU that is charged with organizing the conference. For example, the Director of the Creative and Innovation Economy Center acknowledged "I am very pro-business in developing countries and I'm very pro-multinational companies." William New, *New US Industry, Government-Backed Centre Promotes Industry View in Developing Countries*, INTELL. PROP. WATCH (Apr. 3, 2006), http://www.ip-watch.org/weblog/2006/04/03/new-us-industry-government-backed-centre-promotes-industry-view-in-developing-countries/.

[94] *E.g.*, Letter from Civil Society Groups to Shri Anand Sharma, Honorable Minister of Commerce and Industry in India (Feb. 26, 2010), *available at* http://www.phmovement.org/en/node/2800.

[95] Letter from Pfizer CEO and Professor John Barton to Honorable Max Baucus (Aug. 13, 2008), *available at* http://media.pfizer.com/files/news/kindler_baucus_letter_081308.pdf.

[96] *Id.* at 2. The pricing and reimbursement policies that are recommended for all nations would make those provisions of the US-Australia FTA and US-Korea FTA apply to all developed countries.

[97] "We believe it appropriate that the global and national funds purchasing for these countries pay competitive prices but also believe that these prices should cover an appropriate share of research costs for new products whose primary value is in developing nations." *Id.*

strong patent rights necessarily promote long-term social welfare—and that prospective long-term gains should have priority over short-term desires for access to drugs.

B. PRIVILEGE

This section aims to provide an example of patents as privilege, as well as how that view has been embraced at least in part by some major international organizations, including WHO and WIPO. Although these proposals tend to reflect a pure privilege view such that they are unlikely to be adopted, they may nonetheless provide a useful counterpoint to the uber-right narrative, and possibly also nudge the uber-right toward a middle ground. This section first explains the solutions from a privilege perspective, followed by a discussion of how those with a privilege view have had some success in fostering this view at WHO and WIPO.

1. Privilege View of TRIPS, Doha, and Waiver

Not surprisingly, those with the patents-as-privilege perspective have a very different view of the Doha Public Health Declaration as well as TRIPS–Plus rules. First, as noted earlier, those with the privilege view see the Declaration as ensuring that public health is prioritized over patent rights—even though the Declaration does not explicitly support such a view. This view may explain why some have suggested that TRIPS-Plus rules are either inconsistent or in fact violate the Declaration.[98] The Declaration is an appropriate interpretive tool for TRIPS. However, it only applies to TRIPS, and does not prevent member states from implementing more restrictive measures. The TRIPS framework is notably a *minimum* standard agreement that generally permits countries to provide more protection.

The view of patents-as-privilege also reflects a different perception of developing countries and their sympathizers with respect to the Waiver of the domestic use requirement for compulsory licenses. The Waiver has been strongly criticized for imposing too many barriers to promoting public health.[99] The barriers are seen as betraying the perceived

---

[98] *E.g.*, OXFAM INT'L, PATENTS VERSUS PATIENTS: FIVE YEARS AFTER THE DOHA DECLARATION 13 (2006) (suggesting TRIPS-Plus rules enacted by the United States in FTAs "violate US commitments under the Doha Declaration and prevent developing countries from using safeguards to protect public health"); Oxfam & HAI, *Trading Away Access to Medicines: How the European Union's Trade Agenda Has Taken a Wrong Turn* 12 (2009) (noting that EU efforts to create TRIPS-Plus rules are "directly counter to the spirit and intent of the Doha Declaration").

[99] *E.g.*, Frederick M. Abbott, *The WTO Medicines Decision: World Pharmaceutical Trade and the Protection of Public Health*, 99 AM. J. INT'L L. 317, 339, 342, 347-52 (2005); Brook Baker, *Arthritic Flexibilities for Accessing Medicines: Analysis of WTO Action Regarding Paragraph 6 of the Doha Declaration on the TRIPS Agreement and Public Health*, 14 IND. INT'L. & COMP. L. REV. 613 (2004); Oxfam, WTO Patent Rules Will Still Deny Medicines to the Poor (Aug. 27, 2003) (noting "the agreement is wrapped in so much red tape that it becomes largely unworkable").

"pledge" in Doha that public health is prioritized over patent rights.[100] Given the view that patent rights should give way to access to medicine, a "solution" that is difficult to use is seen as unworkable. As discussed in Chapter 7, some view the single use of the Waiver in five years as evidence that it is ineffective.

2. Solutions Steeped in a View of Patents as Privilege

There have been several international proposals that advocate a radical change to international norms in a manner consistent with the idea that patents are a privilege subservient to other social goals and interests. A Proposed Treaty on Access to Knowledge (A2K) was first proposed in 2005, followed by a Paris Accord in 2005, as well as a proposed Medical Research Treaty. Although these proposals are too radical to be embraced by those with an uber-right view, they highlight the view of patents as a privilege, which helps to provide balance to global discussion.

In 2005, the proposed A2K treaty suggested that rather than require countries to provide minimum levels of patent (and other) protection, there should be minimum standards of *access*.[101] As one example, the proposal suggested that there be a safe harbor from infringement for "compassionate use" of medicine, which although undefined, suggests that access to medicine should be prioritized over patent rights.[102] This is totally opposite to the existing framework of TRIPS and unlikely to ever be seriously considered. However, it does clearly espouse the view that patents are a privilege that can and should be limited as necessary to promote access.

The proposed A2K treaty led to the 2006 Paris Accord, concluded at a meeting by the Transatlantic Consumer Dialogue (TACD) that involved sixty US and EU consumer organizations that similarly espoused views consistent with the notion that patents are a privilege.[103] For example, the proposal seems to suggest that data exclusivity should not be

[100] *E.g*, Oxfam, *supra* note 99 (suggesting that if the Waiver becomes official, it would "be a betrayal of the pledge made in the Doha Declaration to put public health before patent rights."); *see also* World Intellectual Property Organization, Inter-Sessional Intergovernmental Meeting on a Development Agenda for WIPO, *Proposal to Establish a Development Agenda for WIPO: An Elaboration of Issues Raised in Document WO/GA/31/11*, IIM/1/4, at 5 (Apr. 6, 2005) (submitted by Group of Friends of Development) (suggesting the Doha Declaration reflects international agreement that TRIPS should always be supportive of public health objectives).

[101] Draft Treaty on Access to Knowledge [hereinafter Draft A2K Treaty], arts. 1–2, May 9, 2005, *available at* http://www.cptech.org/a2k/a2k_treaty_may9.pdf; *see also* Tim Hubbard & James Love, *A New Trade Framework for Global Healthcare R&D*, 2 PLoS BIOLOGY 147 (2004). Not surprisingly, this proposal has not been greeted with open arms. *E.g.*, J.A. DiMasi & H.G. Grabowski, *Should the Patent System for New Medicines be Abolished*, 82 CLINICAL PHARMACOLOGY & THERAPEUTICS 488 (2007).

[102] Draft A2K Treaty, art. 4-1(b).

[103] Draft Paris Accord, June 17, 2006, *available at* http://www.cptech.org/a2k/pa/ParisAccord-june17draft.pdf. The TACD is comprised of over sixty US and EU consumer organizations that aim to propose joint recommendations to their respective governments. *See* About TACD, TACD, http://tacd.org/index.php?option=com_content&task=view&id=25&Itemid=39.

required, or at least can be negated to ensure that drugs are available at affordable prices. In particular, the proposal states "methods of protecting investments in clinical trials for new medicines should not prevent governments from making medicines available at affordable prices or require unethical or unnecessary replication of human experiments."[104] Without using the term "data exclusivity," this language seems to suggest that it should not be used as a way to protect the financial investments of companies—at least not when there are negative ramifications for access to medicine. In addition, the proposal advocates that the existing business model of financing drug development through a patent premium on drugs be replaced by one that would promote better health outcomes rather than simply more profitable drugs [105] The suggestion to abandon the existing model of research that is based on drugs sold at a patent premium expressly recognizes that patents inherently conflict with providing immediate access to medicine.

An even more radical proposal is the Medical Research and Development Treaty, first set forth in 2005.[106] This proposal provides an alternative method of funding drug development beyond patent profits—a direct response to a desire advocated by the Paris Accord, but for which no details were provided. Although the proposal does not require the elimination of patents, it seems to clearly contemplate that they will be relied upon less as a source to promote revenue for research. In addition, the proposal suggests that members actually forego patent protection and even data protection or exclusivity as the proposed treaty is seen as providing "an alternative and superior framework for supporting innovation."[107] In other words, the proposal takes as a given that patents inherently conflict with immediate access to medicine, such that the proposal attempts to provide a solution that does not involve patent rights. Rather, the proposal adopts the privilege view that patents are simply a tool to promote innovation, but not necessarily the only or best tool. The privilege view thus may be responsible for the proposal that members agree to forego dispute resolution on patent and data exclusivity issues under TRIPS, as well as bilateral and regional agreements.[108] Interestingly, the proposal seeks to nullify the provisions of TRIPS that are seen as intruding on access to health by requiring members to forego the right to bring disputes. Although this is clearly a dramatic challenge to the existing TRIPS and TRIPS-Plus framework, some elements

[104] Draft Paris Accord, at 2.

[105] *Id.* at 2 (suggesting that "when possible and appropriate" the current system of stimulating research and development through high prices "should be replaced with new systems that reward developers for improved health care outcomes").

[106] Medical Research and Development Treaty, Discussion Draft, Feb. 7, 2005, at 1, *available at* http://www.cptech.org/workingdrafts/rndtreaty4.pdf.

[107] *Id.* § 2.3.

[108] *Id.* §§ 2.3; 16(d).

of this privilege-based view are now under discussion at WHO, as will be discussed below.[109]

Although all of these proposals are likely too radical to ever be adopted, they help to both demonstrate the privilege view of patents and set the stage for more moderate proposals that might be embraced. In particular, some of the possible solutions discussed in the next section can be seen as evolving from these more radical solutions.

3. WHO

Those who view patents as a privilege have had some success in propagating their view at WHO. Since the Doha Public Health Declaration, WHO has played an increasingly important role in promoting discussion of issues at the intersection of patents and access to medicine. WHO has provided an official voice promoting the view that patents are a privilege by making specific suggestions against TRIPS-Plus measures. Although all of the WHO resolutions are nonbinding, they may nonetheless provide support to countries desiring to follow one or more of them. In addition, WHO activities and resolutions—especially those that are agreed to by countries that tend to promote an uber-right view (such as the United States)—may help promote a more moderate development of law.

Since 2006, when WHO first released a comprehensive report on the intersection of public health, innovation, and intellectual property rights,[110] it has taken a leadership role in further studying this issue through the creation of working groups to develop near-term strategies.[111] WHO has made a number of specific recommendations that promote a privilege view, pursuant to which patents should be subject to limitations to promote public health. Not only has WHO recommended that FTAs not include TRIPS-Plus measures, but WHO has recommended actions that promote use of TRIPS flexibilities.[112]

---

[109] *See* Catherine Saez et al., *NGOs, Industry Weigh Alternatives for R&D Financing, Counterfeiting at WHO*, INTELL. PROP. WATCH (May 21, 2010), http://www.ip-watch.org/weblog/2010/05/21/ngos-industry-weigh-alternatives-for-innovation-financing-counterfeiting-at-who/ (noting that some countries proposed that WHO consider a R&D Treaty). Although discussion at WHO is still a far cry from actual ratification, the inclusion of a proposal that strongly embraces patents as privilege in a forum with a wide range of stakeholders is still notable.

[110] COMM'N ON INTELLECTUAL PROP. RIGHTS, INNOVATION, & PUB. HEALTH, WORLD HEALTH ORG., PUBLIC HEALTH, INNOVATION AND INTELLECTUAL PROPERTY RIGHTS: REPORT OF THE COMMISSION ON INTELLECTUAL PROPERTY RIGHTS, INNOVATION AND PUBLIC HEALTH (2006) [hereinafter WHO REPORT OF IPR COMMISSION]. This report was the result of a commission established in 2003. *Id.* at ix.

[111] *Public Health, Innovation, Essential Health Research and Intellectual Property Rights: Towards a Global Strategy and Plan of Action*, WORLD HEALTH ASSEMBLY, Resolution WHA59.24 (May 27, 2006); *Global Strategy and Plan of Action on Public Health, Innovation and Intellectual Property*, WORLD HEALTH ASSEMBLY, Resolution WHA61.21 (May 24, 2008); *Establishment of a Consultative Expert Working Group on Research and Development: Financing and Coordination*, WORLD HEALTH ASSEMBLY, WHA63.28 (May 21, 2010). WHO efforts to promote innovation beyond patents reflects a privilege view. In contrast, holders of the uber-right view would assume that strong patent rights inevitably lead to more innovation.

[112] WHO REPORT OF IPR COMMISSION, *supra* note 110, at § 4.26.

For example, WHO has specifically recommended that countries use compulsory licenses as well as parallel trade to lower the prices of medicine.[113] In addition, WHO has recommended that countries not adopt data exclusivity to prevent an unnecessary barrier to generic drugs.[114] WHO also recommended that countries evaluate domestic patent rules to ensure no undue barriers to legitimate competition, citing India's patent law as an example of how to avoid patents on products that are not sufficiently innovative.[115]

Although the WHO activities are generally consistent with a view of patents as privilege, they generally promote a more moderate view, such that those who hold a more strident view of patents as a privilege have criticized WHO. For example, in formulating the 2006 Global Strategy on Public Health (which outlined practical strategies to promote innovation in areas of concern to developing countries), the final Resolution rejected the most extreme view of patents as privilege set forth by LDCs, which suggested the resolution should affirmatively assert that the right to health should come before commercial interests.[116] Similarly, the same resolution refused to include provisions on data exclusivity, as a consensus could not be reached.[117]

4. WIPO

The privilege view of patents has increasingly played a role in a new forum since the Doha Public Health Declaration. In particular, whereas WHO would be considered a natural supporter of the importance of public health, the World Intellectual Property Organization (WIPO) has traditionally had a minimal role in international policy and generally only functioned as a forum for negotiating international agreements to protect intellectual property rights.[118] However, beginning in 2004, at the request of developing countries Argentina and Brazil, WIPO's focus started to embrace the interests of developing countries through a "Development Agenda."[119] The immediate impact was that it forestalled action on a proposed Patent Law Treaty (desired by developed countries) that would have imposed uniform TRIPS-Plus standards for all WIPO members. In 2007,

---

[113] *Id.* §§ 2.10 (compulsory licensing), 4.19 (parallel trade).

[114] *Id.* § 4.20.

[115] *Id.* § 4.27.

[116] William New, *Debate Surrounds Final WHO Global Strategy, Action Plan on IP and Health*, INTELL. PROP. WATCH (May 24, 2008), http://www.ip-watch.org/weblog/2008/05/24/debate-surrounds-final-who-global-strategy-action-plan-on-ip-and-health/.

[117] *Id.*

[118] In fact, some would view WIPO as being established to support the interests of intellectual property owners. *See* Convention for Establishing the World Intellectual Property Organization, opened for signature July 14, 1967, 848 U.N.T.S. 3 (noting that WIPO was established to promote intellectual property protection).

[119] World Intell. Prop. Org., *Proposal by Argentina and Brazil for the Establishment of a Development Agenda*, WO/GA/31/11 (Aug. 27, 2004). For additional detail concerning the development of the WIPO Development Agenda, see IMPLEMENTING THE WORLD INTELLECTUAL PROPERTY ORGANIZATION'S DEVELOPMENT AGENDA (Jeremy de Beer, ed. 2009).

WIPO adopted a Development Agenda and continued to work on the issue. A new group has also been formed within WIPO to transform it into an agency that is more reflective of the interests of developing countries. The Development Agenda is conceived as a shift from "viewing IP as an end in itself to viewing it as a means to serve larger public goals of social, economic and cultural development."[120]

V. Conclusion

The impact of competing patent perspectives on the development of international law remains to be seen. Clearly, the view of patents as privilege has had some impact on moderated positions by major organizations such as WHO and WIPO. In addition, although some institutions of the United States and EU have issued statements that present more moderate positions, these countries have simultaneously taken positions and made statements that present the uber-right approach. For example, although the US Congress and EU Parliament have both formally acknowledged that increased patent rights may pose a barrier to access to medicine,[121] the United States and EU also take uber-right positions and have not ceased to negotiate agreements that impose TRIPS-Plus rules.[122]

---

[120] World Intell. Prop. Org., Committee on Development and Intellectual Property, Fifth Session, *Statement by Egypt on Behalf of Development Group Agenda*, at 2 (April 26–30, 2010), *available at* http://www.ip-watch.org/weblog/wp-content/uploads/2010/04/Development-Agenda-Group.pdf.

[121] S. Res. 241, 110th Cong (2007); H.S. Res. 525, 110th Cong. (2007); European Parliament Resolution of July 12, 2007 on the TRIPS Agreement and Access to Medicines, PA_TA(2007)0353, *available at* http://www.europarl.europa.eu/sides/getDoc.do?Type=TA&Reference=P6-TA-2007-.

[122] In addition, the United States continues to issue its Special 301 Watch list, although perhaps in response to a 2007 Congressional resolution to cease using this list as a tool to force countries to enact stronger intellectual property rights, the 2010 list takes a less aggressive approach toward mandating data exclusivity. Similarly, in 2010, intellectual property owners were not the only ones able to provide input.

12 Looking Toward the Future

This chapter focuses on possible new approaches to dealing with the long-standing problem of how patents may impact short-term access to medicine. To emphasize why new approaches are necessary, the chapter begins with an overview of current and forthcoming tensions in the global debate concerning access to medicine. The chapter then focuses on a few approaches that have been proposed thus far. Although there are many possible solutions to the problem of providing more low-cost medicines, this chapter focuses on a few that help highlight other themes in the book. In particular, this chapter explains how each proposal reflects a certain perspective that may have an impact not only on its possible adoption, but also to counter-proposals.

I. Introduction 356
II. Considering New Approaches 359
A. From Patent Pools to Price Reductions and Free Licenses 359
1. A Privilege-Based Solution: UNITAID Patent Pool 359
2. Solutions from the Uber-Right 364
B. Health Impact Fund 367
C. Other Proposals 370
1. Publicly Funded Clinical Trials 370
2. Drug Development "Corporation" 371
III. Conclusion 372

I. Introduction

The issue of how to promote access to medicine while providing patent rights is not a new one. Nonetheless, promoting such access may soon be more challenging due to a variety of factors.

First, the requirement to patent drugs under TRIPS prevents developing countries such as India and Brazil from continuing to produce a wide range of generic drugs for their own needs and for export.[1] Even for the few drugs that can be made generically, there are some serious challenges. For example, as discussed in Chapter 10, a drug that is generic at both its point of origin and destination may be detained in transit for allegedly infringing local patent rights. In addition, as discussed in the same chapter, there is international discussion of modifying laws concerning what constitutes a counterfeit that could block legitimate generics while creating an improper perception among consumers that generics are unsafe.

Second, a pending crisis threatens to unravel much of the recent success in providing more HIV drugs that have halted unnecessary deaths and opportunistic infections. The cost of HIV treatments is expected to rise dramatically. Not only do HIV patients commonly develop resistance to older treatments, but newer drugs are recommended as initial treatments because they are less toxic than the older ones. As a result of TRIPS, new drugs—whether for initial ("first-line")[2] or subsequent ("second-line") treatment—are

---

[1] India in particular is a major provider of global drugs, currently producing nearly a quarter of global generic drugs. *E.g.*, William Greene, *The Emergence of India's Pharmaceutical Industry and Implications for the U.S. Generic Drug Market* 1, 19 (U.S. Int'l Trade Commission—Off. of Econ., Working Paper No. 2007-05-A, 2007); Indus View, Special Report: Opportunities for India in Generic Drug space, *available at* http://www.theindusview.com/vol2Issue8/pdf/Vol2Issue8Special_report_NA.pdf.

[2] The currently recommended treatment by the WHO (Tenofovir-based combination) costs $200 to nearly $1000 per patient per year (depending on patent protection) versus less than $100 for the previously

generally patented and sold at a patent premium.[3] Although India has long been a key source of low-cost generic HIV drugs, its new laws require it to now patent drugs that it never would have previously patented.[4]

Third, access problems extend far beyond HIV. Although drug companies tend to focus on HIV as the only condition for which developing countries need more access, a majority of deaths in developing countries are actually from noncommunicable conditions, such as heart disease and cancer.[5] Because aids groups tend to focus predominantly on providing low-cost or free drugs that treat communicable diseases such as HIV, malaria, and tuberculosis, many citizens may lack access to drugs for noncommunicable diseases that become death sentences. This poses a major problem for countries that consider right to health care important (either as a constitutional right, or through a public health system that guarantees universal access to medicine). Countries can theoretically issue compulsory licenses on patented drugs, but there are obviously risks in doing so, as discussed in Chapter 5.

A further complication is that some newer drugs called *biologics* made by a biological process (versus a chemical process for conventional drugs)[6] are typically much more expensive than conventional drugs, yet of increasing therapeutic importance.[7] Not only are the original versions of such drugs expensive, but the traditional solution to access

---

recommended (Stavudine) treatment. *E.g.*, ALL PARTY PARLIAMENTARY GROUP ON AIDS, THE TREATMENT TIME BOMB 10 (2009), *available at* http://www.aidsportal.org/repos/APPGTimebomb091.pdf.

[3] Some suggest that the lowest price of a second-line regime of HIV drugs is about $600 per patient per year, which is seven times more expensive than the cheapest first-line drugs. *Id.* at 12. However, the price differential may be much worse in countries where the drugs are patented, or for middle-income countries—some report that the price of second-line treatment can be up to *seventeen times* the cost of first-line drugs. *Id.* MSF quotes even higher figures: in Guatemala, a second-line treatment costs $6,500 per patient per year, which is twenty-eight times that of the first-line treatment ($200). MÉDECINS SANS FRONTIÈRES, TOO LITTLE FOR TOO FEW: CHALLENGES FOR EFFECTIVE AND ACCESSIBLE ANTIRETROVIRAL THERAPY 7 (2006), *available at* http://www.aidsportal.org/repos/REP_IACbriefingdoc_en.pdf.

[4] As discussed in Chapter 4, since 2005, India no longer excludes all drug products from patentability. Although India may continue to make generic HIV drugs that were generic before its 2005 amended patent law came into effect, those lower-cost drugs will likely be of decreasing relevance as patients become resistant to them. In addition, although its present laws (assuming they are not struck down by the WTO or mooted through an FTA) may preclude patents on some drugs, they are unlikely to bar patents on *all* new HIV treatments.

[5] World Health Organization, *Prevention and Control of Noncommunicable Diseases: Implementation of the Global Strategy*, Report by the Secretariat, at 1, EB120/22 (Jan. 8, 2007), *available at* http://www.who.int/gb/ebwha/pdf_files/EB120/b120_22-en.pdf.

[6] Technically, biologics include anything made with a biologic process, which could include vaccines and gene therapies. However, when the term *biologic* is used to refer to a treatment that is an alternative to conventional drugs, it usually involves something made from recombinant DNA (i.e., cells that have been genetically engineered to produce desired compounds, such as synthetic versions of human growth hormone or custom-designed antibodies).

[7] *E.g.*, Judith Johnson, FDA Regulation of Follow-On Biologics, Cong. Research Service Report for Congress 1-2 (Feb. 24, 2009); AARP Public Policy Institute, Biologics in Perspective; Expanded Clinical Options Amid Greater Cost Scrutiny, Fact Sheet, 3 (June 2007), *available at* http://assets.aarp.org/rgcenter/health/fs136_

issues—namely, manufacture of generic versions—is elusive with respect to biologics. Some suggest that it is generally difficult to create an identical biologic such that the usual regulatory pathway to approve generic drugs premised on bioequivalence is inapplicable.[8] Although bioequivalence may not be possible, companies can create a somewhat similar, even if not therapeutically equivalent drug, that could allow some price competition; this similar, but not identical version would be a *biosimilar* or *follow-on* biologic.[9] However, there may be scientific challenges in creating the biosimilar because the manufacturing process is likely proprietary and much more difficult to develop than that of manufacturing a traditional drug.[10] In addition, assuming that scientific obstacles can be overcome, nations are only beginning to establish legal pathways to permit approval of safe and effective biosimilars for sale.[11] In countries that have such pathways, there are usually only a few biosimilars approved for an innovator biologic; as there are relatively few new entrants, the cost savings from competition will not be as deep as with conventional drugs.

Although there are no ready solutions to the above access problems, there may be a silver lining. In particular, as access problems become more exacerbated, that may prompt increased debate and perhaps action on a more dramatic scale than has been seen thus far. This is especially true for biologics because the usual tools of modifying patent standards and issuing compulsory licenses may not work. The next section considers how some currently proposed solutions might be relevant to a new world that

---

biologics.pdf; Gina Kolata & Andrew Pollack, *Costly Cancer Drug Offers Hope and Also a Dilemma*, N.Y. TIMES, A1, July 6, 2008 (noting annual cost of $10,000 for Avastin).

[8] *E.g.*, Simon D. Roger, *Biosimilars: How Similar or Dissimilar Are They?*, 11 NEPHROLOGY 341 (2006). However, recent technical advances are making the process easier. In addition, there are some instances where companies are capable of creating bioequivalent biologics. For example, Insmed developed a bioequivalent for Amgen's biologic sold as Neupogen.

[9] *E.g.*, Bruce S. Manheim et al., *Follow-On Biologics: Ensuring Continued Innovation in the Biotechnology Industry*, HEALTH AFFAIRS, 397-98 (Mar. /Apr. 2006); David. M. Dudzinski, *Reflections on Historical, Scientific and Legal Issues Relevant to Designing Approval Pathways for Generic Versions of Recombinant Protein-Based Therapeutics and Monoclonal Antibodies*, 60 FOOD & DRUG L. J. 143, 144, 206 (2005).

[10] *E.g.*, Jeremiah J. Kelly & Michael David, *No Longer "If," but "When": The Coming Abbreviated Approval Pathway for Follow-on Biologics*, 64 FOOD & DRUG L. J. 115, 119–21 (2009); Bryan A. Liang, *Regulating Follow-On Biologics*, 44 HARV. J. LEGIS. 363, 398, 418 (2007). All or some of the process of making the biologic is likely to be protected as a trade secret. Unlike patent protection, a trade secret can potentially last forever—as long as it is kept reasonably secret.

[11] For example, although the EU approved its first biosimilar in 2006, the United States only recently passed legislation to approve such drugs. *E.g.*, European Generic Medicines Association, Press Release, Biosimilar Products: A Reality in Europe, May 18, 2006, *available at* htm http://www.egagenerics.com/pr-2006-05-18.htm; Patient Protection and Affordable Care Act, Pub. L. No. 111–148, 124 STAT. 119 (2010). The United States has previously asserted that it lacks authority to approve an abbreviated application for most biologicals. Janet Woodcock, Deputy Commissioner, Chief Medical Officer FDA, Testimony before the Committee on Oversight and Government Reform, U.S. House of Representatives (Mar. 25, 2007).

provides greater access to medicine—whether that medicine consists of conventional drugs or biologics.

II. Considering New Approaches

This section considers some proposed solutions to promote access to medicine. Competing patent perspectives are revisited to show how they play a role in both proposed solutions, as well as the likelihood of whether those solutions will be adopted. Three groups of proposed solutions are discussed here. The first section discusses a series of proposals stemming from an original privilege-based proposal to promote access to HIV drugs in developing countries. The second section discusses a proposal that aims to promote innovation and access to a wide range of drugs that are socially beneficial; although this proposal attempts to appeal to patent owners, it still reflects a privilege view and has other logistical challenges that make it unlikely to be adopted. The last section discusses some dramatic proposals whose aim is to promote access to medicine by altering how drugs are developed and marketed; these proposals also face challenges for adoption, but may nonetheless be useful to consider in light of forthcoming access issues that have no ready solutions.

A. FROM PATENT POOLS TO PRICE REDUCTIONS AND FREE LICENSES

This section discusses solutions from both privilege and uber-right views. The section begins with a proposal that predominantly reflects a view of patents as privilege and shows how that proposal may have prompted alternative solutions from patent owners. Although the patent owner proposals reflect more of an uber-right view of patents and are unlikely to satisfy those with a privilege view, they may nonetheless have the potential to promote access to medicine. These solutions can thus illustrate how the intersection of competing patent perspectives may result in modest gains.

1. A Privilege-Based Solution: UNITAID Patent Pool

One intriguing possibility is a proposal by UNITAID[12] to promote access to HIV medicines through a "patent pool." The pool is designed to benefit patients in developing countries where access is usually limited due to inadequate funds to buy medicine at retail prices. UNITAID wants to promote fixed dose combinations (FDCs) to treat HIV as well as pediatric treatments; in both cases, patents from multiple companies are involved.[13]

---

[12] UNITAID purchases drugs to promote increased access to major epidemic killers, including HIV.

[13] The proposal aims to promote not only greater use of drugs for patients, but also greater innovation in these areas. However, because the focus of this book is on access to medicine rather than innovation, this discussion focuses on the access part of the proposal.

Before discussing UNITAID's specific proposal, this section explains how a patent pool operates.

a. Patent Pool Basics

A patent pool is essentially a clearinghouse that promotes licensing of patents that are voluntarily contributed to the pool. Notably, although patents are "volunteered" to be included in the pool, that does not mean patent owners will not be compensated. Typically, a patent pool is managed by an entity that negotiates royalty rates acceptable to patent owners, collects agreed-upon fees from licensees, and transfers the money to patent owners.

A patent pool may exist for any type of technology and for multiple reasons, although generally the patent owner must have some practical interest in participating in the pool.[14] For example, patent owners may elect to create a pool of patents, rather than litigate rights to the patents. Alternatively, there may be an incentive to develop a pool in conjunction with a universal standard adopted for a particular technology that requires use of multiple patents, such as 3-G phones or DVD players. There is one example where patent owners received no financial benefit from a pool but incurred no financial loss and gained positive publicity: a "golden rice" pool of patents was created to help make and distribute genetically engineered rice with enhanced vitamin A to subsistence farmers in developing countries where the patents were not commercially viable. Patent owners volunteered their patents for free for such use, but maintained their full rights to use and enforce the patents in wealthy countries where the patents were valuable.[15]

An effective patent pool can create a similar and even preferable outcome to a series of compulsory licenses involving the same patents. As noted earlier, TRIPS permits countries to issue compulsory licenses on any patent, and some countries may be inclined to do so to reduce costs. However, a compulsory license will generally only permit low-cost manufacture of a drug for domestic use—at least without utilizing the cumbersome procedure for export. A patent pool can easily provide the same type of low-cost use of a patented drug as a compulsory license, but for multiple countries. Moreover, a patent pool could avoid the major drawback of compulsory licenses: potential retaliation from patent owners and countries.

---

[14] *See, e.g.*, David Serafino, *Survey of Patent Pools Demonstrates Variety of Purposes and Management Structures*, KEI Research Note (2007), http://www.keionline.org/misc-docs/ds-patentpools.pdf; Birgit Verbeure et al., *Patent Pools and Diagnostic Testing*, 24 Trends in Biotechnology 115 (2006).

[15] Although patent-owning companies had no financial disincentive, the patent pool owes its existence to the scientists who diligently pursued this goal for developing countries. *See, e.g.*, Simon Robinson & J. Madeleine Nash, *Grains of Hope*, Time, July 31, 2000.

b. An Example: Patent Pool for HIV Drugs

Although a patent pool may be highly desirable as an alternative to compulsory licenses, creating an effective one presents some challenges. In particular, an effective patent pool to promote generic FDCs of HIV drugs depends on a number of factors, including obtaining an adequate number of patents and ensuring that the license conditions are sufficiently attractive to promote generic manufacture.

A key issue is that the pool must contain all of the relevant patents. If all relevant patents are in the pool, the interested user can engage in "one stop shopping" from the patent pool. This could be very helpful if a user wanted to manufacture FDC of HIV drugs without contacting each individual patent owner; after all, FDC typically involve multiple patents owned by different companies. Without a patent pool, any single patent owner could defeat an effort to create a generic FDC. Accordingly, it is critical to have cooperation from all patent owners.

Another important issue in creating an effective patent pool is ensuring that the license rate for pooled patents be financially attractive to potential generic manufacturers. First, it is critical that pooled patents be licensed for a large enough market that potential generic manufacturers will be interested in using the pool. For example, if the pooled patents were restricted to use in only LDCs, there may be inadequate incentive for generic manufacturers to obtain a license from the pool as profits would be slim to nonexistent. However, if the pooled patents were available for use in all developing countries, manufacturers would have more of a profit incentive. Second, it is also important that the license fees for the pooled patents be low enough to enable companies to make a profit—a key issue especially with respect to FDCs as those inevitably involve the payment of royalties to multiple companies.[16] On the other hand, royalties cannot be so low that patent owners are dissuaded from contributing to the pool.

c. Evaluating the Proposal through the Lens of Competing Perspectives

Although the UNITAID patent pool is premised on a laudable goal of promoting broader access to low-cost drugs, its success depends strongly on the voluntary participation of patent owners.[17] UNITAID has heralded the first patent contribution (from the U.S. National Institutes of Health).[18] However, whether the pool will ultimately be successful depends on the number of other relevant patents contributed.

---

[16] UNITAID is reported to have discussed giving companies a 5% royalty. Andrew Jack, *Pharmaceutical Groups Set Up Generics Initiative*, FT.COM, July 21, 2010, *available at* http://www.ft.com/cms/s/0/5f6e1c54-94fb-11df-af3b-00144feab49a.html.

[17] Patent owners may fear that governments will require or pressure them to participate. *E.g.*, Alec Van Gelder & Philip Stevens, *What Purpose Unitaid's Patent Pool?*, at 12 (June 2010), *available at* http://www.policynetwork.net/sites/default/files/EmptyPatentPool.pdf. However, UNITAID has stated that the pool is voluntary.

[18] *E.g.*, UNITAID, Press Release, US National Institutes of Health (NIH) First to Share Patents with Medicines Patent Pool, Sept. 13, 2010, available at http://www.unitaid.eu/en/20100922290/News/US-National-

Whether other patent owners will participate raises a recurring theme in this book: what will be the result of a clash of competing patent perspectives? As discussed in other chapters, divergent perspectives may result in political pressure, as well as posturing. The UNITAID patent pool strongly reflects a view of patents as a privilege, which may create a substantial, if not impossible barrier to obtaining the cooperation of companies that tend to view patents as an uber-right. Those who view patents as a privilege see no problem in limiting that privilege to promote access to medicine; accordingly, they see the pool as an "important and innovative approach to tackling access issues."[19] However, a patent pool is inherently inconsistent with the uber-right view of patents: that holders must be in complete control of the patent so as to maximize profits and innovation. Notably, the single contributor to the patent pool thus far was not a drug company, but a governmental organization. However, most of the relevant patent owners for the UNITAID pool are corporations.

The UNITAID proposal misses the fundamental view of holders of the uber-right perspective that patents are simply not a problem with respect to access to medicine. As stated in a policy paper that seems to reflect an uber-right view, "dilapidated health infrastructure and lack of trained personnel . . . are the biggest barriers to access to medicines," not patents.[20] The policy paper reflects an uber-right view in suggesting there is a "prima facie conflict" between reducing cost of drugs and increasing R&D. An uber-right perspective seems to believe that higher prices will inevitably yield more money to invest in research that will necessarily yield more results.[21] Accordingly, any proposal to trade in patent profits for modest royalties should be seen as inherently incompatible with the uber-right view and thus unlikely to succeed.

Those who believe that political and public pressure will induce patent owners to voluntarily contribute their patents may be highly influenced by a privilege view of patents—and fail to see why those with a different vision of patents would resist. For example, the characterization of the patent pool as a "win-win" situation seems to be

---

Institutes-of-Health-NIH-First-to-Share-Patents-with-Medicines-Patent-Pool.html. The patents are licensed for not only low, but also middle-income countries at a zero percent royalty. Public Health Service Non-exclusive License Agreement, http://www.unitaid.eu/images/news/patentpool/mppf%20patent%20license%20full%20executed%20%28sept%202010%29-ns.pdf.

[19] Médecins Ssans Frontières, MSF Briefing Note, Taking the Plunge: The UNITAID Patent Pool—A New Tool for Achieving Access to Medicine (Feb. 2009). Similarly, MSF suggests that the pool should be attractive to patent owners because it "offers a predictable way to address global access issues while minimising the economic and reputational risks of IP driven controversies . . . ." *Id.* However, as noted above, the uber-right does not believe there is an access problem. Moreover, patent-owning companies are not truly vulnerable to economic and reputational risks when they have the right to exclude all competitors—even if they face sometimes negative publicity.

[20] Van Gelder & Stevens, *supra* note 17, at 3.

[21] This uber-right belief is maintained in the face of recent evidence that increasing research and development has not promoted more innovation.

based on an inherent misunderstanding that a royalty from a patent pool could be deemed adequate compensation to those with an uber-right view.[22] As seen with the South Africa controversy, public pressure may induce pharmaceutical companies to change their positions. However, the patent pool presents a greater challenge. Although patent-owning companies eventually dropped their lawsuit against South Africa, they retained the ability to re-instigate suit or exert other pressure to force South Africa (or other countries) not to issue compulsory licenses. In contrast, a company that willingly contributed its patent would give up this leverage.

Although holders of an uber-right view of patents would consider the pool anathematic to their interests, several companies are purportedly talking to UNITAID about potential licenses.[23] It is possible that additional revenue alone would persuade patent owners to be interested, but that has generally not been the view of large companies faced with compulsory licenses. Any large companies negotiating with UNITAID could have a more moderate view of patents. Alternatively, it is possible these companies maintain an uber-right view, but are nonetheless interested in participating to address public sentiment—if participation can be secured on appropriate terms. For example, even a company with an uber-right view might be willing to license its HIV patents for manufacture in an LDC. After all, companies do not always even obtain patents in such countries because there is no profit to be made.

d. Pool Impact on Innovation

Even if some companies decide to contribute their patents, what that may mean for the future of HIV drug development (as well as development of all drugs) is unclear. In particular, the companies that contributed the patents may decline to engage in further research in the area of HIV if they believe there will be public pressure to forego patent profits.[24] HIV advocates tend to suggest that because companies primarily make their profits on HIV drugs in the developed world, there is no need to fear that innovation will end. However, as companies have a financial interest in developing the most profitable drugs, a company could logically elect to pursue cancer (for which there is currently less pressure to provide drugs at low cost to consumers) instead of HIV treatments.

---

[22] *See* Editorial, *Political Activism Needed for Patent Pool for HIV Drugs*, 374 LANCET 66 (July 25, 2009).

[23] *E.g.*, Kate Kelland, *Some Drug Firms Keen, Others Reluctant on AIDS Pool*, REUTERS, July 1, 2010, at http://www.reuters.com/article/idUSTRE66K2ZP20100721 (noting that UNITAID has been in discussions with Merck, Tibotec, and Gilead); *see also* Ellen 't Hoen, *Pharmaceutical Companies and the UNITAID Patent Pool*, 375 LANCET 30 (Jan. 2, 2010) (representing that "several large pharmaceutical firms" have expressed public interest).

[24] HIV research may have already decreased in response to increasing public pressure to provide drugs at low cost to poor countries. *E.g.*, Kristen Kresge, *Drug Pipelines May Flourish, but Not for HIV*, 5 HIV AIDS Treatment Insider (Nov. 2004) at 1, 2.

A potentially even greater downside to the UNITAID patent pool is that patent-owning companies may fear public pressure to volunteer patents for additional pools. This backdrop could make companies more aggressively police and protect other rights. For example, companies might both promote new laws and retaliate against countries that are perceived to inadequately protect their rights. Although those who view patents as a privilege may counter that patents are necessarily an incomplete solution to promoting innovation, as only large companies currently have the resources to conduct the clinical tests necessary to market a drug, their views have a substantial impact on drug development.

2. Solutions from the Uber-Right

Although companies tend to suggest that patents are not a problem, or at least not a significant one to access to medicine, they sometimes are inspired to suggest solutions that seem to in part concede that patents may impact such access. The inspiration for these solutions may be prompted by privilege-based perspectives that are considered completely unacceptable. This section presents two proposals that each in part acknowledge patents can be a barrier to access to medicine. The first proposal discussed is from Glaxo Smith Kline (GSK); it includes its own patent pool in combination with price cuts and financial investment. The second proposal suggests a royalty-free HIV license for LDCs.

a. GSK Pool, Price Cuts, and Healthcare Investment

Roughly six months after UNITAID announced its patent pool for HIV medicines, GSK, a major pharmaceutical company, announced the creation of its own patent pool for neglected diseases in conjunction with a proposal to reduce costs and invest in health care of poor countries.

At first glance, GSK's announcement would seem to provide a multifaceted and enlightened approach to improving access to medicine. However, this three-part approach reflects a "solution" that is in line with an uber-right view of patents. As discussed earlier, those with an uber-right perspective view control over patent rights as sacrosanct. Accordingly, GSK's voluntarily initiation of a patent pool would seem to be an extreme sacrifice. Indeed, its CEO suggested that the pharmaceutical industry would accuse him of selling the family silver by offering to share patent rights.[25] However, closer analyses of the details of the patent pool suggest that GSK's pool is more consistent with an uber-right view of patents.

Although GSK's pool might seem to endorse patent pools to promote access of medicine, there are some major caveats. First of all, GSK elected not to participate in the

[25] Sarah Boseley, *Drug Giant GlaxoSmithKline Pledges Cheap Medicines for World's Poor*, GUARDIAN, Feb. 13, 2009, http://www.guardian.co.uk/business/2009/feb/13/glaxo-smith-kline-cheap-medicine. The CEO of Glaxo suggested that Glaxo could be a leader in providing a "more balanced view" of promoting innovation among global companies. *Id.*

UNITAID pool,[26] and its own pool excludes commercially valuable HIV treatments.[27] Second, neglected diseases are by definition unprofitable so that contributing a patent on such a disease does not require companies to forego any expected profits. Third, although patent pools can generally promote innovation by providing access to more researchers, any theoretical innovation is unlikely to result in commercial development because there is still no viable market for drugs treating diseases that only impact poor counties.

b. Lowered Prices

GSK announced that it was cutting the prices for some drugs in the fifty least developed countries to no more than 25 percent of their sales price in the UK and United States and that they would also make drugs "more affordable" in middle-income countries. The drugs included relate to HIV, as well as malaria, tuberculosis, and hepatitis.

GSK's price cut can be viewed as representing an uber-right view of patents; at a minimum, it is not consistent with the view of patents as privilege—those with such a view strongly criticized GSK's price cuts as being inadequate for people who may only earn $1–2/day.[28] However, to those who view patents as providing their owner with the right to charge a maximum price in each market, any reduction in price would seem a concession. Of course, it is unclear whether the 25 percent is an actual reduction in price or only a public relations effort to suggest a price reduction. However, an uber-right view would be consistent with either action.

In addition, the differentiation of price cuts for LDCs versus middle-income countries similarly reflects an uber-right view of seeking the maximum price in each market. Although those with a privilege view would treat both groups as developing countries that clearly need access, that is not the view of those with the uber-right perspective. In addition, holders of the uber-right view may fear losing control over segregated markets.

Voluntary price cuts are also more consistent with an uber-right view of patents than compulsory licenses.[29] As discussed earlier, patent owners are strongly opposed to compulsory licenses—even where permitted under national and international laws.

---

[26] Some cynics suggest that GSK's patent pool is an affirmative effort to undermine the success of the UNITAID patent pool by withholding GSK patents from that pool. Glaxo's price cut has also been criticized as not being steep enough as the company would likely net a $20 million profit—which is nearly half of its sales to LDCs. *See e.g.*, Posting of Brook Baker to Health GAP Global Access Project, http://www.healthgap.org/baker-onGSK.htm, *GSK Access to Medicines: The Good, The Bad, and the Illusory* (Feb. 15, 2009) [hereinafter Posting of Baker, *GSK Access to Medicines*].

[27] Glaxo suggested it was not appropriate to include HIV drugs within the pool because the pool was intended to focus on diseases with a "severe lack of treatments" and that the issue of greater access to HIV drugs was "being addressed in other ways." Lisa Richwine, *Glaxo Proposes Patent Pool for Neglected Diseases*, REUTERS, Feb. 13, 2009, http://www.reuters.com/article/idUSN1353591120090213.

[28] *E.g.*, Posting of Baker, *GSK Access to Medicines*, *supra* note 26.

[29] In fact, one criticism of GSK's price cuts is that they might give companies more ammunition to protest compulsory licenses as being unnecessary or unreasonable.

Compulsory licenses always require that patent owners be provided compensation. However, having no control over when a compulsory license issues (as well as limited, if any control over the compensation) makes compulsory licenses inconsistent with the uber-right view of the patent owner's sanctity to control patent rights.

c. Investing in Healthare Infrastructure?

GSK also announced that it would reinvest 20 percent of any profits made from its drugs in LDCs in hospitals, clinics, and staff. GSK's promise to invest part of the profits in infrastructure is also consistent with the uber-right view that an inadequate healthcare infrastructure rather than patents is a major barrier to access to medicine. Although some may criticize GSK's contribution to infrastructure as inadequate or even de minimis, the more important fact to emphasize is that any investment in infrastructure emphasizes the uber-right view that patents do not play a predominant role in limiting access to medicine.[30] By voluntarily contributing to the infrastructure of developing countries, GSK can emphasize its view that issues other than patents are causing access problems.

d. Free Licenses of HIV Drugs–ViiV

Another solution that reflects the uber-right view is a royalty-free license of HIV-related patents from specialist HIV company ViiV, which was actually founded by pharmaceutical giants GSK and Pfizer, to provide treatment for the poorest countries.[31] Similar to the GSK proposal, the ViiV solution seems to have been prompted by development of the UNITAID pool. In particular, the announcement came just as the UNITAID patent pool was declared to be "ready for business"—and after the pool had initiated discussions with major companies such as Merck, Gilead, and Abbott.[32]

Although the royalty-free license seems very generous, it is nonetheless consistent with an uber-right view of patents—or at least it is more consistent with an uber-right view than participation in the UNITAID patent pool. Importantly, the free licenses are

[30] Currently, sales—not profits—in LDCs are roughly $40 million versus Glaxo's overall $40 *billion* in overall profits. Profits are a small fraction such that the actual number that is projected to be invested is $1–3 million. Although this may seem a significant sum, some have suggested that the amount is still negligible as it must be divided by fifty countries, which would result in an investment of about $50,000 per country. *E.g.*, Christian Barry & Matt Peterson, *Shallow cuts: GSK's Voluntary Price Reductions and Patent Pooling Are Not Enough*, Moreover, the amount could be even lower as the price cuts will likely also lead to reduced profits.

[31] The potential patents includes established HIV treatments such as Combivir (by GSK), as well as more recent treatments such as Kivexa and Celsentri. In total, there are reportedly ten current HIV medicines available, with another seven in the pipeline. Andrew Clarke, *GSK Joins Forces with Pfizer to Develop HIV/AIDS Drugs*, GUARDIAN, Apr. 16, 2009, http://www.guardian.co.uk/business/2009/apr/16/gsk-pfizer-hiv-aids; Press Release, ViiV Healthcare Announces Further Initiatives to Improve Access to HIV Medications For People Living in the Least Developed Countries, July 16, 2010, http://www.viivhealthcare.com/media-room/press-releases/2010-07-16.aspx [hereinafter Viiv Press Release].

[32] Andrew Jack, *Pharmaceutical Groups Set Up Generics Initiative*, FT.COM, July 21, 2010, http://www.ft.com/cms/s/0/5f6e1c54-94fb-11df-af3b-00144feab49a.html#axzz14fQktIuu.

limited primarily to LDCs, whereas the UNITAID pool aimed to make HIV patents available to all developing countries to provide generic manufacturers with the greatest possibility of obtaining a profit. In addition, it is unclear whether generic drugs will in fact be manufactured even with the free licenses. After all, some have suggested that generic manufacturers not only need low-cost access to patent rights, but the economies of scale that come with including middle-income developing countries can pay more than LDCs. Regardless of whether any generic manufacturers are able to take advantage of the free HIV licenses, patent owner ViiV can publicly claim that it is responsive to the needs of developing countries.[33]

Regardless of the eventual outcome, the licenses serve as an example of how the privilege view of patents may prompt alternative solutions that, although not completely satisfying those with a privilege view, could nonetheless promote more access to medicine. In particular, given that access to HIV has been a long-standing problem recognized by all (including drug companies), the timing of the ViiV proposal suggests that it may have been prompted by the UNITAID proposal. In other words, even though the UNITAD proposal itself may not be adopted, it may prompt other important developments.

Those with a privilege view would find the ViiV licenses to be inadequate in not covering all developing countries and for not participating in the most-preferred solution of the UNITAID patent pool. However, if each interested party rejected all proposals except those that completely aligned with its interests, no proposal would have any chance of adoption. In addition, even modest progress should not be overlooked. Considering the longer-term landscape, a proposal to provide even some HIV drug patents at no cost—even if only to a limited group of countries—should still be seen as progress in promoting access to medicine.[34]

B. HEALTH IMPACT FUND

An alternative proposal to promote access that attempts to recognize the perspective of the patent owners (but still reflects a somewhat privilege-based view) is the so-called Health Impact Fund.[35] Essentially, this fund aims to promote innovation as well as access by offering prize money to companies that are willing to give up their patent rights to specific inventions. The fund targets socially valuable treatments that may not be

---

[33] For example, in the press release announcing the free licenses, ViiV proclaimed that it was providing support for "people living with HIV/AIDS in the poorest and most vulnerable countries" that account for 80% of people living with HIV. ViiV Press Release, *supra* note 31.

[34] This assumes that the proposal includes some unexpired patents as to LDCs—an issue unclear in the initial press release.

[35] *See* AIDEN HOLLIS & THOMAS POGGE, THE HEALTH IMPACT FUND: MAKING NEW MEDICINES ACCESSIBLE FOR ALL (2008).

commercially valuable such that companies would likely be willing to forego patent rights to such treatments in return for the prize money.

Although the idea of offering prizes for innovation is not new, the Health Impact Fund (HIF) deserves special attention in the context of discussing whether proposals are likely to be adopted by those with differing perspectives of patents. The HIF provides a more moderate view of patents than the UNITAID patent pool. However, as will be discussed, the HIF's core assumptions may still reflect more of a privilege view. Before discussing how the HIF fits into the scheme of competing patent privileges, a brief overview of how the HIF works may be helpful.

The HIF offers companies that develop socially desirable treatments the opportunity to apply for prize money given over the first ten years the treatment is sold. In exchange, a company must agree to sell the drug at a low cost and freely license it to others for global manufacture and sale. No patent owners are required to seek prize money, and the HIF does not envision that companies will cease their efforts to develop and sell blockbuster drugs. However, it does aim to fill a perceived gap in the current system that promotes only the most profitable drugs regardless of their social utility.

At first glance, the HIF proposal seems to be a creative way to encourage valuable innovation and access that would be acceptable to patent owners because it does not require a fundamental change in business practices. Despite its clear attempt to cater to patent owners, the proposal may still reflect a privilege view that presents a hurdle to adoption beyond obvious logistical hurdles of persuading countries to provide financing[36] and difficulties with assessing which treatments have the most impact and thus deserve the most reward.[37] In particular, under the privilege view. Patents are tools intended to promote the HIF proposal is consistent with this assumption and assumes that if it provides a monetary incentive for desired treatments, those treatments will be developed. However, an uber-right view of patents may be that they are a strong property right, they promote profits, and will inevitably lead to appropriate innovation. Accordingly, the holder of an uber-right view may still elect to only develop drugs that maximize profits rather than ones that provide more modest profits—especially when even modest profits might be limited if they come from financially strapped countries. Theoretically, the HIF could address this issue by increasing the reward. However, it seems impossible that the reward could ever approximate the desired profits of the blockbuster drugs that companies seek to develop. For example, the proposed funding for all rewards as well as administrative costs in a single year of the HIF equals the yearly sales of

[36] The HIF assumes substantial government commitment to pay for the fund—an estimated $6 billion a year is required for a minimum of twelve years. HOLLIS & POGGE, *supra* note 35, at 43-44.

[37] The proposal aims to provide more money to treatments that have higher social value based on Quality Adjusted Life Years (QALY). Although this metric is common, it may still be difficult to assess.

a *single* blockbuster drug.[38] Although the money could be attractive to small companies, they are generally not in a position to conduct the clinical trials required to obtain marketing approval—which the HIF proposal assumes as a prerequisite. In addition, a company must apply for the prize money, and there is no guarantee that its new treatment will be deemed valuable enough to gain a prize or what the amount of prize money might be.[39]

The HIF's more moderate view of patent privileges has provoked the ire of those with a strong view of patents as privilege. The HIF has been criticized as undermining proposals that aim to promote more generic drugs, such as the UNITAID patent pool.[40] It has also been criticized as a flawed variation of a prize fund for failing to provide a completely free license—even though that would never have support from patent owners. Fundamentally, those with a strong view of patents as privilege seem opposed to a proposal that permits patent owners to continue to benefit from any patent rights. The HIF is also criticized for creating disincentives not only for production of generics, but for issuance of compulsory licenses.[41] Holders of the view of patents as privilege seem to believe that not only are compulsory licenses permitted, but that proposals that might reduce incentives to use such licenses should be rejected.

The HIF has not been openly embraced by patent-owning companies—even while those with a strong view of patents as privilege seem to suggest that the HIF creators are collaborating with companies to minimize generic drugs. In contrast to the UNITAID patent pool, there is relatively little public press on the reaction of patent owners to HIF. However, there is one published "debate" between HIF creator Thomas Pogge and an executive of Pfizer.[42] In it, the Pfizer executive suggested that the HIF is flawed because part of the document explaining the case for the HIF noted strong patents necessarily limit access. In the view of the Pfizer executive, the history of HIV/AIDS does not support this assertion.[43] This view reflects the traditional uber-right view that patents do not limit access to medicine. In addition, it suggests holders of the uber-right view may be so resistant to the idea that patents minimize access that any reference to this—even if not central to the core of a proposal—may make a proposal inherently less acceptable.

---

[38] Posting of Brook Baker to Health Gap Global Access Project, http://www.healthgap.org/trips/bakeronHIF.htm, *Skepticism about the Health Impact Fund*, Nov. 24, 2008 [hereinafter Posting of Baker, *Skepticism about the Health Impact Fund*].

[39] The proposal suggests that the prize money will be roughly proportional to the health impact of the treatment—in addition to health impact of other proposals. In other words, in a year where there are many other socially valuable treatments, a given company might receive less money.

[40] *E.g.*, Posting of Baker, *Skepticism about the Health Impact Fund*, *supra* note 38.

[41] *Id.*

[42] William Schulz, *Providing Medicines for the Poor*, CHEM. & ENGINEERING NEWS, Dec. 1, 2008.

[43] In particular, he suggests that the barriers to HIV treatment have little to do with the cost as most first-line treatments are less than 5% of the initial price—although neglecting to state that these cheap treatments are made from generic companies and may no longer be recommended treatments. *Id.*

C. OTHER PROPOSALS

Another possible approach to increasing access to drugs is to dramatically reconceptualize how drugs are created and marketed. For example, if the price of drugs was not directly linked to money available for research, the argument that high prices are necessary would disappear, and drugs (at least theoretically) would be cheaper and thus more accessible. There are a variety of different ways to achieve this goal, although all would seem to involve a major change to how drugs are developed and sold—one that might be premised on jettisoning the existing and thus far successful business model of major pharmaceutical companies. Although such change seems unlikely at present, the looming access problems discussed earlier in the Chapter may eventually prompt evaluation of some of these proposals.

1. Publicly Funded Clinical Trials

Although there is a recognized human right to health, the creation of drugs that help promote it is predominantly a commercial operation. Although some may blame companies for prioritizing patents over patients, a corporation has a fundamental mission to its shareholders to make money. Theoretically, corporations could develop more of a humanitarian mission. However, private companies may perhaps not be the appropriate party to develop drugs, or at least not *all* drugs. It is true that private companies are not entirely responsible—in many countries, government and academia play a role in important basic research that may lead to drug discoveries. However, private companies remain the dominant source for engaging in the expensive and lengthy clinical trials necessary to bring a drug to market.

As noted in Chapter 1, drug companies lack neutrality in performing clinical trials of their own drugs. In addition, companies have been known to suppress results that are unflattering to their drugs. Accordingly, one idea is to remove the self-interested drug company from the clinical testing process and instead have tests conducted in a manner similar to publication of scientific articles (i.e., with peer review).[44] Those who previously have been involved in clinical tests could be hired to work for an independent testing agency to conduct the tests. Governments could fund the tests as a public good that benefits all, with possible additional financial assistance from the drug companies. However, a major hurdle is obviously that this is not only a novel idea—albeit one backed by noted academics—but a very costly one.[45]

Although this idea seems to directly address the cost of drugs, as the expense of clinical studies comprises a substantial part of overall drug discovery costs, this proposal may

---

[44] *E.g.*, Tracy R. Lewis et al., *The Case for Public Funding and Public Oversight of Clinical Trials*, ECONOMISTS VOICE (2007).

[45] The proposal is suggested by Professors Lewis (Economics), Reichman (Law), and So (Public Policy). *Id.*

nonetheless reflect more of a privilege view. For example, the proposal assumes that drug companies will reduce prices in proportion to cost savings.[46] Holders of a privilege view would consider this proposal reasonable in that even if a drug were patented, a company should not be entitled to maximize its profits, or at least not when it unduly impacts access to needed medicine. However, corporations that patented the original compounds may feel entitled to reap the entire patent reward, even if that means drugs have a higher retail price. After all, under the uber-right view, neither patents nor drug prices are a problem, as higher costs simply provide fuel for more research that benefits all.

2. Drug Development "Corporation"

Another proposed solution to separate the cost of drug prices from money for research attempts to use existing market incentives to promote innovation while simultaneously lowering costs. The solution requires the creation of a new independent and public agency or nonprofit corporation, referred to as the Drug Development Corporation (DDC) that would play a key role in separating drug research from clinical development and marketing. The idea is that the DDC would purchase (patent) rights to developed drugs and then hold auctions to sell exclusive licenses to companies that would focus solely on making and selling the drug, with the price to consumers being capped.[47]

Although this innovative solution does indeed separate drug prices from the cost of research, it still reflects a privilege-based approach. In particular, this solution addresses what holders of the uber-right view think is a nonexistent problem: the price of drugs. To them, the costs are necessary to promote more innovation such that any proposals to decrease drug prices would be viewed with great skepticism. In addition, multiple parts of this solution strongly reflect a privilege view that would be inconsistent with an uber-right one. For example, although some companies will be paid for basic research, and others have the right to profit from commercial sales, all companies will have only limited abilities to profit. The proposal envisions paying the companies that do basic research a "fair return." However, in the context of compulsory licenses, companies have repeatedly rejected "reasonable" royalty rates and insisted that every drug is entitled to a maximum profit. Similarly, imposition of a cap on final prices that permits only a "reasonable return" would be inconsistent with the profit-maximizing view of patents.

Perhaps recognizing that their solution may not be well received by drug companies, the authors of the proposal explain that it can nonetheless work by force in two ways. First, they suggest that multinational drug companies would be "de-regulated" in a manner similar to telecommunications so that today's drugs companies would no longer exist; rather, in their vision, a company would either develop or manufacture drugs.

---

[46] *Id.* at 2- 3.

[47] Stan Finkelstein & Peter Temin, Reasonable Rx: Solving the Drug Price Crisis (2008).

Second, they suggest that a national regulatory agency reject applications from companies that do both research and market drugs to ensure companies that do both discovery and sales no longer exist. Although mandating change theoretically does not require those of the uber-right view to change their perspective, this may not work in reality, as drug companies have a huge lobbying force that could present a formidable challenge to proposed changes in the law.

III. Conclusion

Although none of the above proposals provides an instant solution to every problem, they show some promise. For example, regardless of whether the UNITAID patent pool comes to fruition, it has had an impact in prompting some unprecedented action from patent owners. Although other proposals may currently seem radical and unlikely to be of any value, it is possible that continuing tensions will require revisiting these solutions, and will perhaps help generate others. This book does not attempt to provide a magic bullet to the problem of access to medicine. Indeed, a premise of the book is that there is no magic bullet—especially as discussion and proposals often involve competing and contrary perspectives. However, hopefully the above discussion at least helps to clarify the importance of perspectives in reception of solutions, as that can hopefully promote realistic and workable solutions. Given the existence of competing perspectives, it may be necessary to concede that an acceptable solution to all will necessarily require compromise that will make it less than ideal to those with more extreme views. Though change may be difficult and slower than desired, movement toward promoting mutually satisfying goals—even if only in part—can still be heralded as a significant step forward from talking past each other.

Appendices

Appendix A: Selected TRIPS Provisions

Note:

The TRIPS Agreement is authored and copyrighted by the WTO. The full text may be found online at www.wto.org. Footnote numbers replicate the numbering online .

This abbreviated version is provided to aid the discussion of TRIPS provisions. Specific terms given discussion in the book are noted through the use of bold text. Similarly, small annotations to clarify references or omission of material are noted in italics, surrounded by square brackets. These annotations are not part of the original text, but provided for ease of reference.

Article 1

NATURE AND SCOPE OF OBLIGATIONS

1. Members shall give effect to the provisions of this Agreement. Members **may**, but shall not be obliged to, **implement in their law more extensive protection than is required** by this Agreement, **provided that such protection does not contravene the provisions of this Agreement.** Members shall be free to determine the appropriate method of implementing the provisions of this Agreement within their own legal system and practice.

2-3. *Omitted*

Article 2

INTELLECTUAL PROPERTY CONVENTIONS

1. In respect of Parts II, III and IV of this Agreement [*standards, enforcement, and acquisition of rights*], Members shall comply with Articles 1 through 12 [*including Art. 5 regarding compulsory licenses for "failure to work" a patented invention*], and Article 19, of the Paris Convention (1967).

Article 3

NATIONAL TREATMENT

1. Each Member shall accord to the nationals of other Members treatment no less favourable than that it accords to its own nationals with regard to the protection[3] of intellectual property, ...

Article 4

MOST-FAVOURED-NATION TREATMENT

With regard to the protection of intellectual property, any advantage, favour, privilege or immunity granted by a Member to the nationals of any other country shall be accorded immediately and unconditionally to the nationals of all other Members. [*Exemptions omitted*]

Article 6

EXHAUSTION

For the purposes of dispute settlement under this Agreement, subject to the provisions of Articles 3 and 4 nothing in this Agreement shall be used to address the issue of exhaustion of intellectual property rights.

Article 7

OBJECTIVES

The protection and enforcement of intellectual property rights should contribute to the promotion of technological innovation and to the transfer and dissemination of technology, to the

[3] For the purposes of Articles 3 & 4, "protection" shall include matters affecting the availability, acquisition, scope, maintenance and enforcement of intellectual property rights, as well as those matters affecting the use of intellectual property rights specifically addressed in this Agreement.

mutual advantage of producers and users of technological knowledge and in a manner conducive to social and economic welfare, and to a balance of rights and obligations.

Article 8

PRINCIPLES

1. Members may, in formulating or amending their laws and regulations, adopt measures necessary to protect public health and nutrition, and to promote the public interest in sectors of vital importance to their socio-economic and technological development, provided that such measures are consistent with the provisions of this Agreement.

2. [*Omitted*]

Article 27

PATENTABLE SUBJECT MATTER

1. Subject to the provisions of paragraphs 2 and 3, patents shall be available for any inventions, whether products or processes, in all fields of technology, provided that they are new, involve an inventive step and are capable of industrial application.[5] Subject to paragraph 4 of Article 65 [*permissible delay of product patents for some developing countries*], paragraph 8 of Article 70 [*transitional requirements, such as "mailbox provision" for countries that do not immediately provide product patents*], and paragraph 3 of this Article patents shall be available and patent rights enjoyable without discrimination as to the place of invention, the field of technology and whether products are imported or locally produced.

2. [*Omitted*]

3. [*Omitted*]

Article 28

RIGHTS CONFERRED

1. A patent shall confer on its owner the following exclusive rights:

(a) where the subject matter of a patent is a product, to prevent third parties not having the owner's consent from the acts of: making, using, offering for sale, selling, or importing[6] for these purposes that product;

---

[5] For the Purposes of this Article, the terms "inventive step" and "capable of industrial application" may be deemed by a Member to be synonymous with the terms "non-obvious" and "useful" respectively.

[6] This right, like all other rights conferred under this Agreement in respect of the use, sale, importation or other distribution of goods, is subject to the provisions of Article 6.

(b) [*Omitted – Process Patent Rights*]

2. [*Omitted*]

Article 30

EXCEPTIONS TO RIGHTS CONFERRED

Members may provide **limited exceptions** to the exclusive rights conferred by a patent, provided that such exceptions do not unreasonably conflict with a normal exploitation of the patent and do not unreasonably prejudice the legitimate interests of the patent owner, taking account of the legitimate interests of third parties.

Article 31

OTHER USE WITHOUT AUTHORIZATION OF THE RIGHT HOLDER

[*Compulsory Licenses*]

Where the law of a Member allows for other use[7] of the subject matter of a patent without the authorization of the right holder, including use by the government or third parties authorized by the government, the following provisions shall be respected:

a. authorization of such use shall be considered **on its individual merits**;
b. such use may only be permitted if, **prior to such use, the proposed user has made efforts to obtain authorization from the right holder on reasonable commercial terms** and conditions and that such efforts have not been successful within a reasonable period of time. **This requirement may be waived** by a Member **in the case of a national emergency or other circumstances of extreme urgency or in cases of public non-commercial use.** [*additional requirements concerning public non-commercial use excluded*];
c. the scope and duration of such **use shall be limited to the purpose for which it was authorized**, . . . ;
d. such use shall be non-exclusive;
e. such use shall be non-assignable, . . . ;
f. any such use shall be **authorized predominantly for the supply of the domestic market** of the Member authorizing such use [*subject to 2003 Waiver, and Article 31bis amendment*];
g. authorization for such use shall be liable, . . . to be terminated if and when the circumstances which led to it cease to exist and are unlikely to recur. The competent authority shall have the authority to review, upon motivated request, the continued existence of these circumstances;

[7] "Other use" refers to use other than that allowed under Art. 30.

h. the **right holder shall be paid adequate remuneration** in the circumstances of each case, taking into account the economic value of the authorization;
i. the legal validity of any decision relating to the authorization of such use shall be subject to judicial review or other independent review . . . ;
j. any decision relating to the remuneration provided in respect of such use shall be subject to judicial review or other independent review . . . ;
k. [*more limited obligations when compulsory license as a remedy for anti-competitive use*] Members are not obliged to apply the conditions set forth in sub-paragraphs (b) and (f) where such use is permitted to remedy a practice determined after judicial or administrative process to be anticompetitive. The need to correct anti-competitive practices may be taken into account in determining the amount of remuneration in such cases. Competent authorities shall have the authority to refuse termination of authorization if and when the conditions which led to such authorization are likely to recur;
l. [*additional conditions for "blocking patents"*] where such use is authorized to permit the exploitation of a patent ("the second patent") which cannot be exploited without infringing another patent ("the first patent"), the following additional conditions shall apply:
    i. the invention claimed in the second patent shall involve an important technical advance of considerable economic significance in relation to the invention claimed in the first patent;
    ii. the owner of the first patent shall be entitled to a cross-license on reasonable terms to use the invention claimed in the second patent; and
    iii. the use authorized in respect of the first patent shall be nonassignable except with the assignment of the second patent.

Article 33

Term of Protection

The term of protection available shall not end before the expiration of a period of twenty years counted from the filing date. [*footnote omitted*]

Article 39

1-2 [*Omitted*]

3. Members, when requiring, as a condition of approving the marketing of pharmaceutical or of agricultural chemical products which utilize new chemical entities, the submission of undisclosed test or other data, the origination of which involves a considerable effort, **shall protect such data against unfair commercial use.** In addition, Members shall protect such data against disclosure, except where necessary to protect the public, or unless steps are taken to ensure that the data are protected against unfair commercial use.

Article 41

1. Members shall ensure that **enforcement procedures** as specified in this Part are available under their law so as to permit effective action against any act of infringement of intellectual property rights covered by this Agreement, including expeditious remedies to prevent infringements and remedies which constitute a deterrent to further infringements. These procedures **shall be applied in such a manner as to avoid the creation of barriers to legitimate trade** and to provide for safeguards against their abuse.
2. Procedures concerning the enforcement of intellectual property rights shall be fair and equitable. They shall not be unnecessarily complicated or costly, or entail unreasonable time-limits or unwarranted delays.

3-5 [*Omitted*]

Article 51

SUSPENSION OF RELEASE BY CUSTOMS AUTHORITIES

Members shall, in conformity with the provisions set out below, adopt **procedures**[13] **to enable a right holder**, who has valid grounds for suspecting that the importation of counterfeit trademark or pirated copyright goods[14] may take place, **to lodge an application** in writing with competent authorities, administrative or judicial, **for the suspension by the customs authorities** of the release into free circulation of such goods. Members may enable such an application to be made in respect of goods which involve other infringements of intellectual property rights, provided that the requirements of this Section are met. Members may also provide for corresponding procedures concerning the suspension by the customs authorities of the release of infringing goods destined for exportation from their territories

[13] It is understood that **there shall be no obligation to apply such procedures** to imports of goods put on the market in another country by or with the consent of the right holder, or **to goods in transit.**

[14] For the purposes of this Agreement: (a) "counterfeit trademark goods" shall mean any goods, including packaging, bearing without authorization a trademark which is identical to the trademark validly registered in respect of such goods, or which cannot be distinguished in its essential aspects from such a trademark, and which thereby infringes the rights of the owner of the trademark in question under the law of the country of importation; (b) "pirated copyright goods" shall mean any goods which are copies made without the consent of the right holder or person duly authorized by the right holder in the country of production and which are made directly or indirectly from an article where the making of that copy would have constituted an infringement of a copyright or a related right under the law of the country of importation.

Article 52

APPLICATION

Any right holder initiating the procedures under Article 51 shall be required to provide adequate evidence to satisfy the competent authorities that, **under the laws of the country of importation, there is *prima facie* an infringement** of the right holder's intellectual property right and to supply a sufficiently detailed description of the goods to make them readily recognizable by the customs authorities. The competent authorities shall inform the applicant within a reasonable period whether they have accepted the application and, where determined by the competent authorities, the period for which the customs authorities will take action.

Appendix B: Declaration on the TRIPS Agreement and Public Health

(Adopted on 14 November 2001)
Ministerial Conference, Fourth Session
WT/MIN(01)/DEC/2

1. We recognize the gravity of the public health problems afflicting many developing and least-developed countries, especially those resulting from HIV/AIDS, tuberculosis, malaria and other epidemics.
2. We stress the need for the WTO Agreement on Trade-Related Aspects of Intellectual Property Rights (TRIPS Agreement) to be part of the wider national and international action to address these problems.
3. We recognize that intellectual property protection is important for the development of new medicines. We also recognize the concerns about its effects on prices.
4. We agree that the TRIPS Agreement does not and should not prevent Members from taking measures to protect public health. Accordingly, while reiterating our commitment to the TRIPS Agreement, we affirm that the Agreement can and should be interpreted and implemented in a manner supportive of WTO Members' right to protect public health and, in particular, to promote access to medicines for all.

In this connection, we reaffirm the right of WTO Members to use, to the full, the provisions in the TRIPS Agreement, which provide flexibility for this purpose.

5. Accordingly and in the light of paragraph 4 above, while maintaining our commitments in the TRIPS Agreement, we recognize that these flexibilities include:
   (a) In applying the customary rules of interpretation of public international law, each provision of the TRIPS Agreement shall be read in the light of the object

and purpose of the Agreement as expressed, in particular, in its objectives and principles.

(b) Each Member has the right to grant compulsory licences and the freedom to determine the grounds upon which such licences are granted.

(c) Each Member has the right to determine what constitutes a national emergency or other circumstances of extreme urgency, it being understood that public health crises, including those relating to HIV/AIDS, tuberculosis, malaria and other epidemics, can represent a national emergency or other circumstances of extreme urgency.

(d) The effect of the provisions in the TRIPS Agreement that are relevant to the exhaustion of intellectual property rights is to leave each Member free to establish its own regime for such exhaustion without challenge, subject to the MFN and national treatment provisions of Articles 3 and 4.

6. We recognize that WTO Members with insufficient or no manufacturing capacities in the pharmaceutical sector could face difficulties in making effective use of compulsory licensing under the TRIPS Agreement. We instruct the Council for TRIPS to find an expeditious solution to this problem and to report to the General Council before the end of 2002.

7. We reaffirm the commitment of developed-country Members to provide incentives to their enterprises and institutions to promote and encourage technology transfer to least-developed country Members pursuant to Article 66.2. We also agree that the least-developed country Members will not be obliged, with respect to pharmaceutical products, to implement or apply Sections 5 and 7 of Part II of the TRIPS Agreement or to enforce rights provided for under these Sections until 1 January 2016, without prejudice to the right of least-developed country Members to seek other extensions of the transition periods as provided for in Article 66.1 of the TRIPS Agreement. We instruct the Council for TRIPS to take the necessary action to give effect to this pursuant to Article 66.1 of the TRIPS Agreement.

Appendix C: Draft Ministerial Declaration

Proposal from a group of developed countries.

IP/C/W/313
4 October 2001
(01-4779)

Council for Trade-Related Aspects of Intellectual Property Rights

Preambular language for ministerial declaration

Contribution from Australia, Canada, Japan, Switzerland and the United States

During the special discussion of the TRIPS Council on 19 September 2001, the delegations of Australia, Canada, Japan, Switzerland and the United States communicated the following text to the Secretariat and have requested it to be circulated as a TRIPS Council document.

Access to medicines for hiv/aids and other pandemics

We Members of the WTO, recognize that access to medicines for treatment of HIV/AIDS and other pandemics, such as malaria and tuberculosis, especially by the poorest populations of the globe, is one of the major challenges for the global community and for its sustainable development;

> *recognize* that an effective response to this challenge requires a mix of complementary social, economic, health policies and practices, including education and prevention programmes;
>
> *recognize* that it is, therefore, the common responsibility of international organizations, governments, non-governmental organizations and private actors, through their areas of

responsibility, to contribute to the promotion of the most favourable conditions for improving access to medicines for treatment of HIV/AIDS and other pandemics;

recognize that among the determinant factors for improving access to medicines are efficient infrastructure to distribute, deliver and monitor drug usage and provide necessary information and education; increased research and development particularly targeted at the major communicable diseases of relevance for developing countries; mechanisms to finance drug purchases, and affordable pharmaceuticals; and the implementation of effective and sustainable healthcare systems;

recognize that strong, effective and balanced protection for intellectual property is a necessary incentive for research and development of life-saving drugs and, therefore, recognize that intellectual property contributes to public health objectives globally.

Therefore, we Members of the WTO reaffirm that the TRIPS Agreement contributes to the availability of medicines and reaffirm our commitment to the TRIPS Agreement and its implementation;

reaffirm the appropriateness of Members using the flexibility afforded by the Agreement to ensure that medicines for treatment of HIV/AIDS and other pandemics are available to their citizens who need them, particularly those who are unable to afford basic medical care;

take note of discussions held by the Council for TRIPS that have clarified Member's views of the flexibility provided under the Agreement;

encourage Members, whatever the exhaustion regime that they may have chosen, to take measures to prevent pharmaceuticals provided to the poorest populations of the globe under discounted pricing schemes or supplied under aid-schemes from being diverted from those for whom they were destined to markets for which they were not intended;

pledge, in the context of the new global fund, to work with the private sector and with affected countries to facilitate the broadest possible provision of drugs in an affordable, medically effective, and WTO-consistent manner; and

reaffirm the importance for developing and least-developed country Members of technical assistance for implementing their obligations under the TRIPS Agreement, while taking into account health concerns.

Clarification language

(To be provided.)*

* *Editor's Note*: No additional language provided.

Appendix D: Draft Ministerial Declaration

Proposal from a group of developing countries.

IP/C/W/312
WT/GC/W/450
4 October 2001
(01-4803)
General Council
Council for Trade-Related Aspects of Intellectual Property Rights

Proposal by the African Group, Bangladesh, Barbados, Bolivia, Brazil, Cuba, Dominican Republic, Ecuador, Haiti, Honduras, India, Indonesia, Jamaica, Pakistan, Paraguay, Philippines, Peru, Sri Lanka, Thailand and Venezuela

During the special discussion of the TRIPS Council on 19 September 2001, the following proposal was communicated to the Secretariat for circulation among Members of the Council by Zimbabwe on behalf of the above-mentioned delegations. When submitting the text, the delegations in question indicated that this was without prejudice to individual country positions and their right to submit additional proposals.

Ministerial declaration on the TRIPS agreement and public health

Ministers,

affirming that the protection and promotion of public health and nutrition is a fundamental obligation and prerogative of the State and that Members retain their sovereign power in this regard;

realizing that the inability of large segments of the population to obtain medicines and treatment at prices they can afford threatens the vital interest of States in protecting and promoting public welfare, preserving law and order, and maintaining social cohesion;

discharging the obligation to protect and promote the fundamental human rights to life and the enjoyment of the highest attainable standard of physical and mental health, including the prevention, treatment and control of epidemic, endemic, occupational and other diseases and the creation of conditions which would assure to all medical service and medical attention in the event of sickness, as affirmed in the International Covenant on Economic, Social and Cultural Rights;

cognizant of the concerns expressed by non-governmental organizations, public health advocates and the worldwide public regarding potential implications of the Agreement on Trade-Related Aspects of Intellectual Property Rights (the TRIPS Agreement) on the availability and affordability of needed medicines and other healthcare products;

concerned about the lack of adequate research and development on medicines for the prevention and treatment of diseases predominantly affecting people in developing and least-developed countries;

emphasizing that the protection of intellectual property rights, in particular patent protection, should encourage the development of new medicines and the international transfer of and access to technology to promote the development and maintenance of sustainable domestic manufacturing capacities for medicines and other healthcare products;

recognizing that in implementing domestic health policies, especially as regards the availability and affordability of medicines and other healthcare products, both the research-based and the generics pharmaceutical industries have important and complementary roles to perform, particularly in developing and least-developed countries;

stressing the importance of the participation of public health officials in discussions and decision-making on intellectual property rules that may have an effect on the availability of and access to healthcare products;

recalling the Preamble of the TRIPS Agreement, which, among others, prescribes that measures and procedures to enforce intellectual property rights should not themselves become barriers to legitimate trade and recognizes the special needs of the least-developed country Members in respect of maximum flexibility in the domestic implementation of laws and regulations in order to enable them to create a sound and viable technological base;

recalling further Article XI:2 of the Marrakesh Agreement Establishing the World Trade Organization and the Decision on Measures in Favour of Least-Developed Countries adopted on 15 December 1993;

reaffirming the General Council decision of 7-8 February 2000 (WT/GC/M/53) that the mandated review of the TRIPS Agreement, among others, should address the impact of the agreement on the trade and development prospects of developing countries;

acknowledging the vulnerability of developing and least-developed country Members to the imposition or the threat of imposition of sanctions and to the prospect of being deprived of incentives or other benefits, including those imposed or offered, as the case may be, beyond the framework of the WTO;

recognizing that challenges within the WTO dispute settlement system may in themselves inhibit or curtail the ability of Members to formulate and implement measures to protect and promote public health;

noting the ongoing examination by the Council for TRIPS on the scope and modalities for the possible application of subparagraphs 1(b) and 1(c) of Article XXIII of GATT 1994 to the settlement of disputes under the TRIPS Agreement;

recognizing that public health crises of unprecedented consequences, of which HIV/AIDS is a most dramatic example, afflict developing countries;

anticipating that drawing attention to and reaffirming the context of the TRIPS Agreement and certain provisions thereof as an initial concrete step will further encourage Members, particularly developing and least-developed country Members, towards considering every possible policy option for the protection and promotion of public health;

emphasizing the fundamental importance of the objectives and principles of the TRIPS Agreement.

Ministers declare that:

1. Nothing in the TRIPS Agreement shall prevent Members from taking measures to protect public health.
2. Each Member retains the right to establish its own policy and rules regarding the exhaustion of intellectual property rights.
3. Each Member has the right to allow other use[1] of the subject-matter of a patent without the authorization of the right holder, including use by the government or third parties authorized by the government, and to determine the grounds upon which such use is allowed.
4. In the case of a national emergency or other circumstances of extreme urgency or in cases of public non-commercial use, Members may grant compulsory licences without prior efforts on the part of the user to obtain authorization from the right holder.
5. A compulsory licence issued by a Member may be given effect by another Member. Such other Member may authorize a supplier within its territory to make and export the product covered by the licence predominantly for the supply of the domestic market of the Member granting the licence. Production and export under these conditions do not infringe the rights of the patent holder.
6. Members are not obliged to apply the conditions set forth in subparagraphs (b) and (f) of Article 31 of the TRIPS Agreement where use of the subject-matter of a patent is permitted to remedy a practice determined after judicial or administrative process to be anti-competitive.
7. Nothing in the TRIPS Agreement shall prevent Members from establishing or maintaining marketing approval procedures for generic medicines and other healthcare products, or applying summary or abbreviated marketing approval procedures based on marketing approvals granted earlier for equivalent products.
8. Nothing in the TRIPS Agreement shall prevent Members from disclosing or using information held by its authorities or the patent holder where it is so required for

[1] "Other use" refers to use other than that allowed under Article 30 of the TRIPS Agreement

reasons of public interest, including where such disclosure or use is necessary to implement effectively any compulsory licences or other measures adopted by public authorities in the public interest.

9. Under Article 30 of the TRIPS Agreement, Members may, among others, authorize the production and export of medicines by persons other than holders of patents on those medicines to address public health needs in importing Members.
10. Each Member shall, within or beyond the framework of the WTO, refrain from imposing or threatening to impose sanctions and refrain from employing the grant of incentives or other benefits in a manner which could curtail the ability of developing and least-developed country Members to avail themselves of every possible policy option to protect and promote public health.
11. Members shall exercise utmost restraint in initiating and pursuing dispute settlement proceedings relating to measures adopted or implemented, particularly by developing and least-developed country Members, to protect and promote public health.
12. In its examination of the scope and modalities for the possible application of subparagraphs 1(b) and 1(c) of Article XXIII of GATT 1994 to the settlement of disputes under the TRIPS Agreement, and without prejudice to recommendations that the Council for TRIPS may adopt and submit to the Ministerial Conference on other relevant aspects, in no event shall such subparagraphs be rendered applicable to measures adopted and implemented by Members, particularly developing and least-developed country Members, to protect and promote public health.
13. In view of the special needs and requirements of developing and least-developed country Members, their economic, financial and administrative constraints, and their need for flexibility to create a viable technological base, the transition period provided for their benefit under Articles 65.4 and 66.1 of the TRIPS Agreement shall be extended for another period of five (5) years from the expiration of the transition periods provided thereunder, particularly in respect of the obligation to render available patent protection on products or processes relating to public health, without prejudice to further extensions.
14. The TRIPS Council shall monitor and evaluate on an ongoing basis, in collaboration with relevant international organizations, the effects of the TRIPS Agreement on health, with particular emphasis on access to medicines and research and development on medicines for the prevention and treatment of diseases predominantly affecting people in developing and least-developed countries.

Appendix E: Glossary of Terms

ACTA: This stands for "Anti-Counterfeiting Trade Agreement," which is currently under negotiation by a number of countries including the United States as well as the EU. The focus of this agreement is on enforcement of intellectual property rights and would establish *TRIPS-Plus* standards for member countries. The agreement may permit member countries to suspend in-transit goods for alleged patent infringement and also pose other barriers to global trade in generic drugs.

ARTICLE 31BIS: The title of a proposed new article of TRIPS once a sufficient number of members agree to adopt it as a formal amendment. In the meantime, the substance of article 31bis is still in effect through an official *Waiver* of the TRIPS domestic use obligation. In essence, the usual TRIPS requirement that compulsory licenses be for the domestic market is waived—but only in limited circumstances and if many additional procedural requirements are met beyond the basic ones that apply to all compulsory licenses.

ARV: This is the commonly used abbreviation for an antiretroviral drug used to treat HIV and AIDS. There are different types of ARVs, with fixed dose combinations (FDC) being the preferred treatment for patients.

BIOEQUIVALENCE: This is the standard that a *generic* drug must establish before it can be approved as a generic to a previously approved drug. Technically, this term refers to the fact that the proposed generic drug results in the same blood concentrations of the active ingredient in the previously approved drug when the same dosage is used.

BIOLOGICS: A new class of treatments created by living cells as opposed to chemicals. They are currently a small percentage of the overall drug market but are becoming increasingly important.

BIOSIMILAR: A drug that is not bioequivalent, but is similar to a biologic compound such that it is an alternative, but not a therapeutic substitute.

BLOCKBUSTER: A drug that has annual sales of at least US $1 billion.

BLOCKING PATENT: A patent that blocks the use of subsequent patents. Although subsequent patents may be granted for inventions that meet the patentability standard, they may nonetheless violate another patent if a prior patent is broad in scope. Some nations permit compulsory licenses to provide for use of the later patent despite the blocking patent.

BOLAR (REGULATORY REVIEW) EXCEPTION: An exception to patent rights that permits a third party to make the patented product during the patent term without authorization of the patent owner for the purpose of obtaining approval to sell the product (as a generic drug) after patent expiry. This is a common exception in patent laws.

COMPULSORY LICENSE: A government-imposed license on a patent that restricts the patent owner's usual ability to exclude all others from the invention, subject only to providing the patent owner a government-determined royalty for the use. Compulsory licenses are often suggested as a tool to provide lower-cost drugs. Compulsory license may be granted pursuant to TRIPS requirements. Nonetheless, use of compulsory licenses is highly controversial.

COUNTERFEIT: This is a highly contentious term. Under trademark laws, a counterfeit is an article that improperly uses the trademark of another, such as a "knock-off" product. More generally, the term is used to refer to illegal products, including drugs, that are sold without prior approval. However, the term is sometimes confusingly (and inappropriately) used to also refer to generic drugs, which are in fact approved for sale by national agencies.

DATA PROTECTION: This term is confusingly used interchangeably with *data exclusivity*. However, as used in this book, data protection refers to the minimal obligation to protect clinical data consistent TRIPS article 39.3.

DATA EXCLUSIVITY: A rule in some countries that bars a generic applicant from relying on the clinical data of a previously approved drug to help establish that its proposed generic is as safe and effective as the original. Essentially, in countries that provide data exclusivity, the entry of generic drugs is delayed for the period of data exclusivity.

DEPENDENT PATENT: A patent whose use depends on obtaining authorization regarding a prior patent that otherwise "blocks" the newer patent's legitimate use. Some nations will consider granting a compulsory license to enable the owner of a dependent patent to use its invention.

DIFFERENTIAL PRICING: The practice of charging different prices for the same product in different markets. In the context of drugs, the preferred business model of large pharmaceutical companies is to charge the maximum price in each market to maximize profits, while also segregating the markets.

DISPUTE SETTLEMENT UNDERSTANDING (DSU): This provides a quasi-judicial process by which *TRIPS* (and other WTO agreements) can be challenged for noncompliance. Member states can bring a dispute that will be heard by a WTO panel and culminate in a panel report—if the parties do not come to a mutually agreeable decision.

DOHA PUBLIC HEALTH DECLARATION: An important declaration concluded at the 2001 Doha Ministerial Conference that should be considered part of the relevant interpretative context for TRIPS. This declaration clarifies the TRIPS requirements for *compulsory licenses* and also clarifies that member states have the right to determine their desired international exhaustion regime. The declaration also contains language that some believe can support broad exceptions to patent rights in the interest of promoting access to medicine. The full text of the declaration, as well as two proposals are reprinted as Appendices.

DOMESTIC EXHAUSTION: The first unrestricted and authorized sale by a patent owner (or its authorized licensee) in the patent-granting country "exhausts" the owner's rights in further use or distribution of the patented product in that country. This is a generally recognized principle.

EFFECTIVE PATENT TERM: The time that a patented drug is both patented and approved for sale; generally, this period is shorter than the actual patent term since the regulatory approval process for a new drug usually happens after a patent is granted.

EVERGREENING: A term used to refer to the common practice of multinational companies of developing incremental variations of existing drugs. The term is usually used to indicate criticism of such companies in keeping their patents "ever green" with respect to providing continued profits. The drug companies refer to the same practice as *life cycle management.*

EXCLUSIVE MARKETING RIGHTS (EMR): Patent-like rights enabling owners to exclude commercial uses of an invention; developing countries were required to provide these in some instances if they took advantage of the TRIPS transitional period before providing full patent rights for drugs.

EXHAUSTION: See Domestic Exhaustion; International Exhaustion.

FREE TRADE AGREEMENT (FTA): A bilateral or multilateral agreement that aims to promote greater trade between its signatories. Most of the FTAs discussed in this book also require intellectual property standards, such as patents that impose higher requirements than TRIPS. Thus, many FTAs are referred to as *TRIPS-Plus.*

GATT: This stands for the General Agreement of Trade and Tariffs—the international agreement governing global trade before the WTO. GATT did not require any members to provide intellectual property protection.

GENERIC DRUG: A drug that is *bioequivalent* to a previously approved drug. A drug deemed generic by a national agency is considered equally safe and effective as its bioequivalent. Generic drugs are often inexpensive because they are generally no longer patented. The term is sometimes used more loosely to refer to an unpatented drug.

GRAY (MARKET) GOOD: This is synonymous with a *parallel import good.* In essence, the good was first sold by the patent owner (or under its authorization) in another country, but its importation is not authorized by the patent owner. Such goods are considered undesirable by patent owners because they are typically sold for less than the patent owner's desired price in a given country.

INTERNATIONAL EXHAUSTION: A legal principle that considers the first global sale of a patented product to "exhaust" the patent owner's rights regarding subsequent global distribution of that product. If a country recognizes international exhaustion, a patent owner has no legal recourse under patent laws to block importation of patented products it first sold in another country. If, however, a country does *not* recognize international exhaustion, the same importation would violate the patent owner's rights and could be stopped as an infringing import.

INDUSTRIAL APPLICATION: This is a requirement for patentability in most countries (although the US uses the roughly similar standard of being *useful* for a credible purpose). Essentially, an invention must be capable of being used in an industry. Drugs easily meet this standard.

INFRINGEMENT: see Patent *Infringement*

INJUNCTION: A legal remedy that a patent owner may obtain (if the court grants it) to prevent violation of its patent rights. An injunction may be *preliminary* if it is granted before trial, or *permanent* if it is granted after a trial that finds the accused entity to have violated patent rights. In either case, a court often considers equitable factors, which could include the impact of an injunction on access to affordable medicine.

INVENTION: This term is used in TRIPS, but there is no definition of what constitutes an invention. Many would consider a drug to be an invention, although India considers slight modifications of prior drugs to not necessarily qualify.

INVENTIVE STEP: This is a requirement for patentability; it attempts to ensure that routine inventions that could have easily been discovered are not patentable. An alternative requirement is that the invention be *nonobvious*.

IP: This stands for "intellectual property", which includes, but is not limited to patents. Other types of intellectual property include trademark (brands), as well as copyright (protection over creative expression).

LEAST DEVELOPED COUNTRY (LDC): A country that is considered among the poorest according to UN classifications. This designation is important under *TRIPS* because LDCs are the only WTO member countries that do not yet have to provide patent protection on drugs.

LIFE CYCLE MANAGEMENT: A term that the pharmaceutical industry (and their sympathizers) use to refer to the same practice that critics call *evergreening*. In essence, the idea is to maximize profits by the sequential filing of patents on a drug to prolong the period of exclusivity and profitability. All the steps taken are consistent with existing laws, such that the industry considers it appropriate management of the "life cycle" of a drug.

LOCAL WORKING REQUIREMENT: A requirement in many national laws that the patent owner make the patented product locally to promote the transfer of technology. The classic definition of local working does not consider importation of the patented drug to suffice. Historically, a patent owner that failed to work a patent locally would be subject to a *compulsory license* or even complete forfeiture; a compulsory license is considered a less onerous imposition than forfeiture.

MAILBOX APPLICATION: This is an unofficial term frequently used to refer to a TRIPS requirement for developing countries. The country must provide a method of receiving applications for drug and drug compounds during the TRIPS transitional period. Importantly, when eventually examined, the original date that the application was submitted is used as the relevant date in comparing the invention to *prior art*.

MARKETING APPROVAL: National approval to sell a drug; it can only be granted after a national agency determines the drug to be safe and effective. This is alternatively referred to as *regulatory approval*.

MINISTERIAL CONFERENCE: Officially, this refers to the highest-level decision-making body in the WTO; it is comprised of all WTO member states. The Ministerial Conference convenes every two years in a new location to address any matter under a WTO agreement. Declarations are decisions that are sometimes made at the Ministerial Conference, such as the *Doha Public Health Declaration*.

MSF: This stands for *Médecins Sans Frontières* (also referred to as Doctors Without Borders). MSF is a nonprofit humanitarian organization that helps provide medical aid worldwide.

Since 1999, MSF has played an active role in international advocacy for laws that would promote greater access to low-cost drugs.

NCE: This stands for "New Chemical Entity" and is often used interchangeably with NME. However, the single TRIPS provision that deals explicitly with drug approval uses this term, rather than NME.

NEW: A requirement of patentability that must be evaluated with respect to what existed before the invention in question (more commonly referred to as *prior art*). This is sometimes alternatively referred to as the *novelty* requirement.

NME: This stands for "New Molecular Entity", which is a new compound that has not been previously approved for sale by the pertinent national agency. This is often used interchangeably with NCE.

NONOBVIOUS: A requirement of patentability; this term is used in the United States and is considered roughly equivalent to the more widely used term *inventive step*.

OPPOSITION PROCEEDING: A proceeding that enables third parties to challenge either a patent application or a patent an opposition of a patent is referred to as *pre-grant* because the patent is not yet granted whereas an opposition to an issued patent is considered a *post-grant* opposition. Usually, the third parties assert that the patent application or patent fails to meet one or more of the patentability requirements. Oppositions are permitted, but not required under TRIPS and some TRIPS-plus agreements limit their scope.

ORIGINATOR: This term is used in contrast to a generic. For example, the first company to develop a drug is an originator company.

PARALLEL IMPORT: The unauthorized importation of a patented product first sold by the patent owner in another country. Whether this importation violates domestic patent laws (of the importing country) depends on whether the country recognizes or rejects the principle of *international exhaustion*.

PARIS CONVENTION: An international agreement on patents that preceded TRIPS. Under this agreement, members were free to grant or not grant patents—so long as they did not provide preferential treatment to their own citizens. In addition, although the Paris Convention did not require patents, it did provide some rules for countries that provide patents. The compulsory license provision of the Paris Convention is incorporated into TRIPS and thus binding on WTO countries.

PATENT: A legal document issued by a country that generally provides the right to exclude others from the patented invention within that country. A patent is only granted if a filed application is found to meet the requirements of patentability (i.e., that the invention has industrial applicability, is new, and has an inventive step).

PATENT INFRINGEMENT: Violation of patent rights. For a patented drug, patent infringement will occur if the patented drug is made, sold, offered for sale, or imported (subject to international exhaustion) without authorization from the patent owner. This is subject to any domestic exception, such as the *Bolar (Regulatory) Review and International Exhaustion*.

PATENT LINKAGE: A controversial requirement in some nations' laws that limits approval of generic drugs when one or more patents might be infringed by the proposed generic. This law is generally enforced by a regulatory agency responsible for approving drugs. The agency does not typically conduct an independent investigation of patent issues, but instead relies on assertions from interested patent owners.

PATENT POOL: This is essentially a clearinghouse that contains patents volunteered by their owners; the pool may expedite licensing of patents as it can serve as "one-stop shopping" thus permitting potential licensees to avoid contacting each individual patent owner. The patent pool negotiates royalty rates to be paid to the patent owners and transfers money from the licensees to the patent owners. A patent pool may be used for any type of subject matter, but the pool's existence and utility depends upon the cooperation of patent owners.

PATENT TERM: The length of time the patent is in effect; after the patent term ends, the patented invention is free for anyone to use without liability. TRIPS provides that the patent term ends twenty years from the filing date of the corollary patent application. However, because TRIPS does not specify when the patent term begins and most nations do not provide any rights while a patent application is pending, the term of a patent is usually less than twenty years

PATENT TERM EXTENSION: A lengthening of the standard patent term that is provided in some countries for inventions that undergo lengthy patent examination or marketing approval. This is a common requirement in *FTAs*.

PATENTABLE SUBJECT MATTER: This is the largest possible scope of what may be considered for patentability. However, even an invention that is patentable subject matter may be denied a patent if it fails other requirements, such as that it be *new*, or have an *inventive step*.

PCT: This stands for Patent Cooperation Treaty. The PCT provides a streamlined process for applying for patents in multiple countries. WTO members are not required to follow the PCT, but some *FTAs* do require this.

PHRMA: This stands for Pharmaceutical Research and Manufacturers of America a US-based group that lobbies on behalf of large pharmaceutical companies. However, it is often used to refer to multinational companies that make and sell drugs the international arena this book follows this convention.

PIONEER: A synonym for *originator.* A pioneer company is one that creates new, as opposed to generic, drugs. Similarly, a pioneer drug is a new (not generic) drug.

PRIOR ART: This is an important term in assessing whether an invention is patentable; essentially, public knowledge that existed before the applicant's invention date (including, but not limited to patents, published patent applications, and scientific articles) is considered prior art. Prior art is used to evaluate whether an invention is new and has an *inventive step*.

PRIORITY FOREIGN COUNTRY: Designation on the annual Special 301 Report that is created by the USTR. This is the top designation and refers to a country with the most onerous and egregious policies. Countries with this designation are most vulnerable for imposition of unilateral trade sanctions from the United States.

PRIORITY WATCH LIST: Designation on the annual Special 301 Report that is created by the USTR. This is the middle level designation and refers to countries the USTR considers to inadequately protect intellectual property rights.

REGULATORY APPROVAL: A synonym for marketing approval (mandatory requirement to sell a drug) as regulations often apply to requirements to obtain such approval.

REVERSE ENGINEERING: The process of figuring out how something was made or done, starting only with the end product. In the context of drugs, a generic company may need to reverse engineer how to make the product for commercial production as that information would not be disclosed in public documents such as patents (which usually only discuss production of limited amounts of material).

REVOCATION PROCEEDING: A proceeding in which a patent may be revoked, and thus no longer exist.

SUPPLEMENTAL PROTECTION CERTIFICATE (SPC): Patent-like protection granted in European countries after the patent expires. A SPC essentially provides the same protection as an extension of the patent term. The SPC is generally available in European countries that do not provide patent term extension.

SPECIAL 301 REPORT: An annual report prepared by the USTR pursuant to US law that requires a report of countries whose protection of US intellectual property owners' rights is inadequate or ineffective. A country that is listed on this report may be subject to withdrawal of trade benefits or other trade sanctions if it does not modify its laws in accordance with the US request. Importantly, a country can be listed even if it is in full compliance with TRIPS. There are different categories of countries on this list, ranging from watch list, priority watch list, and priority foreign country, with *priority foreign country* being used to refer to a country whose laws are most troubling.

TRADEMARK: Anything that is used in commerce to identify its source, such as a brand name (e.g., Coca-Cola) or the name of a drug (e.g., Lipitor). Typically a trademark is a word, phrase, or logo, but can also include color and packaging.

TRADE SECRET: Confidential and proprietary information that is valuable if it is kept "reasonably" secret. Unlike patent protection, trade secret protection does not provide a remedy against those who independently discover the information. However, trade secret protection can potentially last indefinitely—unless the secret is revealed. This is generally considered poor protection for drug compositions (which can be reverse engineered), but may be useful for processes of making drugs and especially *biologics* if the process cannot be detected from the end product.

TRANSITIONAL PROVISIONS: Provisions that applied to developing countries who were original members of TRIPS. Under these provisions, developing countries that did not previously provide patents on drugs and drug compounds were entitled to a five-year transitional period before having to do so. However, during the interim, developing countries were required to provide a *mailbox application*, as well as *exclusive marketing rights*.

TRIPS: This stands for the "Agreement on Trade Related Aspects of Intellectual Property Rights," although the full name is rarely used. TRIPS is an international agreement that set forth the first minimum standards of intellectual property protection, including patent law; this agreement requires all WTO members to grant patents, including patents on drugs and drug compounds.

TRIPS COUNCIL: A body created pursuant to TRIPS that is intended to ensure countries comply with TRIPS and provide assistance to members to resolve issues outside the formal Dispute Settlement Process.

TRIPS-PLUS: This is an informal term commonly used to refer to any requirement that goes beyond what is required in TRIPS.

UNITAID: This organization helps purchase drugs for poor countries and has recently proposed an HIV *patent pool* to promote greater access to lower-cost treatments for developing countries.

USEFUL: This is a patentability requirement that is roughly synonymous with *industrial application*. This term is used in the United States.

USTR: This stands for the Office of the "United States Trade Representative." The USTR is a government agency that is responsible for overseeing and negotiating US trade interests, which have historically reflected the interests of major corporations. By law, the USTR is required to issue an annual report called the *Special 301 Report*, which lists countries that have inadequate intellectual property standards.

VIENNA CONVENTION: International agreement that is used for customary interpretation of international treaties, such as TRIPS.

WAIVER: In this book, this term is used to refer to a complicated waiver of the TRIPS requirement that compulsory licenses be predominantly for domestic use, in limited circumstances and pursuant to many procedural requirements. This has been called different things, including the "paragraph 6 solution" (referring to the paragraph in the Doha Public Health Declaration that requests the TRIPS Council to find a solution to the problem of countries with inadequate infrastructure to use a compulsory license). It has also been called the "August 30 solution" in reference to the August 30 date when the TRIPS Council provided a solution. See *Article 31bis.*

WATCH LIST: Designation on the annual *Special 301 Report* that is created by the USTR. This is the lowest of three possible designations. A country on the watch list is subject to greater scrutiny by the United States, but no formal proceedings need occur. However, countries on the Watch List may nonetheless be subject to pressure to adopt a free trade agreement with the United States that contains intellectual property provisions that "remedy" perceived deficiencies.

WIPO: This stands for "World Intellectual Property Organization." WIPO is a specialized UN agency that administers many international agreements related to intellectual property. Before TRIPS, the only existing international agreements related to patents were WIPO treaties—the Patent Cooperation Treaty (PCT) and the Paris Convention.

WTO: This stands for "World Trade Organization." The WTO is an organization that was created by a 1994 WTO Agreement with the overall goal of liberalizing global trade. The WTO supervises a number of related agreements, some of which have a less direct connection to liberalizing trade, such as patent standards under TRIPS. The WTO has a broad membership of countries from different levels of economic development that overall represent most of the world's trade.

WTO APPELLATE BODY: A standing three-member body that reviews decisions of WTO panels. Its decision is binding on the parties before the WTO Body, and its decisions are often considered persuasive in subsequent WTO disputes.

WTO REQUEST FOR CONSULTATION: This is the first step toward initiating a proceeding under the WTO Dispute Settlement Process. However, the request may not result in a formal proceeding (that culminates in a panel report) if member states are able to come to a mutually agreeable solution.

WTO PANEL REPORT: A report of a three-person panel after the conclusion of a quasi-judicial process that considers input from countries directly involved in the dispute, as well as other countries with an interest. The Report is automatically adopted unless one of the parties requests an appeal to the WTO Appellate Body. Parties are given a short time frame to comply with a report (or decision of the Appellate Body). Failure to do so may result in sanctions pursuant to the *DSU.*

Index

Abbott, Frederick M., 318–19
Abbott Laboratories, 19, 44, 178, 366
  decision to withdraw drugs from Thai market, 178
  Kaletra and, 150, 154, 158*n*1, 168–69, 171
Abbreviated new drug application procedure (ANDA), 260
Access to Medicine Campaign (1999), 332
Active ingredient, 10–12, 17–18, 19, 25, 30, 93, 104, 116, 233–34
Adelman, Martin, 164
Advertising, of prescription drugs, 32, 249
Agreement on Trade Related Aspects of Intellectual Property Rights. *See* TRIPS
"Aide Memoire" (from Switzerland), 169
American Enterprise International (AEI), 181
Amoxicillin, 300
Anthrax, 211
Anti-Counterfeiting Trade Agreement (ACTA), 288
Apo-Triavir, 214–15
Apotex, 214–17
Appellate Body, WTO, 58, 58*n*8, 68, 79
Astra Zeneca, 32
AZT, 11, 233

Baker, Brook, 322
Bate, Roger, 181
Bayer Inc v. Canada, 77*n*55
Bayh-Dole Act, 1980, 8*n*13
Bioequivalence, 13, 13*n*23, 15–16, 15*n*31, 24, 150, 256, 259, 358
  approval based on, 150
Bioequivalent drug, 26, 256
Biologics, 4, 357–58, 358*n*10
Biosimilar, 358
Black market, 43
Blockbuster drug, 27–29, 31, 368–69
Blocking patent, 139–40, 141, 379
Boehringer Ingelheim, 215
Bolar (regulatory review) exception, 72, 260
Boyce, Ralph, 169
Brand, 16, 20–21. *See also* Trademark
Brazil
  in-transit goods position, 318–21
  local working requirement, 333–35
  pipeline protection, 229–30
Bristol Myers Squibb (BMS), 267, 333
Brooktree Corp. v. Advanced Micro Devices, Inc., 103*n*36
Brown v. Duschene, 323*n*145

Canada, 203
  FTAs, 226
  Patent linkage, 281
  use of Waiver, 212–15
  WTO Panel Report, Patent Protection of Pharmaceutical Products, 71–74, 95, 106*nn*44, 46, 162*n*33, 236*n*28, 310*n*102
  WTO Panel Report, Term of Patent Protection, 68

Canada Patent Act, 22*n*44, 103*n*35, 212*n*30
compulsory licenses for drugs, 130, 153. *See also* Waiver
regulatory review provision, 106, 106*n*45
stockpiling exception, 72–73
Cancer drugs, 17, 29
EU seizure of, 298–99
Thailand's compulsory licenses, 169–70, 170*n*48
Ciprofloxacin (Cipro), 173, 211
Clinical Data (of new drugs)
Creation of, 7, 259, 264
Protection. *See also* Data exclusivity; Data protection
Reliance on (by generic companies), 15, 256, 259–61, 267. *See also* Bioequivalence
Submission of (to regulatory authorities), 256
Clinical testing, 5–6, 8 (Fig 1), 9, 25, 31, 370–71
Clinton, Bill, 172, 332
Clinton Foundation, 172
Combivir, 366*n*31
Compulsory licenses, 70–71. *See also* India Patent Act, Thailand's compulsory licenses, South African controversy
anticompetitive use (remedy for), 140
to Apotex, 214–17
automatic, 115, 130
blanket, 180
blocking patents, 132–33, 139–40
Canada (termination of), 153
challenging of, 141
dependent patents. *See* Compulsory licenses – blocking patents
domestic use (TRIPS requirement), 135. *See also* Waiver
duration, 136, 145–46
economic implications, 151–52
emergency (national), 132*n*18, 134–35, 137*n*26, 141, 155, 237, 337*n*54, 382, 388. *See also* Thailand compulsory licenses, India Patent Act – compulsory licenses; Waiver
failure to work. *See* Compulsory licenses – local working (lack of)
local working (lack of), 130–32, 333–34
grounds for, 130–33
innovation (impact on), 152–54, 186–87
and middle-income countries, 183
myths related to, 180–84
prior negotiation, 133–34, 182, 186–87
public noncommercial use (and TRIPS), 184–86
parallel imports, relation to, 190*n*147
Paris Convention, 132
patent owner rights and recourse, 137–38, 141
procedural requirements (under TRIPS), 133–36
remuneration (under TRIPS), 138
scope and duration (under TRIPS), 136–37, 180
termination (under TRIPS), 136
TRIPS requirements, 127–41, 179–80, 181
"TRIPS-Plus" limitations, 237–38
retaliation, 149–52
Contributory infringement, 104
Conventional drugs, 4, 357
Copyrights, 43, 61
Correa, Carlos, 317
Counterfeit trademark, 75
Cross-border sales, 51

Data exclusivity, 21, 25–26, 30, 255–56, 350
Rationales challenged, 263–68
vs cost sharing, 270–71
vs data protection, 118–19
and developing countries, 268–69
duration of, 257–58, 272
as FTA requirement, 242
minimizing harm from, 269–73
in EU Member States, 261
genesis of, 260–62
global, 262–73
overview, 256–60
and patented drugs, 258–60
PhRMA's view, 266–68
free riding rationale, 265–66
and TRIPS, 79–80, 346
in United States, 262
Data protection (under TRIPS)
obligations of member states, 80–82
term of, 82–83
Declaration on the TRIPS Agreement and Public Health. *See* Doha Public Health Declaration
Dependent patent, 139–40

Developing countries, generic drug application in, 15
approval based on bioequivalence, 150
Development process, of drugs. *See* Generic drug, New drugs
Differential pricing, 27
Disease mongering, 11*n*19
Dispute Settlement Understanding (DSU), 58, 60, 61, 151
Request for Consultation, 77*n*57, 287*n*1, 290*n*14, 302*n*78
Docetaxel, 170*n*48
Doctors without Borders. *See Médecins sans Frontiers*
Doha Paragraph Six Problem. *See* Waiver
Doha Public Health Declaration, 327, 382–83
and compulsory licenses, 129, 134–35, 135*n*22, 171–72, 175, 175*n*86, 179, 185, 187, 192
genesis (and proposals), 335–39, 384–89
and international exhaustion, 108
as interpretive tool, 60, 96, 179, 179*n*106. *See also* In-transit generic drugs (suspension of)
and LDC, 87, 87*n*77–78
post-Doha, 339–53
pre-Doha, 327–35
privilege view of patents, 336–37, 348–53
proposed language, 384–89
TRIPS-Plus obligations and, 341–43
uber-right view of patents, 337–38, 340–48
Waiver of domestic use requirement, 198, 340–41
Domestic exhaustion, of patent rights, 38–39, 44
Drug
commercialization, 5. *See also* Sales strategy
conventional versus biologic, 4, 357
Development cost, 7–8, 263
Development process, 6–13
generic. *See* Generic drug
legal protection of, 16–24
marketing approval process, 13–16
new. *See* New drugs
manufacturers, 5–6
sales. *See* Sales strategy
stages of development, 6
types, 4–5
Drug Development Corporation (DDC), 371–72
Drug Development process
abbreviated development, 10–12, 14–15
clinical testing of chemical compound, 9
identifying the chemical compound, 9
new combinations, 11
new dosage or delivery system for an existing drug, 10
new uses, 11
new variations, 11–12, 11*n*20
purer drug, 12*n*21
Drug Price Competition and Patent Term Restoration Act of 1984 (US), 260*n*11

Efavirenz, 168
Epstein, Richard, 182
Erlotinib, 111, 170*n*48
EU Member States, 85, 149, 203
Council Regulation 1383/2003, 289, 295, 298
data exclusivity in, 261
patent linkage (lack of), 278
EU-Korea FTA, 237*n*35
EU Regulation No. 1383/2003. *See also* In-Transit generic drugs (suspension of)
ACTA, 288
and counterfeit drugs, 287, 294, 299, 300
as barrier to trade in generic drugs, 290, 297–98
and customs, 288, 290, 297, 300
and detainment of drugs, 287, 289
and destruction of drugs, 287
EU view, 294–97
history, 289
and similar laws, 288, 301–2
overview, 288–89
patent infringement of in-transit goods, 289
patent owner view, 292–94
privilege view, 297–302
TRIPS (compliance with), 302–14
as TRIPS-Plus standard, 287
Evergreening, 10, 20, 30*n*65, 93, 96, 275
Exclusive marketing rights (EMRs), 85–87, 230
mailbox applications, 87
time period of, 86–87
Exhaustion of patent rights. *See* Domestic exhaustion, of patent rights; International exhaustion, of patent rights

F. Hoffmann-La Roche AG v. Cipla Ltd., 110–11, 111*n*68, 72
F. Hoffmann-La Roche AG v. Wockhardt Ltd, 102*n*31
Failure to work a patent. *See* Local working requirement
Federal Food, Drug, and Cosmetic Act 21 U.S.C. § 355 (b)(1) (2006), 14*n*28
Fixed dose combinations (FDCs), of HIV drugs, 359–61
Forum shifting, 227, 227*n*7, 325
Free trade agreement (FTA), 225, 258
  Access to medicine, impact on, 243–46
  bilateral or multilateral, 226
  Central America-Dominican Republic-United States, 236*n*31
  compulsory licenses (limits on), 237–38
  data exclusivity (requirement of), 242. *See also* Data exclusivity
  and developing countries, 226
  direct-to-consumer advertising of prescription medicine, 249
  drug costs (impact on), 225, 248
  EU-Korea, 237*n*35
  negotiations, 227–28
  patent revocation (limitations on), 239*n*41
  parallel import (limitations on), 240–41
  "Medicines Working Group," 247
  patent linkage (requirement), 242–43, 275. *See also* Patent linkage
  patent term extension, 235–37
  and price control (limits), 246–49
  and power disparity, 227
  and US modification of, 237, 244
  U.S.-Australia, 226, 240*n*47, 247, 247*n*63, 247*n*65, 249, 275*n*59
  U.S.-Chile, 242*n*51
  U.S.-Jordan, 225, 229, 236*n*29, 240*n*44, 249–50
  U.S.-Morocco, 239*n*41, 240*n*46
  U.S.-Singapore, 238n37
  U.S.-Singapore FTAs, 242*n*49–50
  U.S.-South Korea, 247
  U.S.–Pan. Trade Promotion Agreement, 233*n*24
  WHO recommendations, 351
Froehner, Juliano, 181, 188

General Agreement on Tariffs and Trade (GATT), 59, 59*n*12, 137*n*27, 302*n*79, 317*n*125, 387, 389
Generalized System of Preferences, 151
Generic drug, 5. *See also* Compulsory license, in-transit good (suspension of)
  barriers to regulatory approval of, 15–16, 25–26. *See also* Data exclusivity; Patent linkage
  and biologics (challenges to), 358
  vs brand, 20
  development process of, 12–16
  as "legitimate trade," 311
  manufacturers, 6, 32–34
  patent barriers, 24–25
  regulatory approval, 15–16. *See also* Bioequivalence
  sales strategy, 33–34
Gervais, Daniel, 131*n*14
Gilead, 366
Glaxo SmithKline, 5, 182
  price cuts, and healthcare investment, 364–66
Gleevec, 230, 347
Global Fund, 299
Gore, Al, 332
Gray market goods, 43–44
GWU India Project, 346–47

Hassan, Fred, 164, 164*n*20
Hatch-Waxman Act, 260, 277
Health Action International and Consumer Project for Technology, 331
Health Impact Fund, 367–69
HIV/AIDS treatment
  Abbott's HIV drug Kaletra, 150, 158*n*1, 168–69, 171
  Brazil, 334–35
  and Doha Public Health Declaration, 338
  fixed dose combinations (FDCs), 359, 361
  free licensing, 366–67
  patent pool for, 361
  in-transit goods (suspension of), 292
  South African controversy, on compulsory licensing, 332
  and Thailand's compulsory licenses, 150, 152, 158*n*1, 168, 170
  and TRIPS impact, 356–57
  and uber-right proposal, 364–66

Hong Kong, 203
HOPE for Africa Bill, 331*n*21

Iceland, 203
Imitinab, 170*n*48
Imports. *See* International exhaustion rights
In dubio mitius, principle of, 79
In-transit generic drugs (suspension of). *See also* EU Regulation No. 1383/2003
  counterfeit drugs, 294–95
  "country of importation" (determining), 307–8, 316*n*119
  and Doha Public Health Declaration, 313–14, 318–19
  domestic laws (misinterpreted), 322–23
  drugs suspended, 290
  Eli Lilly letter to Cipla, 294*n*23
  EU perspective, 287–88, 294–97
  fake medicines, 296
  and GATT, 302*n*79, 317*n*125
  and HIV treatments, 290, 292
  of Indian companies, 292–93, 295
  "legitimate trade," 309–10
  patent infringements of, 289, 304–6
  patent perspectives, relevance of, 291–302
  privilege view, 290–91, 297–302, 314–23
  territoriality, 310, 317, 319–20
  TRIPS article 7, 311–12, 319
  TRIPS article 8, 311–13, 319
  TRIPS article 41, 308–12
  TRIPS article 51, 303–6, 315–16
  TRIPS article 52, 304, 306–8, 316–17
  TRIPS consistency, 302–14
  uber-right (patent owner) view, 291–97
  Waiver (as interpretative tool), 307, 310–11
India, 203
  and data protection, 118–20
  and patent linkage, 120–21
  as generic drug supplier, 121–23, 357
  patent laws as model, 123
  Drugs and Cosmetics Rules, 1945, 119*n*88
India Patents Act
  § 3d, 92–5, 101, 111
  § 48, 104*n*39
  § 85, 109*n*62
  § 107A(a), 105*n*42
  §§ 64.1(j), 109*n*60
  § 64.1(p), 109*n*60
  § 107A(b), 107*n*49
  compulsory licenses, 106
    compensation, 146
    grounds for, 142–44
    and national Emergency, 145, 147*n*55, 148*n*61
    Patent owner rights, 146–47
    procedural requirements, to compulsory licenses, 144–46
  infringement, 104
  injunctions, 110–12
  and international exhaustion, 107–8
  inventive step, 96–97
  mailbox applications, 84*n*75, 112–18
  oppositions, 100–102
  patentability (limitations on), 91–99
  patent linkage (lack of), 120–21
  patent term, 104–5
  remedies (limited), 104–18
  revocation, 109–10
  and third parties, 100–103
  TRIPS compatibility, 94–96, 101–3, 106, 101*n*25, 107–8, 110, 119–20, 144
Industrial application, 17–18, 63–64, 63*n*27
Industrial Property Act (2001), Kenya, 107*n*48
Injunction, 42, 69, 74, 88, 110–12, 120, 283
  interim, 110
  interlocutory, 110
  permanent, 111, 165, 175*n*80, 188
  preliminary, 19, 75, 110–11, 273–74
Intellectual property rights, 43, 59
  norms, 59
  Special 301 report, 151
  TRIPS articles, 7–8, 311–12
International Covenant on Economic, Social and Cultural Rights (ICESCR), 164
International exhaustion, of patent rights, 67
  arguments against, 46–48
  arguments for, 45–46
  in Canada, 51
  commercial implications, 48–51
  consumers' point of view, 45, 46*n*6, 47–48
  countries recognizing, 48–51
  cross-border sales, 51
  definition, 37, 44
  distinction with domestic exhaustion, 40
  drug companies' point of view, 50
  in EU Member States, 48–49

International exhaustion, of patent rights (*cont'd*)
- and geographic price discrimination, 46
- importing country, relevance of laws, 40–41
- importation right (exception to), 37, 39–43
- India's patent laws, 107–8
- and pricing of drugs, 46–47
- related terms, 43–45
- safety issues, 47–48
- in United States, 48, 51
- unrestricted global sale, 49

Invention, 62–65, 87. *See also* India Patent Act
- exclusion, 92–96
- industrial application, 63–64
- new, 64
- not available at reasonably affordable price (as grounds for compulsory license), 143

Inventive step, for patentability, 18–19, 64–65, 96–97, 101, 232

Israel, 203

Japan, 60, 203
- average period of examination, 102*n*26
- industrial application, 63*n*27

Jordan, 225, 245, 249–50

Kaletra, 150, 158*n*1, 168–69

Kapczinski, Amy, 333

Kieff, Scott, 182

Kivexa, 366*n*31

Knock-offs, 304

Korea, 203

Kuwait, 203

Labeling, 43

Lamivudine, 215

Least developed countries (LDCs), 83–87, 122–23
- Ethiopia, 197
- Glaxo's sale profits, 366*n*30
- and use of Waiver, 202–4

Legal protection of Drugs
- patent, 16–19
- regulatory protection, 21
- trademarks, 20–21

Legitimate trade, 291, 309–10

Letrozole, 170*n*48

Life cycle management, 10, 30

Local working requirement, 130–32, 333–4. *See also* Compulsory licenses

Maco, 203

Mexico, 203

Mailbox application, 85, 85*n*75, 105, 230
- patents, 105, 112–18

Market exclusivity, 30

Marketing approval of new drugs, 13–15

Marrakesh Agreement Establishing the World Trade Organization, 84*n*73

Médecins Sans Frontiers (MSF), 172, 212, 215, 332–33

Medical Research and Development Treaty, 350

Medicare Modernization Act of 2003, 50*n*12

Medicines and Related Substances Control Amendment Act 90 of 1997 (South African), 328*n*5

Mendelssohn, Peter, 169

Merck, 366

Mercurio, Bryan, 309

Metabolite, 12*n*22

Mexico, and pipeline patent protection, 230

Military-governed countries and compulsory licenses, 183–84

Monetary damages, 69

Morton Salt Co. v. G.S. Suppiger Co., 144*n*44

Myths, on compulsory licensing
- Country limitation, 183–84
- emergency limitation, 181–83
- subject matter limitation, 180–83

National emergency and compulsory licensing. *See* Compulsory licenses; Thailand's compulsory licenses

National exhaustion. *See* Domestic exhaustion, of patent rights

Nevirapine, 215

New chemical entity (NCE), 7, 64, 81–82

New drugs, 5
- cost of, 7–8, 263
- development process of, 6–12
- intersection between patent and regulatory approval, 21–24
- marketing approval, 13–14
- sales strategy, 26–32

New molecular entity (NME), 7. *See also* NCE

New Zealand, 203

Nokia v. HMRC, 289*n*8
Noncommunicable conditions, 357
Nondiscrimination requirement, for patent rights (TRIPS), 73–74
Nonobvious, 17–19
Normal exploitation, of a patented drug, 72
North American Free Trade Act, 230*n*14
Norway, 203
Novartis, 169, 230, 347

Opposition proceedings, 99, 101, 234
Optical isomers (mirror images), 29
Optimization, 9*n*16
Originator, 21, 33, 77, 79, 80, 118, 255–57, 257*n*4, 260, 261, 263, 265–66, 269, 277, 279

Panama, 237, 244
Parallel imports, 43–44, 47, 240. *See also* International exhaustion
  adverse impacts, 46–48
  barred, 45, 49
  conflated as counterfeit drugs, 329
  definition, 44
  permitted, 49
Paris Convention for the Protection of Industrial Property, 58*n*5, 59, 131*n*12, 132, 132*n*16, 227, 245*n*59, 350
Patent, requirements to obtain. 17, 23. *See also* Patent application; India Patent Act; Patent rights
  inventive step, 17–18, 64–65, 96–97, 99
  industrial application, 17–18, 63–64
  nonobviousness, 17–19
  new/novelty, 17–18, 64, 99
  patentable subject matter, 17–18
  patentability analysis, 18 (Table 2), 62–65
  and prior art, 18, 99
  TRIPS, 62–65, 377
Patent application
  examination of, 19, 99–100
  position, 100–102
  prior art, comparison to, 19, 99–100
  timing, 21–22
  TRIPS, 65–66
Patent Cooperation Treaty (PCT), 228–29, 231–32
Patent infringement, 25, 26, 40, 75, 120, 193, 205, 243, 259–60, 268, 273–75, 277
  in-transit, 289–91
  under TRIPS, 304–10
Patent Law Treaty, 352
Patent linkage, 21, 118, 120, 242–43, 261
  in Australia, 273, 281, 281*n*74
  countries without, implications, 276
  vs data exclusivity, 282
  and FTAs, 278–79
  generic manufacturers' perspective, 274–75
  genesis of, 277–78
  in global arena, 278–81
  India, 120–21
  legal challenges to, 287–81
  mediating the harm of, 280–81
  overview, 273–76
  patent owner's perspective, 273–74
  vs patent enforcement, 273–74, 279
Patent perspectives theory. *See also* Privilege view of patents; Uber-right view of patents
  Brazil compulsory licensing controversy, impact on, 333–35
  challenges to, 189–93
  Doha Public Health Declaration development, impact on, 336–38
  and intellectual property theories (liability vs property), 160
  overview, 158–66, 290–91
  privilege perspective explained, 161–63
  social science literature, 166–67
  solutions, impact on, 349–51, 359–72
  South Africa compulsory licensing controversy, impact on, 327–30
  Thailand's compulsory licenses, case study, 167–79
  TRIPS compulsory license, impact on, interpretation, 179–89
  TRIPS border measures (and in-transit goods), impact on interpretation, 314–19
  uber-right perspective explained, 163–66
  and waiver, impact on, 340
Patent pool
  basics, 360–61
  vs. compulsory license (360)
  GSK, 364
  for HIV/AIDS treatment, 361
  impact on innovation, 363–64
  UNITAID's pool, patent perspectives on, 359–64

Patent rights. *See also* Patent rights – exceptions to, India patent act
commercial advantage, 16–17
and data exclusivity, 21, 25, 258–59, 283. *See also* Data exclusivity
enforcement of, 19, 74–76
exclusionary rights, 67
and generic drugs (impact on), 24–25
and international trade, 36–37
nondiscrimination requirement (under TRIPS), 63, 73–74
patent examination process, 65–66
and patent linkage, 21, 25, 283. *See also* Patent linkage
provisional enforcement, 75
term of. *See* Patent term
Patent rights – exceptions to
Bolar exception, 72
Compulsory licenses. *See* Compulsory licenses
"experimental use" of a patented invention, 73
international exhaustion of. *See* International exhaustion
legitimate interest (for TRIPS article 30), 72
limited exceptions (under TRIPS). *See* TRIPS article 30
"normal exploitation" of a patented drug (for TRIPS article 30), 72
stockpiling exception, 72–73
TRIPS art. 30, 70, 71, 73, 73*n*44, 114, 197*n*3, 219
TRIPS art. 31, 70, 114, 207
Patent term. *See also* Effective patent term
basic, 16
consistent with TRIPS, 104–5
effective, 22–24, 72, 235–36, 258, 259, 360
extensions, 235–37
in United States, 19, 23
Patents Act of 1970 (India), 103*n*32
Pender, Jon, 182
Peru, 237, 244
Pfizer, 5
Proposal to promote access and innovation, 347
Pharmaceutical Benefit Scheme (PBS), 246–47
Pharmaceutical Research and Manufacturers of America (PhRMA), 132, 266–67, 329, 347
and Thailand's compulsory licenses, 173–79
Pipeline Patent Protection, 229–32
Pioneer companies, 256, 260, 265–66
Pioneer drugs, 256, 277, 282
Pirated copyright goods, 75, 290
Plavix, 168, 177, 185–86
Pre-grant oppositions, 234–35
Preambular language, for ministerial declaration, 384–85
Prequalification of product, 13*n*27
Prilosec, 32
Prima facie infringement, 304, 381
Prior art, 18, 22–23, 64, 85, 87, 99–100, 112*n*75, 121, 123, 232–33
Prior negotiations. *See also* Compulsory licenses
prerequisite of, 133–35
and Thailand's compulsory licenses, 171, 178, 187–88
and Waiver, 211
Priority foreign country, 330
Priority Watch list, 169, 250*n*76, 330
Privilege view of patents, 159, 161–63
Doha Public Health Declaration, 336–37, 348–53
in-transit goods, 290–91, 297–302
and prior negotiations, 187
South African controversy, on compulsory licensing, 328
Thailand's compulsory licenses, 170–73
Promotional activity, 32
Proposed Treaty on Access to Knowledge (A2K), 349
Protection and Utilization of Public Funded Intellectual Property Bill, 2008, 8
Prozac, 10–11, 28, 29, 234
Public noncommercial use and compulsory licensing, 184–86, 191
Publicly funded clinical trial, 370–71
Punta Del Este Declaration, 59*n*13

Qatar, 203

Racemix mixture, 11*n*20, 12*n*21
Reference pricing, 26
Regional exhaustion, 48*n*9
Regulatory approval of drugs, 13–16, 105–6
as barrier to generic drug entry, 25–26
Regulatory protection of drugs, 255. *See also* Data exclusivity; Patent linkage

Reichman, Jerome, 79*n*64
Retroactive damages, 69
Reverse engineering, 150
Revocation proceedings
FTAs limit grounds for, 239
in India, 109
TRIPS, 110
Roche, 5
Ruse-Khan, Grosse, 57*n*4, 303*n*80, 317
Rwanda, use of Waiver, 214–17

Sales strategy
generic drugs, 33–34
new drugs, 26–32, 50–51
Sanofi -Aventis, 168
Schwab, Susan C., 181
Sept. 11, 2001 terrorist attacks, 339
Seuba, Professor, 316–17
Shanker, Daya, 315–16, 323
Singapore, 203
Sosecal v. Sisvel, 289*n*7
South African controversy, 191–92
confluence of perspectives, 330–33
and Gore's campaign, 332–33
and nongovernmental consumer groups, 331–32
privilege view of patents, 328
South African law on parallel imports and compulsory licensing, 327
uber-right view of patents, 328–30
Special 301 list, 330
UN Commission resolution, 331
USTR resolution, 330
WHO resolution, 331
South Korea, 203
Special 301 list, 169, 330
Special 301 report, 151, 151*n*69, 226, 330*n*17
Special Watch list, 151
Stockpiling exception, 72–73
Stocrin, 168
Sub-Saharan Africa
economy, 176
rates of HIV infection in, 176
Substandard drugs, 5
Substantive Patent Law Treaty (SPLT), 226*n*5
Supplementary Protection Certificate (SPC), 23–24
Switzerland, 169, 203
Sykes, Alan, 182
Taiwan, 203
Tarceva, 111
Targets, 9*n*15
Tax benefits, on new drugs, 8
Taxol, 267
Territoriality principle and in-transit goods, 317, 320
Thailand
access to medicine (universal), 167, 170, 176
cancer, 170
defense spending vs. health care, 174, 176
Facts and Evidence on the Ten Burning Issues Relating to Government Use of Patents in Thailand, 167, 167*n*35, 168
heart disease, 177
and HIV, 167–68, 176–77
military government, 173–74
second-line treatment ARVs, 171
Ten Burning Questions on the Government Use of Patents on Four Anti-Cancer Drugs in Thailand, 167n.35
Thai National Health Security Act, 167*n*36
and WHO, 168, 170
Thailand's compulsory licenses. *See also* Myths, on compulsory licensing
and Abbott's decision to withdraw drugs from Thai market, 178
background, 167–70
for cancer drugs, 167*n*35, 169–70, 170*n*48
and confusion regarding TRIPS requirements, 192–93
emergency, confusion regarding relevance, 180–83
criticism of, 169, 173–79
of effavirenz, 168
and EU Trade Commissioner pressure, 169
Government Pharmaceutical Organization (GPO), 176
of Kaletra (Abbott's HIV drug), 150, 158*n*1, 168–69, 171
national emergency *See* Thailand's compulsory licenses – emergency
as new market for pharmaceutical companies, 172
and negotiations with patent owners, 168, 171, 178, 187–88
of Plavix, 168, 177
Priority Watch status, 169
privilege perspective, 170–73

Thailand's compulsory licenses (*cont'd*)
support for, 168, 170–73
and TRIPS compliance, 171–72, 177, 180–89
uber-right perspective, 173–79, 192
US retaliation, 152, 170*n*51, 173, 174*n*80
US Special 301 report, 151
World Bank, 152, 168, 170
Trade-Related Aspects of Intellectual Property Rights (TRIPS). *See* TRIPS
Trade secret, 119, 260, 267, 358*n*10
Trademark, 16, 20, 43, 61, 296
Transatlantic Consumer Dialogue (TACD), 349
Transitional provisions, 84–87
exclusive marketing rights (EMRs), 85–87
mailbox provision, 85
TRIPS, 57*n*1
and anti-property activists, 175
Appellate Body, 58, 58*n*8
article 1, 375
article 2, 376
article 3, 376
article 4, 376
article 6, 107–8, 376
article 7, 148, 311–12, 319, 336*n*51, 376
article 8, 96, 118, 311–13, 319, 377
article 27, 63*n*24, 377
article 28, 67*n*30, 104, 377
article 30, 71–73, 115–17
article 31, 129*n*3, 179–80, 184, 213. *See also* Compulsory licenses
article 31bis, 202–14, 218–20. *See also* Waiver
article 33, 68*n*33, 379. *See also* TRIPS – patent term
article 39(3). *See* Data protection
article 41, 74*n*49, 75*n*53, 234*n*26, 308–12, 380
article 51, 75*n*51, 303–6, 309, 315–16, 318–20, 380
article 52, 303–8, 316–17, 321, 381
border measures, 303–8
enforcement of, 58. *See also* Dispute settlement understanding
compliance dates, 83–85
compulsory licensing. *See* Compulsory licensing
data protection under, 76–83
dispute resolution. *See* Dispute settlement understanding
developing countries, 58–59, 83–87, 356
enforcement of patent rights, 74–76
exceptions to patent rights, 70–74, 105–18, 378
exclusive marketing right (EMR), 85–87
exhaustion of rights, 67, 107–8, 376
FTA restrictions. *See* FTA
and HIV treatment, 356–57
importance of, 57–58
interpretation of, 60–62
least developed countries (LDCs), 83–87
"minimum" standards, 57, 65. *See also* In-transit drugs (suspension of)
"most favored nation" (MFN) requirement, 57*n*3, 376
"national treatment" requirement, 57*n*3, 376
negotiating history, 58–60
non-discrimination against a field of technology, 73–74, 95–96, 117–18
objectives, 382 *See also* TRIPS article 7
patentable inventions, 62–65
principles, 382. *See also* TRIPS article 8
patent examination (and opposition), 65–66
patent rights and remedies, 66–70
patent term, 68–70
provisional measures, 75. *See also* Injunctions
and Thailand's compulsory licenses. *See* Thailand's compulsory licenses
transitional provisions, 84–87
unfair commercial use (protection from), 76–80, 82–83, 379
TRIPS Council, 202, 206, 212, 335
"TRIPS-Plus" standards
adoption of, 226–28
compulsory license (limits), 237–38
controversy, 249–51
data exclusivity, 241–44
parallel imports (limitations), 240–41
revocation (limits), 239
opposition (limits), 234–35
patent linkage, 242–43
patent right exceptions (limited), 239–40
patent term extension, 235–37
patentability, 232–34
PCT, 228–29, 231–32
pipeline protection, 229–32
price controls on drugs (limits), 246–48
new uses of previously known compounds, 233–34

regulatory protection of drugs, 241–46
requirements, 228–49
significance, 225
WIPO (Patent Law Treaty), 352–53
Turkey, 203

UAE, 203
Uber-right view of patents, 327, 362*n*21
compulsory licenses, 183–89
and denials of injunctive relief, 189
Doha Public Health Declaration, 337–38, 340–48
Explanation, 159, 163–66
proposed solutions, impact on, 364–67
in-transit goods, 291–94
South African controversy, on compulsory licensing, 328–30
Thailand's compulsory licenses, 173–79, 192
UN rules, for determining a LDC, 84
UNAIDS, 299
Unfair commercial use, 76–80, 82–83
India's patent laws, 118–19
UNITAID, 299
UNITAID patent pool, 361–64
United States, 60, 203
average period of examination, 102*n*26
data exclusivity in, 262
free trade agreements. *See* FTA
"government use" of patents, 189
negotiation of TRIPS, 227
Patent and Trademark Office (USPTO), 345–46
procedural safeguards of public interest, 281
retaliation against Thailand, 170*n*51
and Special 301 process, 151–52
and Thailand's compulsory licenses, 151
TRIPS compliance dates, 84
and TRIPS-plus standards, 225–26
Universal Declaration of Human Rights, 164, 164*n*22
Unpatented drugs. *See* Generic drug
U.S. Patent Act, 103*n*35
U.S.C. § 271(c), 104*n*40
US Office of Trade Representatives, 226, 330

Venetian Patent Act of 1474, 131*n*12
Vienna Convention on the Law of Treaties, 60*n*15, 60*n*17, 78, 152*n*74
Vivv, 366–67

Waiver of Domestic Use Requirement (for compulsory license consistent with TRIPS)
applicability of, 200–202
TRIPS article 31bis, 199, 209, 212, 219–20
Apotex, 214–17
and national emergency, 203, 208, 211–12
case study (Canada-Rwanda), 214–17
Chairperson Statement, 199
and compulsory license in importing country, 207–8
and compulsory license in exporting country, 208
countries eligible to import, 202–4, 212–13
coverage, 202–6
criticisms, 217–19
products eligible, 200–201, 204, 294, 211–12
Doha Public Health Declaration, 204, 340–41
exporting country, 203–4, 208–9, 211
General Council Chairperson. *See* Chairperson Statement (above)
and generic companies (impact on business model), 219–20
history, 197–200, 219–20, 340–41, 348–49
importing country, 202–3, 212–13
industrialized countries, 203
insufficient manufacturing capacity (relevance of), 207
LDCs, 202–4
Labeling requirements, 205, 206, 208
Legal status, 198–99
national implementation, 209–17
notification to TRIPS Council – exporting country, 208–9
notification to TRIPS Council – importing country, 206–8
process, 206, 209
re-sale of licensed goods, 205
regulatory approval of licensed drug, 213
remuneration, 213–14
and trade diversion concerns, 204–5
TRIPS amendment, 198
WHO prequalification, 213
Watch list, 330*n*17, 332
Priority, 250
Special, 151

WHO, 299, 346, 351
and Thailand's compulsory licenses, 167–68, 171–72, 178
and counterfeit, 345
and generic drugs stopped in transit, 315
prequalification, 213
and South Africa, 330–31
WHO Good Manufacturing Practices, 13*n*27
WHO List of Essential Medicines, 214
WIPO Patent Law Treaty. *See* Patent Law Treaty
WIPO Provisions on Protection Against Unfair Competition, 78*n*63
World Bank, 152
and Thailand's compulsory licenses, 168, 170
World Health Organization (WHO), 13
and Doha Public Health Declaration, 351–52
World Intellectual Property Organization (WIPO), 226*n*5, 352–53
World Trade Organization (WTO), 57
Appellate Body, 58, 58*n*8, 68, 79
dispute settlement, 61, 141
and GATT, 59, 59*n*12
ministerial conference, 197, 336, 382, 384
Ministerial Declaration on the TRIPS Agreement and Public Health. *See* Doha Public Health Declaration
Sanctions, 58. *See also* Dispute settlement understanding
World Trade Organization Panel Report. *See also* Dispute Settlement Understanding – Request for Consultations
Canada – Term of Patent Protection, 68
Canada – Patent Protection of Pharmaceutical Products, 71–74, 95*n*12, 106*n*44, 106*n*46, 162*n*33, 236*n*28, 310*n*102
China – Measures Affecting the Protection and Enforcement of Intellectual Property Rights, 111*n*73
Intellectual property decisions, 61*n*19, 62*n*20

Zidovudine, 215, 233